Single Atom Catalysts

Single Atom Catalysts

Design, Synthesis, Characterization, and Applications in Energy

Edited by

PRASHANTH W. MENEZES

Materials Chemistry Group for Thin Film Catalysis—CatLab, Helmholtz-Zentrum Berlin für Materialien und Energie, Berlin, Germany
Department of Chemistry, Technische Universität Berlin, Berlin, Germany

DEBASISH SARKAR

Department of Physics, Malaviya National Institute of Technology Jaipur, Jaipur, Rajasthan, India

KAMLENDRA AWASTHI

Department of Physics, Malaviya National Institute of Technology Jaipur, Jaipur, Rajasthan, India

ELSEVIER

Elsevier
Radarweg 29, PO Box 211, 1000 AE Amsterdam, Netherlands
125 London Wall, London EC2Y 5AS, United Kingdom
50 Hampshire Street, 5th Floor, Cambridge, MA 02139, United States

ISBN: 978-0-323-95237-8

Publisher: Matthew Deans
Acquisitions Editor: Stephen Jones
Editorial Project Manager: John Leonard
Production Project Manager: Sujithkumar Chandran
Cover Designer: Mark Rogers

Typeset by MPS Limited, Chennai, India

Contents

List of contributors

Priyanka Aggarwal
Department of Physics, Malaviya National Institute of Technology Jaipur, Jaipur, Rajasthan, India

Shahzad Ahmed
Department of Metallurgical and Materials Engineering, Indian Institute of Technology Jodhpur, Jodhpur, Rajasthan, India

Arshiya Ansari
Department of Metallurgical and Materials Engineering, Indian Institute of Technology Jodhpur, Jodhpur, Rajasthan, India

Kamlendra Awasthi
Department of Physics, Malaviya National Institute of Technology Jaipur, Jaipur, Rajasthan, India

Jinli Fan
School of Materials and Chemistry, University of Shanghai for Science and Technology, Shanghai, P.R. China

Chengyu Fu
School of Materials and Chemistry, University of Shanghai for Science and Technology, Shanghai, P.R. China

Minbin Gao
National Engineering Laboratory for Methanol to Olefins, Dalian Institute of Chemical Physics, Chinese Academy of Sciences, Dalian, P.R. China

Yuan Ha
School of Advanced Materials and Nanotechnology, Xidian University, Xi'an, P.R. China

Weiju Hao
School of Materials and Chemistry, University of Shanghai for Science and Technology, Shanghai, P.R. China

Arindam Indra
Department of Chemistry, IIT (BHU), Varanasi, Uttar Pradesh, India

Prashant Kumar
Global Innovative Centre for Advanced Nanomaterials (GICAN), College of Engineering, Science and Environment (CESE), School of Engineering, The University of Newcastle, Callaghan, NSW, Australia

GuiFen Li
Science Center of Energy Material and Chemistry, College of Chemistry and Chemical Engineering, Inner Mongolia University, Hohhot, P.R. China

Haoxuan Liu
State Key Laboratory of Featured Metal Materials and Life-cycle Safety for Composite Structures, MOE Key Laboratory of New Processing Technology for Nonferrous Metals and Materials, School of Resources, Environment and Materials, Guangxi University, Nanning, P.R. China

Xijun Liu
MOE Key Laboratory of New Processing Technology for Non-Ferrous Metals and Materials, and Guangxi Key Laboratory of Processing for Non-Ferrous Metals and Featured Materials, School of Resource, Environments and Materials, Guangxi University, Nanning, P.R. China

Prashanth W. Menezes
Materials Chemistry Group for Thin Film Catalysis—CatLab, Helmholtz-Zentrum Berlin für Materialien und Energie, Berlin, Germany; Department of Chemistry, Technische Universität Berlin, Berlin, Germany

Indranil Mondal
Department of Chemistry, Technische Universität Berlin, Berlin, Germany; School of Chemistry, IISER Thiruvananthapuram, Kerala, India

Qingbao Ni
Science Center of Energy Material and Chemistry, College of Chemistry and Chemical Engineering, Inner Mongolia University, Hohhot, P.R. China

Pranay Ranjan
Department of Metallurgical and Materials Engineering, Indian Institute of Technology Jodhpur, Jodhpur, Rajasthan, India

Rui Ren
Science Center of Energy Material and Chemistry, College of Chemistry and Chemical Engineering, Inner Mongolia University, Hohhot, P.R. China

Debasish Sarkar
Department of Physics, Malaviya National Institute of Technology Jaipur, Jaipur, Rajasthan, India

Subhash Chandra Shit
Department of Chemistry, Gwangju Institute of Science and Technology (GIST), Gwangju, Republic of Korea

Moin Ali Siddiqui
Department of Metallurgical and Materials Engineering, Indian Institute of Technology Jodhpur, Jodhpur, Rajasthan, India

Ajit Kumar Singh
Department of Chemistry, IIT (BHU), Varanasi, Uttar Pradesh, India

Baghendra Singh
Department of Chemistry, IIT (BHU), Varanasi, Uttar Pradesh, India

Jing Sun
Science Center of Energy Material and Chemistry, College of Chemistry and Chemical Engineering, Inner Mongolia University, Hohhot, P.R. China

Carsten Walter
Materials Chemistry Group for Thin Film Catalysis—CatLab, Helmholtz-Zentrum Berlin für Materialien und Energie, Berlin, Germany

Tianwei Wang
State Key Laboratory of Featured Metal Materials and Life-cycle Safety for Composite Structures, MOE Key Laboratory of New Processing Technology for Nonferrous Metals and Materials, School of Resources, Environment and Materials, Guangxi University, Nanning, P.R. China

Dongdong Xue
Science Center of Energy Material and Chemistry, College of Chemistry and Chemical Engineering, Inner Mongolia University, Hohhot, P.R. China

Jiangwei Zhang
Science Center of Energy Material and Chemistry, College of Chemistry and Chemical Engineering, Inner Mongolia University, Hohhot, P.R. China

Yan Zhao
Science Center of Energy Material and Chemistry, College of Chemistry and Chemical Engineering, Inner Mongolia University, Hohhot, P.R. China

CHAPTER 1

Introduction to single-atom catalysts

Priyanka Aggarwal[1], Kamlendra Awasthi[1], Debasish Sarkar[1] and Prashanth W. Menezes[2,3]
[1]Department of Physics, Malaviya National Institute of Technology Jaipur, Jaipur, Rajasthan, India
[2]Materials Chemistry Group for Thin Film Catalysis—CatLab, Helmholtz-Zentrum Berlin für Materialien und Energie, Berlin, Germany
[3]Department of Chemistry, Technische Universität Berlin, Berlin, Germany

1.1 Introduction

Our modern culture faces significant challenges in the areas of both energy and the environment [1−5]. Conventional nonrenewable fossil fuels like coal, petroleum, and natural gas are currently fulfilling the global energy demands and are therefore being depleted at a fast pace in support of the ongoing improvement in the economy and to satisfy the energy needs of the fast-growing population. The increasing use of these nonrenewable resources has also resulted in severe environmental pollution, particularly in recent years. To address the upcoming global energy and environmental crisis, it is indispensable to search for sustainable and green energy, where energy conversion becomes one of the critical aspects. As an example, the electrochemical reduction of carbon dioxide delivers valuable fuels (CH_4, CO, C_2H_4, etc.), which also helps in lowering the CO_2 level in the atmosphere, thus killing two birds with one stone. Unfortunately, due to the extremely stable $C=O$ bond (806 kJ mol^{-1}) in CO_2, the direct conversion of CO_2 to fuels is kinetically slow and not much efficient [6]. Hence, such energy processes are dominantly controlled by the catalysts. They lie at the center of this paradigm and are vital pillars of modern chemistry that are used in more than 90% of industrial processes. This is due to their crucial role in lowering the desired activation energy and boosting the rate, efficiency, and selectivity of a desired specific reaction that provides essential chemicals and renewable fuels to our society [7]. Their use guarantees high-yield transformations into the desired products without producing undesirable by-products. Given their huge potential to contribute to the development of a

Single Atom Catalysts
DOI: https://doi.org/10.1016/B978-0-323-95237-8.00010-0

sustainable future, academic interest in catalysis is on the rise. Significant research has been put into developing and manufacturing low-cost catalysts that are both extremely efficient and selective. Success in catalysis leads to the advancement of science and pushes the limits of human knowledge. Researchers from various fields have entered into the topic of catalysis, including those in science, chemistry, biology, materials science, and engineering, to mention a few. According to the Web of Knowledge, the number of publications on the topic of "catalysis" has skyrocketed over the last decade. Approximately, 20 Nobel Prizes have been awarded for research on catalysis [8]. Two important aspects of efficient and selective catalysis are determined by facile dissociation of reactants and favorable binding energy of intermediates to ease the adsorption of reactants and desorption of product.

In search of efficient catalysts, noble metals, such as Pt, Ru, and Ir, emerge as well-known catalysts that are employed in the majority of energy transformations and chemical reactions [9]. This is due to their unique physicochemical properties enabling them to activate the process with greater efficiency. However, their massive cost and acute shortage prevent them from being used on a big scale. Furthermore, neither of these is the best performing concurrently for various reactions. For example, Pt-based materials are popularly useful for hydrogen evolution reaction (HER) but not for standard oxygen evolution reaction (OER), for which Ir/Ru-based catalysts serve as the benchmark [9,10]. In this direction, significant efforts have been directed toward searching for those catalysts that are not based on noble metals. These include various earth-abundant and nonprecious metals (chalcogenides [11–14], carbides [15], borophosphates [16,17], pnictides [18], borides [19], alloys [20], intermetallics [21–25], phosphonates [26], phosphides [27], etc.) and metal-free-based electrocatalysts [28,29] that have been actively used in the past several years in various catalytic reactions. Sadly, the catalytic performances are indeed lacking in terms of activity, selectivity, and stability compared to noble materials.

1.1.1 Fundamentals of single-atom catalysts

Basically, catalysts can be categorized into two types: homogeneous catalysts (homo = same, geneous = phase) and heterogeneous catalysts (hetero = different). In homogeneous catalysis, catalysts and reactants are in the same phase (often a liquid), whereas in heterogeneous catalysis, catalysts exist in different phases (usually a solid) from the reactants [30,31].

Out of these, heterogeneous catalysts are primarily employed in many significant commercial reactions due to their robustness and easy separation [32]. Supported metal nanostructures in the nano-size range on a high surface area support, a type of heterogeneous catalyst has been extensively researched in numerous industrial reactions owing to their good catalytic performance. However, their performance lags behind homogeneous catalysts. Several parameters were shown to be critical in affecting the performance (activity, selectivity, and longevity) of heterogeneous catalysts based on supported nanoparticles. The size and shape of metal particles, nature of exposed faces, defects sites, step sites, corners, interactions between metal and support, and bimetallic effects (electronic or steric), metal-support interface are just a few of these characteristics [33]. Out of these, the size and shape of metal particles significantly influence the performance of such catalysts [34]. This is because supported metal catalyst often comprises a mixture of metal particles with a wide range of sizes and irregular morphologies. Each metal particle may have several active sites with various functionalities. Such variability causes unwanted side reactions resulting in the decrease in selectivity toward a particular product. Moreover, the only atoms that can act as active sites and are accessible to reactants are those that are exposed to the surface and have an unsaturated coordination environment, while those in the interior are not directly involved in the catalytic process, thus minimizing the utilization of metal active sites in heterogeneous catalysts [35]. The intricacy of heterogeneous systems made it difficult to understand the relationship between structure and activity/reactivity, which is quite important to improve catalytic performance. On the other hand, homogeneous catalysts display almost 100% metal atom utilization, as well as well-defined active sites [36]. Although homogeneous catalysts feature high selectivity and high metal atom utilization compared to heterogeneous catalysts, they are fragile and face thermal stability and product-catalyst separation issues [37].

One obvious approach to tackle the earlier-mentioned drawbacks associated with both types of catalysts is to gradually reduce the size of metal nanoparticles to clusters and ultimately to a single atomic limit. With the size reduction, the fraction of active atoms exposed to a surface with an unsaturated coordination environment (low coordination) increases so that now each and every atom of the supported metal catalyst in the targeted catalytic reaction can be utilized [38]. The final ultimate dispersion of metal atoms leads to an increase in specific activity and also lowers down the cost of catalysts, mainly beneficial for high-cost noble metals (Fig. 1.1A). This approach gives rise to the production of single-atom catalysts (SACs) that are defined as atomically dispersed catalysts with catalytically single metal atoms uniformly dispersed on appropriate support [39]. SACs, in contrast to other heterogeneous catalysts, feature homogenized active centers for catalytic reactions that are similar to those of homogeneous catalysts, integrating the advantages of both homogeneous and heterogeneous catalysts (Fig. 1.1B) [40]. SACs have produced a real sensation in the era of heterogeneous catalysis in barely a decade, resulting in several scientific breakthroughs. The mononuclear nature of active sites in SACs offers nearly 100% metal atom utilization efficiency and exposes more active sites. In addition, they provide well-defined and uniformly dispersed active sites, high selectivity, strong metal-support interactions, recyclability, high catalytic activity, as well as robust stability [41,42]. Moreover, recent advances in both atomic resolution characterization techniques and theoretical modeling and simulations that allow us to directly examine the distribution of single atoms (or even their precise location) on the support surface have sped up the growth of this discipline. Interestingly, there have been more than 1000 publications coming every year on this subject throughout the past few years. SACs are regarded as the most promising next-generation catalysts that have the potential to utilize 100% of the active participating atoms in a reaction, resulting in a significant increase in catalytic efficiency. The field has advanced at a breakneck pace, yielding potential catalytic leads. They

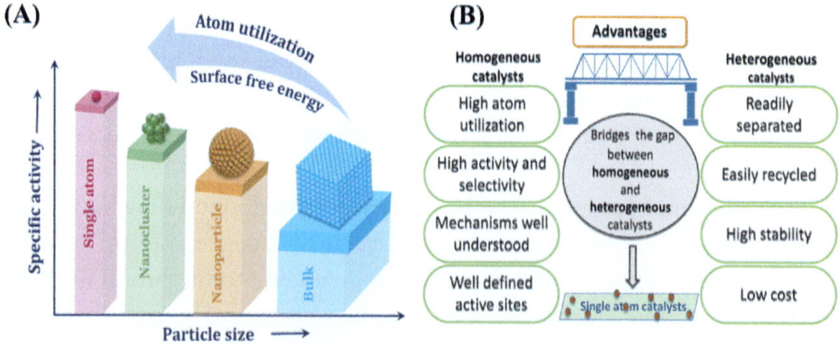

Figure 1.1 (A) Schematic diagram depicting the correlations between specific activity, atom utilization, and surface free energy per atom with particle size. (B) SACs combine the advantageous properties of both homogeneous and heterogeneous catalysts.

outperform nanoparticle/cluster catalysts in terms of activity, stability, as well as selectivity in a rising number of chemical reactions. SACs with well-defined active centers present intriguing opportunities owing to their unique physical and chemical properties [43]. They hold great potential to replace the state-of-the-art noble metals employed in the majority of electrochemical application systems, such as HER, OER, oxygen reduction reaction (ORR), nitrogen reduction reaction (NRR), CO_2 reduction reaction (CO_2RR), etc. [9,44,45]. SACs have since been used in a variety of chemical reactions, including thermocatalytic, photocatalytic, and electrocatalytic reactions. However, their practical applications are limited by their inherent problems, namely, atom diffusion, and aggregation during synthesis and reaction conditions [46,47]. This is because, with the continuous downsizing, from bulk, small nanoparticles, clusters, and ultimately to single atoms, the surface free energy also drastically increases owing to a higher number of unsaturated atoms [48].

Consequently, it is a difficult task to construct SACs. They are commonly synthesized by carefully maintaining a low loading density of metal precursors on appropriate solid support with sufficiently high surface area, including metal oxides, carbon materials, metal sulfides, MXenes, g-C_3N_4, metal-organic frameworks (MOFs), to avoid any aggregation during the construction (reduction) process [49]. This, however, lowers the concentration of supported metal atoms. The support of SACs in some ways mimics the organic ligands used in homogeneous catalysis from the perspective of coordination chemistry. Thus, two obstacles to overcome in the production of highly efficient SACs are (1) to uniformly distribute metal single atoms with higher loadings on the substrate by preventing their aggregation. (2) To regulate the coordination and electronic environment for every single atom to enhance the interaction between a single atom and neighboring support atoms [50]. These two challenges may be overcome by employing advanced synthesis techniques based on defect engineering (introducing vacancies, steps, and edges that bind single atoms strongly), spatial confinement [trapping single atoms into the void spaces of porous materials like MOFs, zeolites, covalent-organic framework (COFs), etc.], coordination design (bonding of single atoms with electron-rich heteroatoms, usually N, O, and S) strategy [51]. Fig. 1.2 presents the overview of SACs, including synthesis methods, properties of supports and their types, and characterization techniques that will be discussed in detail in the present and upcoming chapters of this book.

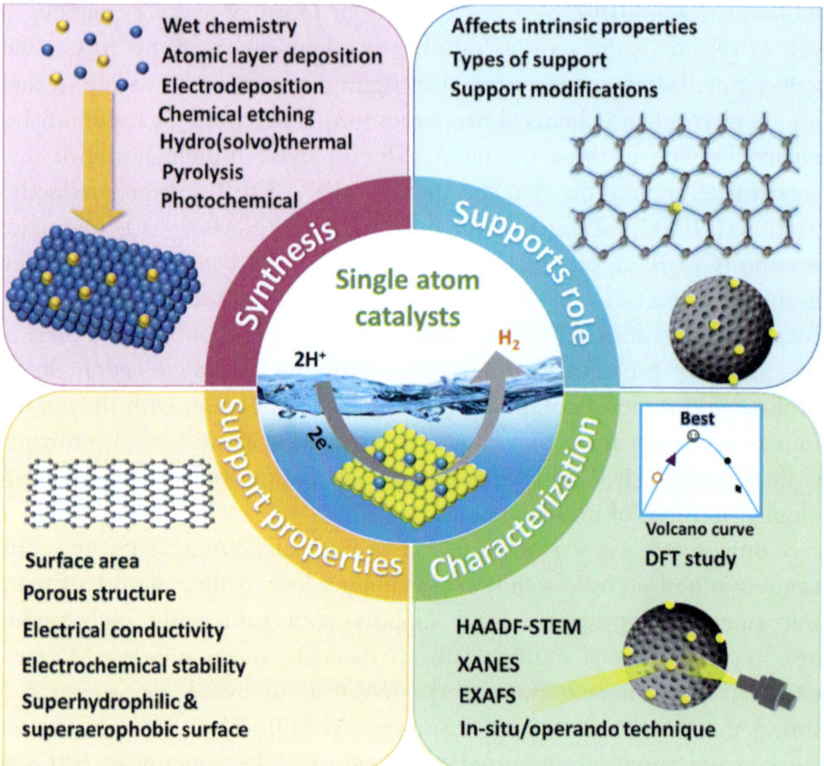

Figure 1.2 Overview of SACs with their synthesis strategies, supports and their properties and characterization techniques employed to uncover the nature of SACs. *Reproduced with permission from P. Aggarwal, D. Sarkar, K. Awasthi, P.W. Menezes, Functional role of single-atom catalysts in electrocatalytic hydrogen evolution: current developments and future challenges, Coord. Chem. Rev. 452 (2022) 214289. Copyright 2022, Elsevier.*

1.1.2 Single-atom catalysis: its beginning and evolution

In the past, the idea of SACs was developed by gradually decreasing the size of nanoparticles/clusters counterparts supported on a solid substrate to the level of single atoms to improve the usage of expensive noble metals (Pt, Pd, Ru, Ir, etc.) and subsequently minimize their cost. Fig. 1.3 illustrates the rapid progress in this field from the past to present. In 1995, the breakthrough work by Thomas and coworkers traced the development of single metal atoms as active centers for heterogeneous catalysis. They prepared oxygen-coordinated Ti single-site heterogeneous catalyst by grafting metallocene complexes on mesoporous silica followed by calcination for

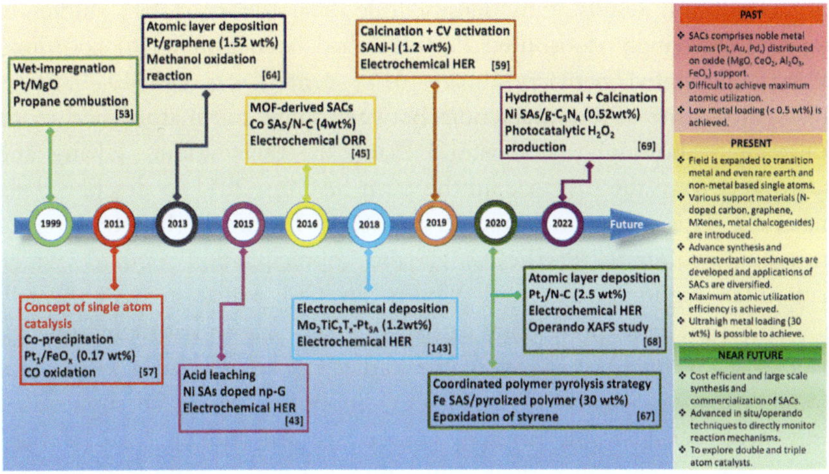

Figure 1.3 Time map to illustrate progress in the field of single-atom catalysis.

the epoxidation of cyclic alkenes [52]. Initial studies of catalysts by Asakura's and teammates in 1999 confirmed that atomically dispersed Pt species on MgO support were as active as Pt particles for propane combustion [53]. Flytzani-Stephanopoulos et al. in early 2003 provided persuasive evidence that the activity observed in the water–gas shift reaction was caused by atomically dispersed gold or platinum species strongly linked with the surface of CeO_2 rather than metal nanoparticles [54]. The group found that even after removing metal nanoparticles from CeO_2 surface by using highly corrosive chemicals, the catalytic activity did not decay and was preserved. This suggests that the single-atom species may actually be the real active sites. Although single atoms have been shown to have a good potential for enhancing catalytic performance during this initial stage, the configuration of single atoms and their interaction with the catalyst support could not be adequately explored in depth. In 2007, Hackett's group demonstrated that atomically distributed Pd sites on mesoporous alumina substrate (meso-Al_2O_3) showed outstanding aerobic oxidation catalytic activity [55]. The group performed high-angle annular dark field (HAADF) scanning transmission electron microscopy (STEM) to directly visualize single Pd species dispersed on meso-Al_2O_3 support. Further, they utilized extended X-ray absorption (XAS) fine structure spectroscopy (EXAFS) to predict the coordination structure of isolated Pd species. Kwak and colleagues in 2009 used the traditional incipient wetness approach to design γ-Al_2O_3 assisted Pt SACs [56]. They combine characterization results with density functional theory (DFT) studies to confirm that singly distributed Pt atoms had been bonded to coordinatively unsaturated pentacoordinate Al^{3+} centers on γ-Al_2O_3 surface. Thus, the nature of the interactions between single metal atoms and oxide substrate is thoroughly understood thanks to their studies. Zhang and coworkers were the first to coin the term SACs in 2011 to describe these materials. The group anchored Pt SACs (Pt loading: 0.17 wt.%) on an iron oxide substrate (Pt_1/FeO_x) via the coprecipitation method, which had much higher activity and stability for CO oxidation than their Pt cluster counterparts [57]. They carried out DFT along with experimental data to study Pt SACs. Most of the earlier studies on SACs focused on noble metal catalysts (Pt, Pd, Au, etc.) supported on oxide support via using mass-selected soft landing, coprecipitation, and incipient wetness approaches mainly. Also, their application is largely confined to organic transformation reactions to increase the reaction rate and activity. But now, SACs have crossed the periodic table in just a decade, and their research field has recently been considerably enlarged owing to their ongoing advancements in support material type, synthesis strategies, characterizations, and applications.

Now, noble-free (Ni, Co, Fe, Cu) and even nonmetal-based single atoms such as P [58], I [59], O [60] are added to the family. In addition, the variety of support materials has also expanded from the initial and most widely used metal oxides to carbon materials, heteroatom doped carbons, MXenes, transition metal dichalcogenides (TMDs), microporous materials, including MOFs, COF materials, zeolites, and others (metal hydroxides, metal phosphides, metals, etc.). Advanced synthetic techniques, including atomic layer deposition (ALD), spatial confinement strategy, defect engineering strategy, coordination design technique, photochemical method, electrochemical deposition, chemical etching, hydro/solvothermal, ball milling, microwave-assisted, and high-temperature atom trapping, have also been developed to prepare SACs [61–63]. Sun' group, in 2013 for the first time, utilized ALD technique to fabricate Pt SACs on graphene support with Pt loading of 1.52 wt.% and investigated its high catalytic activity toward the oxidation of methanol [64]. However, the majority of the reported SACs contain fairly low metal loading amounts (often less than 2 wt.%) to avoid the aggregation of individual metal atoms, which restricts the development and practical application of SACs. Hence, novel synthesis methods along with suitable support materials providing a variety of anchoring sites to capture single atoms with strong metal-support interaction are developed to increase metal loading. Qi et al. in 2019 synthesized Co SACs with a higher Co loading content of 3.54 wt.% on the distorted 1 T phase of MoS_2 (D 1T-MoS_2) by employing an electrochemical leaching approach [65]. The strong interactions between the Co single atoms (SAs) and the D 1T-MoS_2 support owing to the strong Co-S covalent bond are beneficial to achieving high loading of Co SAs without undergoing aggregation. One year later, Ji's group utilized atom-confinement and coordination techniques to stabilize Er SAs on carbon nitride nanotube support and reported metal loading of Er to be as high as 20.1 wt.% [66]. The resulting catalyst exhibits remarkable activity for photocatalytic CO_2RR. Eventually, Xiong et al. synthesized a series of high-loading SACs by using a simple pyrolyzing coordinated polymer strategy out of which metal loading of Fe SACs reached a record value of 30 wt.% [67], opening the door for their practical applications. In the same year, Fang's group carried out operando XAS fine structure (XAFS) technique to examine the structural evolution of single Pt species dispersed on MOF-derived N-doped carbon support during electrochemical HER [68]. They concluded that Pt SAs approach to near-zero valence state as a result of weakened interactions with the surrounding support atoms during the reaction are the true active sites that contribute to high HER activity. Zhang and coworkers in 2022 employed ultrathin g-C_3N_4 nanosheet as support to anchor Ni SAs, and the obtained catalyst exhibits high selectivity of 87.3% toward photocatalytic H_2O_2 production [69]. Earlier SACs were broadly employed for thermocatalytic applications. However, interest in single-atom electrocatalysis has expanded significantly in recent years to become a prominent topic. Single-atom photocatalysis is still in its infancy but already holds great potential for the area [70,71].

Despite these developments, ongoing investigations of novel SACs remain a key area of research in the field of catalysis from the viewpoints of materials synthesis and mechanism interpretation. Hence, it becomes necessary to propose and conduct systematic theoretical studies and operando experimental studies to gain in-depth knowledge about the nature of SACs.

1.2 Properties of single-atom catalysts

The supported single atoms do not form a metal-metal bond in SACs unlike nanoparticles/clusters counterparts because there is no significant interaction between isolated single atoms [72]. However, individual atoms make bonds with the support atoms that can be of different nature, including ionic, covalent, and coordination bonds. This results in isolated single atoms to usually carry a positive (sometimes negative) charge. SACs not only possess maximal atom utilization efficiency but also provide numerous possibilities to tune reaction rates and selectivity. The development of well-defined active metal centers in SACs strengthens our ability for comprehending structure—activity relationships at an atomic and subatomic level [73]. The notably high catalytic performance of SACs is attributed to their advantageous properties as discussed here: (1) large proportions of low coordination metal atoms compared to their bulk/nanoparticle's counterparts. (2) It is easier to load single atoms to substrates due to their lower coordination, leading to their strong interactions with the support. Furthermore, it makes the charge transfer process between singly dispersed metal centers and supports more efficient [74]. (3) Almost 100% atomic dispersion of single atoms on the support accounts for their high specific activity. (4) Single atoms escape the problems caused by particle size effects, which are dominant factor in altering the performance of metal nanoparticles or clusters as discussed in Section 1.1.1. (5) The electronic bands of metal narrow as the number of metal atoms in a particle decrease, finally being replaced by discrete energy levels (widening of HOMO-LUMO gap) on reaching the ultimate atomic level due to confinement of electrons caused by quantum size effect [43]. Features in the density of single atom d-states become broader for the supported single atoms. The size of the broadening is determined by the interaction between the single atom and the nearby atoms of the support. (6) Because single atoms lack a metal surface, their adsorption model differs dramatically from that of metal nanoparticles, regardless of whether the reaction is carried out in the solution phase or gas phase.

All these unique physicochemical features of SACs discussed earlier make them special from those nanoparticles or sub-nanometer clusters. SACs therefore emerged as versatile materials that perform really well as catalysts to advance the catalysis process. For instance, Lee's group developed various types of Pd catalysts with different particle sizes supported on mesoporous Al_2O_3 support out of which the Pd single atoms (Pd SAs) demonstrated the superior catalytic activity towards aerobic oxidation of alcohols [55]. These properties of SACs are drastically affected by the local coordination environment of single atomic species dispersed on support and discussed in detail in the following section.

1.2.1 The crucial role of support in influencing intrinsic properties of SACs and the concept of support-metal charge transfer

Supports, in addition to single atoms, are crucial in the area of SACs since they also contribute to dramatically improving HER performance [48]. It is obvious that an individual atom is unable to stay at the support surface alone without any interaction with the substrate. Support tailors the stability, local geometric environment, and electronic structures of supported single atoms similar to ligands in homogeneous catalysts, and so offering a foundation for modifying heterogeneous catalysts for specific applications [75]. The almost limitless possibilities of single atom-support combinations resulted in fine-tuning of their coordination structures and related properties leading to their key applications in many industrially important reactions [76]. Hence, it becomes important to discuss key support features that will enable us to choose a suitable support material and are as follows: (1) high availability of defects, steps, edges, kicks, or corners on the support surface to stabilize active single atom sites. (2) High electrical conductivity to facilitate quick transfer of electrons. (3) Highly porous structure to easily allow transfer of reactants and electrons. (4) Large electrochemical surface area to anchor high density of SAs and expose abundant active sites to the reactants. (5) Extremely high chemical and electrochemical stability to endure severe electrolyte conditions. (6) Finally, the support surface should possess superhydrophilic and superaerophobic features to promote ion transfer and escape of H_2 bubbles, respectively [77] (Fig. 1.4).

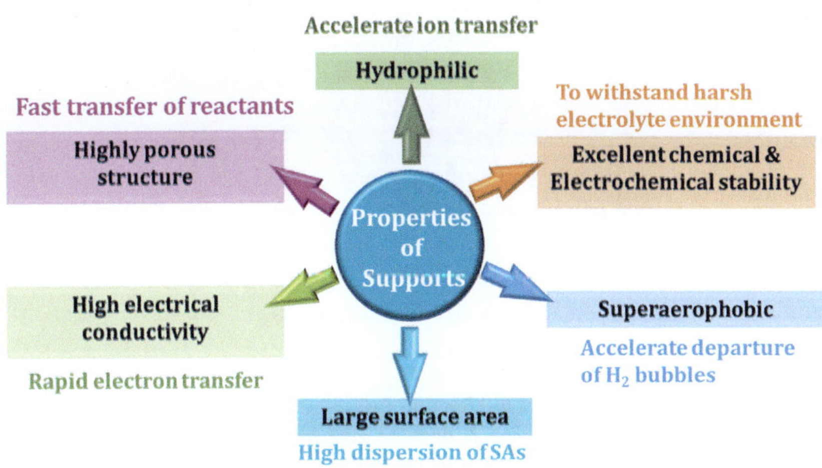

Figure 1.4 Summary of some essential properties of support material.

1.2.1.1 Concept of support-metal charge transfer

As discussed earlier, support is not only used to disperse and stabilize SAs but it also plays a vital role in affecting the geometric and electronic properties of supported isolated SAs as a result of interaction between dispersed metal atoms and support (Fig. 1.5). In 2012, C.T. Campbell coined the term electronic metal—support interaction, EMSI accounting for the related charge transfer caused by the strong bonding through ionic, covalent, or coordinate interactions between active metal center and support [78]. Hence, a single metal atom can be anchored to the support via the strong metal—support interaction to prevent sintering [79]. The stability of SACs refers to thermodynamic and kinetic stability. The thermodynamic stability implies Gibbs free energy of formation of single atoms or their adsorption energy must be more negative compared to cohesive energy of single atoms, implying spontaneous dispersion of nanoparticles to single atoms. On the other hand, kinetic stability refers to a high migration energy barrier to prevent aggregation [80—82]. The EMSI not only helps to stabilize active single atoms but also strongly influences the electronic structure of SACs [50,83]. When a single atom is anchored on a support, the transfer of charge takes place between the active metal site

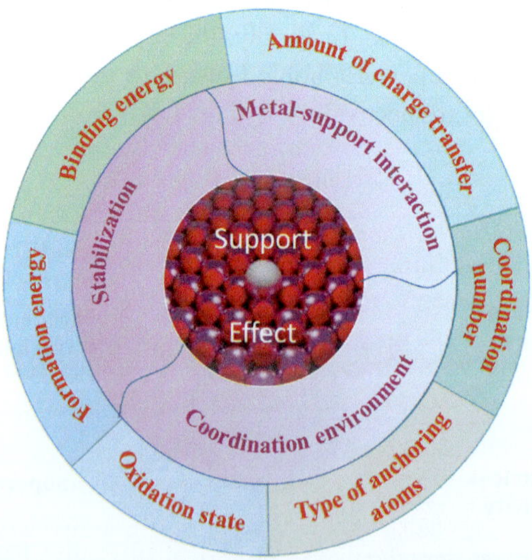

Figure 1.5 Influence of support in affecting properties of SACs. *Reproduced with permission from P. Aggarwal, D. Sarkar, K. Awasthi, P.W. Menezes, Functional role of single-atom catalysts in electrocatalytic hydrogen evolution: current developments and future challenges, Coord. Chem. Rev. 452 (2022) 214289. Copyright 2022, Elsevier.*

and the anchoring sites of the support due to the variation in their Fermi energy levels or/and differences in their electronegativities [84–86]. This causes the active metal site to become electron-deficient or electron-rich, which modifies its electronic configuration by adjusting the position of d-band center and ultimately tunes the adsorption and desorption behavior of reactive species on the active metal site to improve catalytic efficiency [87]. The electrochemical stability of support also influences the activity of SACs. Because a single metal atom is chemically and/or physically bonded to the substrate, any type of chemical phase transition or physical destruction of the support may cause considerable activity loss in SACs. Corrosion of the support surface results in the loss of single-atom species hence negatively influencing the overall electrocatalytic performance. The physical properties of single metal atom and supports, as well as the bonding environment that determine their oxidation states, amount of charge distribution, metal d state characteristics, and the geometry of metal bonding configurations on the substrate, all influence the interaction between single atoms and the support [88–90]. All of these characteristics determine the strength of the EMSI and can be linked to catalytic activity, selectivity, and stability [91,92]. Liu et al. demonstrated the role of local geometric configuration of carbon support in influencing the electronic structure of SACs and hence the overall ORR catalytic performance via DFT computations [93]. For this, they selected three types of FeN_4 active sites with different local carbon structures ($FeN_4{-}C_8$, $FeN_4{-}C_{10}$, and $FeN_4{-}C_{12}$). They concluded that $FeN_4{-}C_8$ site has the lowest activation energy for O_2 dissociation (about 0.2 eV) among the three for boosting the four e^- ORR.

As mentioned earlier, due to the high surface free energy of single atoms, they easily tend to agglomerate to produce nanoparticles/clusters counterparts that ultimately reduce their activity and selectivity. Hence, the loading density of single atoms in SACs is usually kept low. Therefore, recent advancements in synthetic routes (defect engineering, spatial confinement strategy, heteroatom doping) and materials science (nanoengineering, functionalization, porous structure) are required to realize strong metal–support interaction to avoid agglomeration of active single atoms and achieve more controllable and practical synthesis of high density, active and stable SACs [94–98] (Fig. 1.6). Wang et al. synthesized Fe SACs dispersed on ZIF-8 derived mesoporous N-doped carbon framework via a facile benzoate-assisted self-template strategy [99]. The synthesized Fe SACs demonstrated excellent ORR catalytic efficiency as well as superior

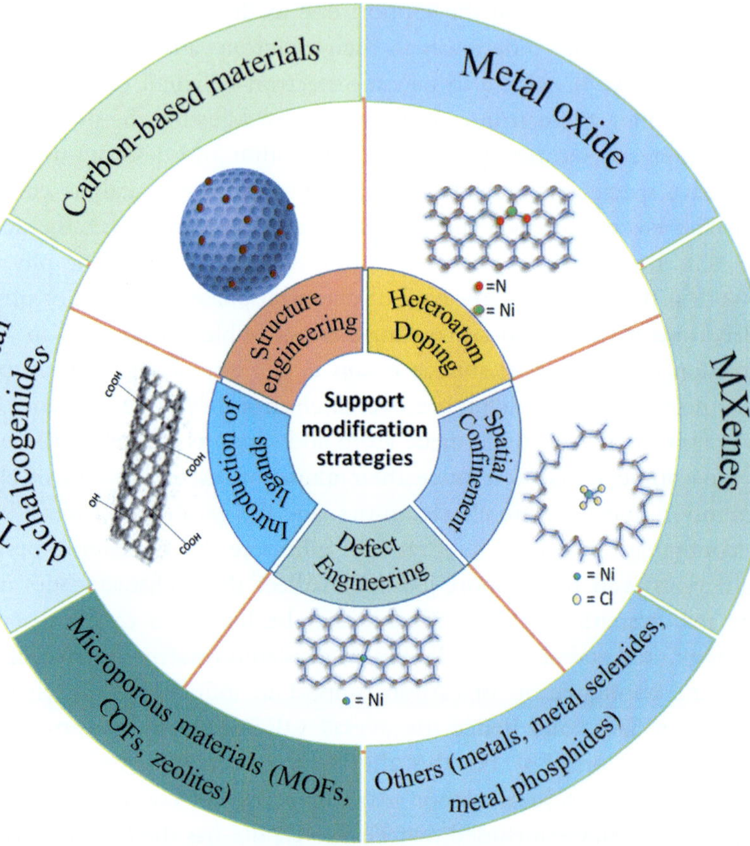

Figure 1.6 Numerous support modification techniques employed to enhance the interaction between an active single atom and the support. *Reproduced with permission from P. Aggarwal, D. Sarkar, K. Awasthi, P.W. Menezes, Functional role of single-atom catalysts in electrocatalytic hydrogen evolution: current developments and future challenges, Coord. Chem. Rev. 452 (2022) 214289. Copyright 2022, Elsevier.*

durability with over 90% activity retained after 90 hrs i-t test owing to the hierarchical porous structure and exposure of a large number of active sites. To understand the interaction between the metal atom and support and develop highly effective SACs, more thorough and in-depth understanding of the geometric and electrical characteristics of the active center at the atomic level are needed. Hence, it becomes important to develop advanced characterization tools or methods with enhanced precision, including HAADF-STEM, ex/in situ/operando XAS, emission and photoelectron spectroscopies, etc., to investigate exceptional physicochemical properties (oxidation state, coordination environment, structure–activity relationships, dynamic electronic/geometric structure evolution, reaction-driven restructuring) of SACs [100]. The in situ/operando characterization technique permits real-time monitoring of surface oxidation states and local atomic structure transformations. [101–103]. In addition, theoretical studies (DFT computations) must go hand in hand with practical study, allowing for a deeper knowledge of the catalytic system's fundamentals.

1.2.2 Types of support

1.2.2.1 Metal oxides

Metal oxides (FeO_x, CeO_2, Al_2O_3, MgO, TiO_2) have long been employed as supports to uniformly disperse single atoms via strong metal–support interactions [104]. The beneficial features of utilizing metal oxides as support include their low cost, easy synthesis, abundant metal or oxygen vacancies, strong metal-support interaction, structural diversity, and ultrahigh stability [105,106]. Qiao' group incorporated Au SAs into the Ce vacancy sites of CeO_2 support and revealed its extremely high activity, selectivity, and stability for preferential oxidation of CO [107]. TiO_2 is on the list of popular semiconductors and serves as an appropriate photochemical support [108]. Bi SAs were supported on TiO_2 to examine its performance for photoelectrochemical (PEC) water splitting by Pang's group [108]. They demonstrated high PEC performance of the prepared catalyst for water splitting, generating ultrahigh photocurrent density that is 4 times higher than pristine TiO_2. However, metal oxides possess poor electrical conductivity, which can be improved by modifying their surface by introducing metal or oxygen vacancies, doping, and structural engineering [105]. Zhou's group introduced 3-D Pt SACs by electrodepositing Pt SAs on NiO/Ni heterostructures grown on Ag nanowires [109]. The resultant catalyst showed superior HER activity and longevity of 30 h in 1 M KOH, thanks to the abundant Ni and O vacancies in NiO.

1.2.2.2 Carbon-based materials

These are one of the most extensively used support in the field of single atom catalysis owing to their fascinating properties, which are ultrahigh electrochemical surface area, exceptional conductivity, ease of accessibility, tunable structural (size, shape, and porosity), geometric, and electronic features, and low cost [110]. Numerous carbon-based materials including graphene, heteroatom-doped carbon, carbon nitride, defective carbon, MOF-based carbon materials, carbon nanosheets, and many more, have been reported to support many transition metals [111,112]. It is observed that interaction between pristine carbon-based materials and single atoms is quite weak and therefore single atoms easily agglomerate on the pristine carbon support. For instance, activation energies (migration barrier) of loaded single atoms on pure graphene are very low, lying in the range of 0.14–0.8 eV [113]. At the same time, activation energy is increased to 2.1–3.6 and ~5 eV after creating monovacancies and divacancies on perfect graphene, respectively [114,115]. It implies single atoms are tightly bonded to the surface of defective graphene compared to pristine ones. The presence of defects acts as trapping sites for single atoms in which single atoms get trapped and can not move, hence preventing their agglomeration and increasing stability. Hence, it becomes prominent to modify pristine carbon material surfaces using one or more procedures such as the introduction of vacancies, defects, heteroatoms (N, P, S, O), functional groups or ligands, and structural engineering to ensure strong metal–support interactions [116]. Surprisingly, metal atom-supported nitrogen-doped carbon support has shown remarkable catalytic performance in a wide range of electrocatalytic reactions (HER, OER, NRR, ORR, CO_2RR) and has the potential to replace state of art Pt [117–119]. For instance, Li's group anchored Fe SAs on N-doped hollow carbon nanospheres connected by carbon nanotubes [120]. The designed electrocatalyst demonstrated an outstanding ORR half-wave potential of 0.84 V in acidic media better than Pt/C owing to its porous structure, abundant active Fe-N_x sites, and high electrical conductivity of CNTs. Besides, carbon-supported SACs have also shown great potential to catalyze a wide range of essential reactions in energy storage devices such as supercapacitors, sodium-sulfur batteries, etc. [121].

1.2.2.3 Transition metal chalcogenides (MoS_2, $MoSe_2$, WS_2, WSe_2)

Layered Transition metal chalcogenides (TMCs), due to their appealing physical and chemical properties (morphological diversity, high stability, and high efficiency) and low cost, have sparked much attention as catalyst support for various energy applications [13,122,123]. TMCs are specifically investigated as effective catalysts for water splitting [124]. MoS_2 is popular for HER due to the similar hydrogen adsorption energy of MoS_2 edge to that of Pt [125]. In an alkaline electrolyte, $CoSe_2$ has been employed as a promising bifunctional catalyst for water electrolysis [126]. Unfortunately, TMCs have poor electrical conductivity and inert basal planes that cover a large proportion of the total surface area [127]. Therefore, it becomes essential to address these unfavorable features of TMCs to realize their commercial applications. In this perspective, researchers have unveiled different modification techniques including phase transformations (thermodynamically stable and semiconducting 2 H phase to metastable and conducting 1 T phase [128]), decoration of TMCs on conductive substrates [129], transition metal atom doping [130], defect and strain engineering [131]. 1 T phase of MoS_2 is beneficial owing to its high electrical conductivity, increased interlayer spacing, active basal planes, and high density of active sites. Doping has been demonstrated to be an effective strategy for enhancing MoS_2 HER performance out of earlier-listed modification methods. A variety of single transition metal atoms have been effectively deposited onto MoS_2 in recent years, greatly enhancing its electrocatalytic activity [132–134]. As an example, Ru SAs supported MoS_2 exhibit significantly enhanced HER activity compared to pristine MoS_2. The introduction of S vacancies, MoS_2 phase transition, and the electronic structural changes of Ru SAs doped MoS_2 were all credited with the improved HER activity [135].

1.2.2.4 MXenes

MXene is a novel type of 2D nanomaterial, including transition metal carbides, nitrides, and carbonitrides, which has attracted interdisciplinary interest due to their intriguing electronic, structural, and catalytic properties [136–138]. Because of MXene's distinctive properties (such as high electrical conductivity, high density of active sites, strong metal–support interaction, tunable band structure, ultrahigh stability, and hydrophilic nature), they have been widely used as solid support and catalyst in numerous electro/photocatalytic reactions [139–141]. Zhang's group constructed Pt SACs by embedding Pt SAs into the Mo vacancies of molybdenum-titanium carbide ($Mo_2TiC_2T_X$) MXene [142]. Benefitting from the strong interaction between Pt SAs and nearby C atoms of MXene and high density of Mo vacancies on the MXene surface, the developed catalyst showed excellent HER performance, outperforming state of art Pt/C as well as long-term stability of 100 h. Despite the inspiring features of MXenes, research in this field is still at its initial stages compared to other 2D-based materials used as support.

1.3 Distinction between single-atom catalysts and grafted organometallic catalysts

Surface organometallic chemistry (SOMC), or more specifically, surface organometallic catalysis (SOMCat), a well-established approach associated with specific techniques, is used for the design, synthesis, and characterization of well-defined single-site heterogeneous catalysts [33]. This approach operates by reacting organometallic/coordination compounds with well-defined surfaces (oxides, metal nanoparticles, carbon, or graphene) to produce the desired surface functionalities. In this approach, the surface is considered as a rigid ligand that binds the organometallic complex, molecular in nature, mainly by covalent (or sometimes ionic) bonds [143,144]. Since this technique relies on grafting organometallic transition metals onto well-defined substrates in a strictly regulated environment, it enables the creation of catalytic sites that are theoretically similar (single-site or single complex). One must enter a catalytic cycle through a presumptive catalytic intermediate made from organometallic compounds to produce well-defined surface organometallic/coordination fragments (SOMFs). These ideas serve as the foundation for the "catalysis by design" strategy ("structure-activity" relationship), aiming to boost catalyst performance by enhancing knowledge of the mechanistic elements of the catalytic process [145]. The majority of organometallic chemistry principles can be transferred to design and rationalize heterogeneous catalysis that effectively erases the existing gap between homogeneous and heterogeneous catalysis [146]. Notably, a transition is taking place in heterogeneous catalysis from "ill-defined materials" to "well-defined single-site catalysis." The single-site approach reduces the possibility of metal sintering since a grafted complex cannot move freely on the surface to undergo leaching or agglomeration. In parallel to the most recent developments in SOMCat, SACs have also recently emerged as a new frontier in the field of heterogeneous catalysis. The fundamental idea behind SAC is to reduce the size of metal/nonmetal nanoparticles until they reach the limit of a single atom leading to maximal dispersion of active atoms, increased activity, and selectivity. In SACs, single atoms are uniformly and highly dispersed on the surface of oxides, nitrides, carbides, phosphides, functionalized carbons, graphene, microporous materials, including zeolites, MOFs, etc. Although single metal atom serves as an active site in both cases, the main difference as shown in Fig. 1.7 lies in the fact that, in SOMCat, the isolated metal atom is bonded to "ligands" (spectator or functional) chosen particularly for the desired reaction. The spectator ligands (oxo, amido, imido, nitride, halogen, etc.) are mostly essential to tune the metal center in SOMC. While on the other hand, in SACs, these ligands are absent and the isolated metal atom must adopt the appropriate coordination environment with the reagents to obtain the target reaction. In SOMCat, the structure of SOMF mostly determines how catalytic reactions turn out. For instance, metal hydrides promote C-H bond activation of alkanes, whereas a metal

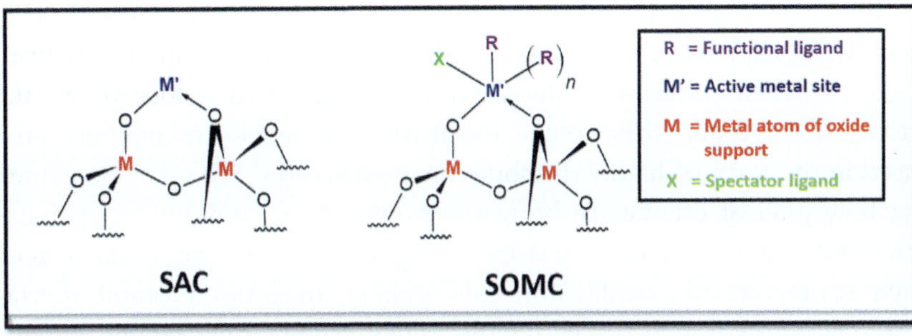

Figure 1.7 The comparison between single-atom catalyst (SAC) and grafted surface organometallic catalyst (SOMC) supported on an oxide support.

carbene is necessary for olefin metathesis. The well-defined nature of an active site in single-atom catalysis makes it easy to characterize their structure by employing advanced tools, such as solid-state nuclear magnetic resonance spectroscopy (SS-NMR), FTIR, XPS, Raman, IR, EXAFS, XANES, diffuse reflectance UV-Vis spectroscopy (DRUV-Vis), dynamic nuclear polarization-surface enhanced NMR spectroscopy (DNP-SENS), etc. Out of these, SS-NMR is one of the most effective tools to characterize well-defined heterogeneous catalysts since it provides incredibly comprehensive information on the chemical structure and dynamics of surface active sites [147]. However, XAS, including XANES and EXAFS, is a potent technique in SACs to identify local geometric (type of active metal center and coordinating atoms, coordination number, bond length) and electronic structure (oxidation state) of active species in SACs. However, in SOMCat, NMR is the highly preferred technique to characterize surface organometallic compounds due to its excellent consistency in structure determination as well as quantification. Additionally, it makes atomic-level characterization easier and provides data on dynamics.

1.4 Applications of single-atom catalysts

SACs have attracted a great deal of attention as recently developed materials in catalysis due to their extraordinary activity from the perspectives of both scientific research and industrial applications. Pt/Ru/Ir-based materials continue to be the most suitable catalysts in the majority of catalytic processes and are regarded as the benchmarks due to their strong tendency to activate the targeted reaction with minimum overpotential, high exchange current density, superior catalytic activity, and high stability in all pH range.

However, the high price and rare availability of these noble metals restrict their widespread usage as a catalyst [148]. An increase in noble material's usage, i.e., their surface-to-volume ratio, is required to reduce their total loading. As a result, these noble metal-based materials are typically produced in the form of highly distributed nanostructures. SACs have emerged as a holy grail of catalysts to boost usage even more and are occupying a significant status in various sectors owing to their unique characteristics (such as unsaturated coordination environment, maximized atomic utilization, quantum size effect, well-defined active centers, tunable, and distinctive electronic/geometric structure). They usually outperform their bulk nanoparticle/cluster counterparts in terms of catalytic performance. Hence, they have been employed in a wide range of applications (see Fig. 1.8) in thermochemical/organic transformations (selective hydrogenation, water gas shift reaction, oxidation reaction, C–C coupling reactions, CH_4 conversion, hydroformylation, methanol steam reforming, etc.), electrochemical applications [HER, OER, ORR, NRR, CO_2RR, etc.], photochemical applications (Water splitting, nitrogen fixation, CO_2 reduction),

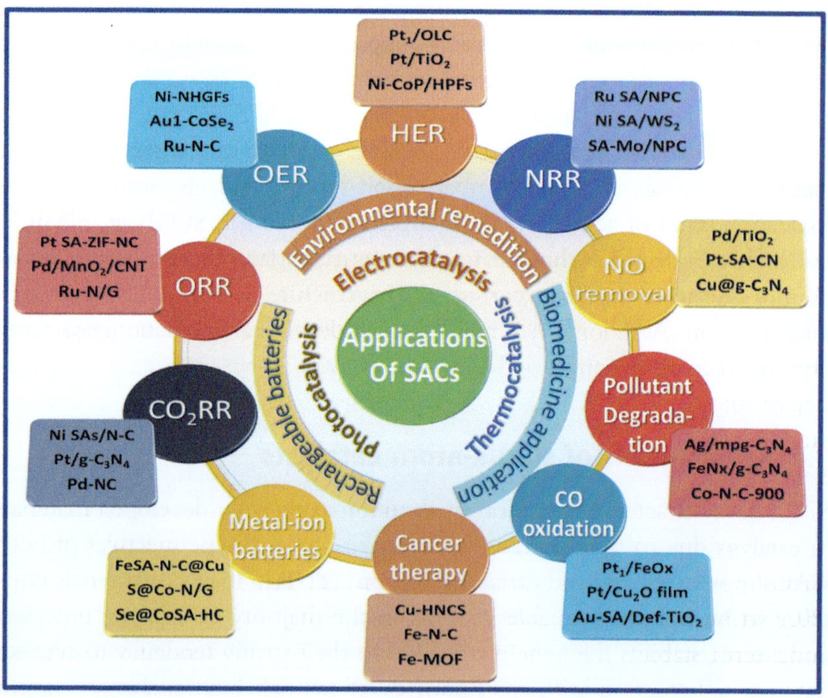

Figure 1.8 A diagram demonstrating potential areas of applications of SACs.

biomedicine (cancer therapy, biosensing technology), and environmental monitoring and remediation (CO_2RR, NO removal, pollutant degradation) [149,150]. With rising concerns over the energy crisis, environmental issues, and threats to human health and the earth's ecosystem, the fixation of gaseous contaminations (CO_2, NO) by converting harmful gases into value-added fuels or chemicals and degradation of water pollutants, such as heavy metals, dyes, volatile organic compounds, pesticides, etc., has gained extensive attention.

In this direction, SACs, including atomically dispersed Pt-group and earth-abundant elements, are playing key roles that can effectively and selectively boost the target reaction kinetics to mitigate these serious issues. Metal-nitrogen-doped carbon (M-N-C) SACs have played the most important role in substituting or lowering noble metal loadings in electrocatalysis [111,151,152]. This is due to the numerous benefits that carbon materials as supports provide, such as their large specific surface area (accommodate enough single-atom sites), high electrical and thermal conductivity (boost charge transfer), superior chemical stability (high resistance to corrosion), tunable porosity, and adaptability with dopants [111]. Commonly, M-N-C SACs are prepared by pyrolyzing metal precursors with heteroatom-doped carbon substrate and subsequent acid leaching. Maximizing the catalytic performance (activity, selectivity, and stability) of SACs is critical to fulfill future commercial/industrial/clinical requirements. Stability issues associated with SACs under varied settings are gradually becoming apparent, obstructing related applied research. The aggregation and leaching phenomenon of SACs has remained a serious issue to be tackled in the practical reaction procedures for SACs stability. Also, preventing aggregation with the increasing loading of single atoms is of utmost importance to achieve higher utilization and hence high activity. It is common knowledge that enhancing the interaction between active single atoms and supports overcomes these issues and improves the stability of SACs. To do so, we will need to select the right metals and supports. For instance, Baek's group choose dicyandiamide (DCD) as a rich nitrogen source precursor to securely trap a large number of isolated Cu atoms and protect them from aggregation. The resultant Cu-N-C SACs possess a higher loading content of Cu single atoms of up to 20.9 wt.% and display ultrahigh oxygen reduction catalytic activity and long-term stability compared to benchmark Pt/C [153]. The conversion rate, production/yield rate, Faradaic efficiency, degradation efficiency, detection limitation, overpotential/onset potential/half-wave potential, and other parameters have been utilized to assess the catalytic abilities of SACs toward certain processes in terms of activity/selectivity [150]. In this regard, appropriate ways can be employed to boost the SACs performance for desired catalytic reactions. The appropriate manipulation of chemical compositions, morphologies, and structures can lead to boosted catalytic performances toward target processes, according to experimental and theoretical investigations of SACs. It allows for the successful deployment of SACs for rapid energy conversion and storage devices, such as water electrolyzers, metal-air batteries, fuel cells, etc., reducing energy crisis and environmental issues to a large extent. Certain biomedical breakthroughs have also been documented in the field of SACs. They have been particularly useful in cancer treatment, wound healing, bioassay, and biosensing applications.

1.5 Summary and outlook

As the global need for environmentally safe energy has surged, there is a strong desire to develop high-performance catalysts at an affordable cost. Presently, catalysts based on noble metals are seen as the benchmark for numerous electrocatalytic reactions. Nevertheless, their widespread use is impeded due to their limited accessibility and high expenses. This situation has prompted the innovation and creation of noble metal-free electrocatalysts with high catalytic activity. Furthermore, research has shown that the specific shape, size, and composition of nanostructured electrocatalysts have notable impact on their catalytic activity. As most of the electrolytic reactions happen mainly at the surface and/or near-surface regions of the catalysts, downsizing catalysts to the ultimate atomic level is highly advantageous toward achieving efficient electrocatalytic activity. Such innovative idea has resulted in the development of SACs, which are actually isolated atoms of catalysts uniformly dispersed over a suitable platform called support. Over the years, SACs have emerged as a fascinating class of materials for achieving maximum catalytic efficiency by tuning the coordination environment and electronic structure of an active single metal atom. Here in this chapter, and the successive ones, design, synthesis, characterization, and several promising areas of applications regarding SACs have been highlighted. It is envisaged that such a detailed overview of SACs would give impetus to further development of cost-effective SACs through novel scalable methods for futuristic energy applications.

Acknowledgments

The author graciously thanks the scientists referred to in this chapter for their exceptional contributions to the research of electrochemical catalysis. PA gratefully acknowledges financial support from the Council of Scientific & Industrial Research (CSIR), Human Resource Development Group (HRDG), India for the Senior Research Fellowship (File No: 09/964(0025)/2019-EMR-I). DS and KA acknowledge financial support from UGC-DAE CSR Indore through projects CRS/2021–22/01/442 and CRS/2021–22/01/445, respectively. Authors are thankful to MRC, MNIT Jaipur for providing characterization facilities.

References

[1] A.K. Singh, D. Sarkar, Enhanced light absorption and charge carrier management in core-shell Fe_2O_3@nickel nanocone photoanodes for photoelectrochemical water splitting, ChemCatChem 11 (2019) 6355–6363. Available from: https://doi.org/10.1002/cctc.201901836.

[2] A.K. Singh, D. Sarkar, A facile approach for preparing densely-packed individual p-NiO/n-Fe_2O_3 heterojunction nanowires for photoelectrochemical water splitting, Nanoscale 10 (2018) 13130–13139. Available from: https://doi.org/10.1039/c8nr02508h.

[3] S. Agarwal, M.J. Ahemad, S. Kumar, D. Van Dung, P. Rai, M. Kumar, et al., Enhanced hydrogen sensing performances of PdO nanoparticles-decorated ZnO flower-like nanostructures, J. Alloy. Compd. 900 (2022) 163545. Available from: https://doi.org/10.1016/J.JALLCOM.2021.163545.

[4] S. Agarwal, S. Kumar, H. Agrawal, M.G. Moinuddin, M. Kumar, S.K. Sharma, et al., An efficient hydrogen gas sensor based on hierarchical Ag/ZnO hollow microstructures, Sens. Actuators B Chem. 346 (2021) 130510. Available from: https://doi.org/10.1016/J.SNB.2021.130510.

[5] N. Pal, N. Saini, M. Agarwal, K. Awasthi, Experimental investigation of natural polysaccharide-based mixed matrix membrane modified with graphene oxide and Pd-nanoparticles for enhanced gas separation performance, Int. J. Hydrog. Energy. (2022). Available from: https://doi.org/10.1016/J.IJHYDENE.2022.03.049.

[6] W. Ren, X. Tan, W. Yang, C. Jia, S. Xu, K. Wang, et al., Isolated diatomic Ni-Fe metal-nitrogen sites for synergistic electroreduction of CO_2, Angew. Chem. Int. Ed. Engl. 58 (2019) 6972–6976. Available from: https://doi.org/10.1002/ANIE.201901575.

[7] Z. Kou, W. Zang, P. Wang, X. Li, J. Wang, Single atom catalysts: a surface heterocompound perspective, Nanoscale Horiz. 5 (2020) 757–764. Available from: https://doi.org/10.1039/d0nh00088d.

[8] C.W. Jones, Another nobel prize for catalysis: Frances Arnold in 2018, ACS Catal. 8 (2018) 10913. Available from: https://doi.org/10.1021/ACSCATAL.8B04266.

[9] C. Zhu, Q. Shi, S. Feng, D. Du, Y. Lin, Single-atom catalysts for electrochemical water splitting, ACS Energy Lett. 3 (2018) 1713–1721. Available from: https://doi.org/10.1021/ACSENERGYLETT.8B00640.

[10] B. Lu, Q. Liu, S. Chen, Electrocatalysis of single-atom sites: impacts of atomic coordination, ACS Catal. 10 (2020) 7584–7618. Available from: https://doi.org/10.1021/ACSCATAL.0C01950.

[11] B. Chakraborty, R. Beltrán-Suito, V. Hlukhyy, J. Schmidt, P.W. Menezes, M. Driess, Crystalline copper selenide as a reliable non-noble electro(pre)catalyst for overall water splitting, ChemSusChem 13 (2020) 3222–3229. Available from: https://doi.org/10.1002/cssc.202000445.

[12] D. Sarkar, D. Das, S. Das, A. Kumar, S. Patil, K.K. Nanda, et al., Expanding interlayer spacing in MoS_2 for realizing an advanced supercapacitor, ACS Energy Lett. 4 (2019) 1602–1609. Available from: https://doi.org/10.1021/ACSENERGYLETT.9B00983.

[13] J.N. Hausmann, P.W. Menezes, Why should transition metal chalcogenides be investigated as water splitting precatalysts even though they transform into (oxyhydr) oxides? Curr. Opin. Electrochem. 34 (2022) 100991. Available from: https://doi.org/10.1016/J.COELEC.2022.100991.

[14] P. Aggarwal, D. Sarkar, P.W. Menezes, K. Awasthi, Boosting electrochemical hydrogen evolution activity of MoS_2 nanosheets via facile decoration of Au overlayer, Int. J. Hydrog. Energy 47 (2022) 41795–41805. Available from: https://doi.org/10.1016/j.ijhydene.2022.08.113.

[15] M. Chen, J. Liu, W. Zhou, J. Lin, Z. Shen, Nitrogen-doped graphene-supported transition-metals carbide electrocatalysts for oxygen reduction reaction, Sci. Rep. 5 (2015) 1–10. Available from: https://doi.org/10.1038/srep10389. 2015 51.

[16] P.W. Menezes, C. Walter, B. Chakraborty, J.N. Hausmann, I. Zaharieva, A. Frick, et al., Combination of highly efficient electrocatalytic water oxidation with selective oxygenation of organic substrates using manganese borophosphates, Adv. Mater. 33 (2021) 2004098. Available from: https://doi.org/10.1002/adma.202004098.

[17] P.W. Menezes, A. Indra, I. Zaharieva, C. Walter, S. Loos, S. Hoffmann, et al., Helical cobalt borophosphates to master durable overall water-splitting, Energy Environ. Sci. 12 (2019) 988–999. Available from: https://doi.org/10.1039/c8ee01669k.

[18] R. Beltrán-Suito, P.W. Menezes, M. Driess, Amorphous outperforms crystalline nanomaterials: surface modifications of molecularly derived CoP electro(pre)catalysts for efficient water-splitting, J. Mater. Chem. A. 7 (2019) 15749–15756. Available from: https://doi.org/10.1039/c9ta04583j.

[19] J. Wu, M. Hou, Z. Chen, W. Hao, X. Pan, H. Yang, et al., Composition engineering of amorphous nickel boride nanoarchitectures enabling highly efficient electrosynthesis of hydrogen peroxide, Adv. Mater. (2022) 2202995. Available from: https://doi.org/10.1002/ADMA.202202995.

[20] C. Panda, P.W. Menezes, S. Yao, J. Schmidt, C. Walter, J.N. Hausmann, et al., Boosting electrocatalytic hydrogen evolution activity with a $NiPt_3$@NiS heteronanostructure evolved from a molecular nickel-platinum precursor, J. Am. Chem. Soc. 141 (2019) 13306–13310. Available from: https://doi.org/10.1021/jacs.9b06530.

[21] P.W. Menezes, C. Panda, S. Garai, C. Walter, A. Guiet, M. Driess, Structurally ordered intermetallic cobalt stannide nanocrystals for high-performance electrocatalytic overall water-splitting, Angew. Chem.—Int. Ed. 57 (2018) 15237–15242. Available from: https://doi.org/10.1002/anie.201809787.

[22] P.W. Menezes, C. Walter, J.N. Hausmann, R. Beltrán-Suito, C. Schlesiger, S. Praetz, et al., Boosting water oxidation through in situ electroconversion of manganese gallide: an intermetallic precursor approach, Angew. Chem.—Int. Ed. 58 (2019) 16569–16574. Available from: https://doi.org/10.1002/anie.201909904.

[23] P.W. Menezes, S. Yao, R. Beltrán-Suito, J.N. Hausmann, P.V. Menezes, M. Driess, Facile access to an active γ-NiOOH electrocatalyst for durable water oxidation derived from an intermetallic nickel germanide precursor, Angew. Chem.—Int. Ed. 60 (2021) 4640–4647. Available from: https://doi.org/10.1002/anie.202014331.

[24] B. Chakraborty, R. Beltrán-Suito, J.N. Hausmann, S. Garai, M. Driess, P.W. Menezes, Enabling iron-based highly effective electrochemical water-splitting and selective oxygenation of organic substrates through in situ surface modification of intermetallic iron stannide precatalyst, Adv. Energy Mater. 10 (2020) 2001377. Available from: https://doi.org/10.1002/aenm.202001377.

[25] I. Mondal, J.N. Hausmann, G. Vijaykumar, S. Mebs, H. Dau, M. Driess, et al., Nanostructured intermetallic nickel silicide (pre)catalyst for anodic oxygen evolution reaction and selective dehydrogenation of primary amines, Adv. Energy Mater. (2022) 2200269. Available from: https://doi.org/10.1002/AENM.202200269.
[26] A. Indra, P.W. Menezes, I. Zaharieva, H. Dau, M. Driess, Detecting structural transformation of cobalt phosphonate to active bifunctional catalysts for electrochemical water-splitting, J. Mater. Chem. A. 8 (2020) 2637–2643. Available from: https://doi.org/10.1039/c9ta09775a.
[27] P.W. Menezes, A. Indra, C. Das, C. Walter, C. Göbel, V. Gutkin, et al., Uncovering the Nature of active species of nickel phosphide catalysts in high-performance electrochemical overall water splitting, ACS Catal. 7 (2017) 103–109. Available from: https://doi.org/10.1021/acscatal.6b02666.
[28] Y. Zheng, Y. Jiao, Y. Zhu, L.H. Li, Y. Han, Y. Chen, et al., Hydrogen evolution by a metal-free electrocatalyst, Nat. Commun. 5 (2014) 1–8. Available from: https://doi.org/10.1038/ncomms4783.
[29] Q. Zhai, Y. Pan, L. Dai, Carbon-based metal-free electrocatalysts: past, present, and future, Acc. Mater. Res. 2 (2021) 1239–1250. Available from: https://doi.org/10.1021/ACCOUNTSMR.1C00190.
[30] R.T. Hannagan, G. Giannakakis, M. Flytzani-Stephanopoulos, E.C.H. Sykes, Single-atom alloy catalysis, Chem. Rev. 120 (2020) 12044–12088. Available from: https://doi.org/10.1021/ACS.CHEMREV.0C00078.
[31] J. Li, M.F. Stephanopoulos, Y. Xia, Introduction: heterogeneous single-atom catalysis, Chem. Rev. 120 (2020) 11699–11702. Available from: https://doi.org/10.1021/acs.chemrev.0c01097.
[32] H. Zhang, G. Liu, L. Shi, J. Ye, Single-atom catalysts: emerging multifunctional materials in heterogeneous catalysis, Adv. Energy Mater. 8 (2018) 1701343. Available from: https://doi.org/10.1002/aenm.201701343.
[33] M.K. Samantaray, V. D'Elia, E. Pump, L. Falivene, M. Harb, S. Ould Chikh, et al., The Comparison between single atom catalysis and surface organometallic catalysis, Chem. Rev. 120 (2020) 734–813. Available from: https://doi.org/10.1021/ACS.CHEMREV.9B00238.
[34] Z. Li, S. Ji, Y. Liu, X. Cao, S. Tian, Y. Chen, et al., Well-defined materials for heterogeneous catalysis: from nanoparticles to isolated single-atom sites, Chem. Rev. 120 (2020) 623–682. Available from: https://doi.org/10.1021/acs.chemrev.9b00311.
[35] C. Zhu, S. Fu, Q. Shi, D. Du, Y. Lin, Single-atom electrocatalysts, Angew. Chem. Int. Ed. 56 (2017) 13944–13960. Available from: https://doi.org/10.1002/ANIE.201703864.
[36] Z. Lei, C. Sathish, Y. Liu, A. Karokoti, J. Wang, L. Qiao, et al., Single metal atoms catalysts—promising candidates for next generation energy storage and conversion devices, EcoMat. (n.d.) e12186. https://doi.org/10.1002/EOM2.12186.
[37] A.Z. Fadhel, P. Pollet, C.L. Liotta, C.A. Eckert, Combining the benefits of homogeneous and heterogeneous catalysis with tunable solvents and nearcritical water, Molecules 15 (2010) 8400–8424. Available from: https://doi.org/10.3390/MOLECULES15118400.
[38] M. Dhiman, V. Polshettiwar, Supported single atom and pseudo-single atom of metals as sustainable heterogeneous nanocatalysts, ChemCatChem 10 (2018) 881–906. Available from: https://doi.org/10.1002/CCTC.201701431.
[39] L. Zhang, Y. Ren, W. Liu, A. Wang, T. Zhang, Single-atom catalyst: a rising star for green synthesis of fine chemicals, Natl. Sci. Rev. 5 (2018) 653–672. Available from: https://doi.org/10.1093/nsr/nwy077.
[40] F. Chen, X. Jiang, L. Zhang, R. Lang, B. Qiao, Single-atom catalysis: bridging the homo- and heterogeneous catalysis, Chin. J. Catal. 39 (2018) 893–898. Available from: https://doi.org/10.1016/S1872-2067(18)63047-5.

[41] Y.S. Wei, M. Zhang, R. Zou, Q. Xu, Metal-organic framework-based catalysts with single metal sites, Chem. Rev. 120 (2020) 12089–12174. Available from: https://doi.org/10.1021/ACS.CHEMREV.9B00757.

[42] S. Mitchell, J. Pérez-Ramírez, Single atom catalysis: a decade of stunning progress and the promise for a bright future, Nat. Commun. 11 (2020) 1–3. Available from: https://doi.org/10.1038/s41467-020-18182-5. 2020 111.

[43] W. Zang, Z. Kou, S.J. Pennycook, J. Wang, Heterogeneous single atom electrocatalysis, where "Singles" are "Married", Adv. Energy Mater. 10 (2020) 1–18. Available from: https://doi.org/10.1002/aenm.201903181.

[44] H.J. Qiu, Y. Ito, W. Cong, Y. Tan, P. Liu, A. Hirata, et al., Nanoporous graphene with single-atom nickel dopants: an efficient and stable catalyst for electrochemical hydrogen production, Angew. Chem.—Int. Ed. 54 (2015) 14031–14035. Available from: https://doi.org/10.1002/anie.201507381.

[45] P. Yin, T. Yao, Y. Wu, L. Zheng, Y. Lin, W. Liu, et al., Single cobalt atoms with precise n-coordination as superior oxygen reduction reaction catalysts, Angew. Chem.—Int. Ed. 55 (2016) 10800–10805. Available from: https://doi.org/10.1002/anie.201604802.

[46] L. Zhang, J. Zhu, X. Li, S. Mu, F. Verpoort, J. Xue, et al., Nurturing the marriages of single atoms with atomic clusters and nanoparticles for better heterogeneous electrocatalysis, Interdiscip. Mater. 1 (2022) 51–87. Available from: https://doi.org/10.1002/IDM2.12011.

[47] H. Xiong, A.K. Datye, Y. Wang, Thermally stable single-atom heterogeneous catalysts, Adv. Mater. 33 (2021) 2004319. Available from: https://doi.org/10.1002/ADMA.202004319.

[48] G.M. Tomboc, T. Kim, S. Jung, H.J. Yoon, K. Lee, Modulating the local coordination environment of single-atom catalysts for enhanced catalytic performance in hydrogen/oxygen evolution reaction, Small (2022) 2105680. Available from: https://doi.org/10.1002/SMLL.202105680.

[49] J. Su, R. Ge, Y. Dong, F. Hao, L. Chen, Recent progress in single-atom electrocatalysts: concept, synthesis, and applications in clean energy conversion, J. Mater. Chem. A. 6 (2018) 14025–14042. Available from: https://doi.org/10.1039/c8ta04064h.

[50] J. Yang, W. Li, D. Wang, Y. Li, Electronic metal–support interaction of single-atom catalysts and applications in electrocatalysis, Adv. Mater. 32 (2020) 2003300. Available from: https://doi.org/10.1002/ADMA.202003300.

[51] Y. Chen, S. Ji, C. Chen, Q. Peng, D. Wang, Y. Li, Single-atom catalysts: synthetic strategies and electrochemical applications, Joule 2 (2018) 1242–1264. Available from: https://doi.org/10.1016/j.joule.2018.06.019.

[52] T. Maschmeyer, F. Rey, G. Sankar, J.M. Thomas, Heterogeneous catalysts obtained by grafting metallocene complexes onto mesoporous silica, Nat. 378 (1995) 159–162. Available from: https://doi.org/10.1038/378159a0.

[53] K. Asakura, H. Nagahiro, N. Ichikuni, Y. Iwasawa, Structure and catalytic combustion activity of atomically dispersed Pt species at MgO surface, Appl. Catal. A Gen. 188 (1999) 313–324. Available from: https://doi.org/10.1016/S0926-860X(99)00247-1.

[54] Q. Fu, H. Saltsburg, M. Flytzani-Stephanopoulos, Active nonmetallic Au and Pt species on ceria-based water-gas shift catalysts, Sci. (80-.) 301 (2003) 935–938. Available from: https://doi.org/10.1126/SCIENCE.1085721.

[55] S.F.J. Hackett, R.M. Brydson, M.H. Gass, I. Harvey, A.D. Newman, K. Wilson, et al., High-activity, single-site mesoporous Pd/Al_2O_3 catalysts for selective aerobic oxidation of allylic alcohols, Angew. Chem. Int. Ed. 46 (2007) 8593–8596. Available from: https://doi.org/10.1002/ANIE.200702534.

[56] J.H. Kwak, J. Hu, D. Mei, C.W. Yi, D.H. Kim, C.H.F. Peden, et al., Coordinatively unsaturated Al^{3+} centers as binding sites for active catalyst phases of platinum on γ-Al_2O_3, Sci. (80-.) 325 (2009) 1670–1673. Available from: https://doi.org/10.1126/SCIENCE.1176745.

[57] Q. Botao, W. Aiqin, Y. Xiaofeng, A. Lawrence F, J. Zheng, C. Yitao, et al., Single-atom catalysis of CO oxidation using Pt_1/FeOx, Nat. Chem. 3 (2011) 634–641. Available from: https://doi.org/10.1038/NCHEM.1095.

[58] W. Fu, Y. Wang, W. Tian, H. Zhang, J. Li, S. Wang, et al., Non-metal single-phosphorus-atom catalysis of hydrogen evolution, Angew. Chem. Int. Ed. 59 (2020) 23791–23799. Available from: https://doi.org/10.1002/ANIE.202011358.

[59] Z. Yongqiang, L. Tao, C. Shuangming, J. Bo, V. Anthony, J. Yan, et al., Non-metal single-iodine-atom electrocatalysts for the hydrogen evolution reaction, Angew. Chem. Int. Ed. Engl. 58 (2019) 12252–12257. Available from: https://doi.org/10.1002/ANIE.201905554.

[60] C. Xiao, R. Sa, Z. Cui, S. Gao, W. Du, X. Sun, et al., Enhancing the hydrogen evolution reaction by non-precious transition metal (Non-metal) atom doping in defective $MoSi_2N_4$ monolayer, Appl. Surf. Sci. 563 (2021) 150388. Available from: https://doi.org/10.1016/J.APSUSC.2021.150388.

[61] S. Ji, Y. Chen, X. Wang, Z. Zhang, D. Wang, Y. Li, Chemical synthesis of single atomic site catalysts, Chem. Rev. 120 (2020) 11900–11955. Available from: https://doi.org/10.1021/ACS.CHEMREV.9B00818.

[62] T. Gan, Q. He, H. Zhang, H. Xiao, Y. Liu, Y. Zhang, et al., Unveiling the kilogram-scale gold single-atom catalysts via ball milling for preferential oxidation of CO in excess hydrogen, Chem. Eng. J. 389 (2020) 124490. Available from: https://doi.org/10.1016/j.cej.2020.124490.

[63] M. Moliner, J.E. Gabay, C.E. Kliewer, R.T. Carr, J. Guzman, G.L. Casty, et al., Reversible transformation of Pt nanoparticles into single atoms inside high-silica chabazite zeolite, J. Am. Chem. Soc. 138 (2016) 15743–15750. Available from: https://doi.org/10.1021/JACS.6B10169.

[64] S. Sun, G. Zhang, N. Gauquelin, N. Chen, J. Zhou, S. Yang, et al., Single-atom catalysis using Pt/graphene achieved through atomic layer deposition, Sci. Rep. 3 (2013) 1775. Available from: https://doi.org/10.1038/srep01775.

[65] K. Qi, X. Cui, L. Gu, S. Yu, X. Fan, M. Luo, et al., Single-atom cobalt array bound to distorted 1T MoS_2 with ensemble effect for hydrogen evolution catalysis, Nat. Commun. 10 (2019) 1–9. Available from: https://doi.org/10.1038/s41467-019-12997-7.

[66] S. Ji, Y. Qu, T. Wang, Y. Chen, G. Wang, X. Li, et al., Rare-earth single erbium atoms for enhanced photocatalytic CO_2 reduction, Angew. Chem. Int. Ed. 59 (2020) 10651–10657. Available from: https://doi.org/10.1002/ANIE.202003623.

[67] Y. Xiong, W. Sun, P. Xin, W. Chen, X. Zheng, W. Yan, et al., Gram-scale synthesis of high-loading single-atomic-site Fe catalysts for effective epoxidation of styrene, Adv. Mater. 32 (2020) 2000896. Available from: https://doi.org/10.1002/ADMA.202000896.

[68] S. Fang, X. Zhu, X. Liu, J. Gu, W. Liu, D. Wang, et al., Uncovering near-free platinum single-atom dynamics during electrochemical hydrogen evolution reaction, Nat. Commun. 11 (2020) 1–8. Available from: https://doi.org/10.1038/s41467-020-14848-2.

[69] Y.Z. Zhang, C. Liang, H.P. Feng, W. Liu, Nickel single atoms anchored on ultrathin carbon nitride for selective hydrogen peroxide generation with enhanced photocatalytic activity, Chem. Eng. J. 446 (2022) 137379. Available from: https://doi.org/10.1016/J.CEJ.2022.137379.

[70] W. Zhang, Q. Fu, Q. Luo, L. Sheng, J. Yang, Understanding single-atom catalysis in view of theory, JACS Au 1 (2021) 2130–2145. Available from: https://doi.org/10.1021/JACSAU.1C00384.

[71] C. Gao, J. Low, R. Long, T. Kong, J. Zhu, Y. Xiong, Heterogeneous single-atom photocatalysts: fundamentals and applications, Chem. Rev. 120 (2020) 12175–12216. Available from: https://doi.org/10.1021/ACS.CHEMREV.9B00840.
[72] J. Liu, Catalysis by supported single metal atoms, ACS Catal. 7 (2017) 34–59. Available from: https://doi.org/10.1021/ACSCATAL.6B01534.
[73] J. Xi, H.S. Jung, Y. Xu, F. Xiao, J.W. Bae, S. Wang, Synthesis strategies, catalytic applications, and performance regulation of single-atom catalysts, Adv. Funct. Mater. 31 (2021) 2008318. Available from: https://doi.org/10.1002/ADFM.202008318.
[74] D. Liu, A. Barbar, T. Najam, M.S. Javed, J. Shen, P. Tsiakaras, et al., Single noble metal atoms doped 2D materials for catalysis, Appl. Catal. B Environ. 297 (2021) 120389. Available from: https://doi.org/10.1016/J.APCATB.2021.120389.
[75] S.K. Kaiser, Z. Chen, D. Faust Akl, S. Mitchell, J. Pérez-Ramírez, Single-atom catalysts across the periodic table, Chem. Rev. 120 (2020) 11703–11809. Available from: https://doi.org/10.1021/ACS.CHEMREV.0C00576.
[76] A. Wang, J. Li, T. Zhang, Heterogeneous single-atom catalysis, Nat. Rev. Chem. 2 (2018) 65–81. Available from: https://doi.org/10.1038/s41570-018-0010-1.
[77] H. Xu, Y. Zhao, G. He, H. Chen, Race on engineering noble metal single-atom electrocatalysts for water splitting, Int. J. Hydrog. Energy. 47 (2022) 14257–14279. Available from: https://doi.org/10.1016/J.IJHYDENE.2022.02.152.
[78] C.T. Campbell, Electronic perturbations, Nat. Chem. 4 (2012) 597–598. Available from: https://doi.org/10.1038/nchem.1412. 2012 48.
[79] L. Li, X. Chang, X. Lin, Z.-J. Zhao, J. Gong, Theoretical insights into single-atom catalysts, Chem. Soc. Rev. 49 (2020) 2869–2885. Available from: https://doi.org/10.1039/D0CS00795A.
[80] J.C. Liu, Y. Tang, Y.G. Wang, T. Zhang, J. Li, Theoretical understanding of the stability of single-atom catalysts, Natl. Sci. Rev. 5 (2018) 638–641. Available from: https://doi.org/10.1093/nsr/nwy094.
[81] W. Yang, M. Zhao, X. Ding, K. Ma, C. Wu, I.D. Gates, et al., The effect of coordination environment on the kinetic and thermodynamic stability of single-atom iron catalysts, Phys. Chem. Chem. Phys. 22 (2020) 3983–3989. Available from: https://doi.org/10.1039/c9cp05349b.
[82] L. Wu, S. Hu, W. Yu, S. Shen, T. Li, Stabilizing mechanism of single-atom catalysts on a defective carbon surface, Npj Comput. Mater. 6 (2020) 23. Available from: https://doi.org/10.1038/s41524-020-0292-y.
[83] N.J. O'Connor, A.S.M. Jonayat, M.J. Janik, T.P. Senftle, Interaction trends between single metal atoms and oxide supports identified with density functional theory and statistical learning, Nat. Catal. 1 (2018) 531–539. Available from: https://doi.org/10.1038/s41929-018-0094-5. 2018 17.
[84] K. Qi, M. Chhowalla, D. Voiry, Single atom is not alone: metal–support interactions in single-atom catalysis, Mater. Today. 40 (2020) 173–192. Available from: https://doi.org/10.1016/J.MATTOD.2020.07.002.
[85] D.N. Sredojević, M.R. Belić, Ž. Šljivančanin, Hydrogen evolution reaction over single-atom catalysts based on metal adatoms at defected graphene and h-BN, J. Phys. Chem. C. 124 (2020) 16860–16867. Available from: https://doi.org/10.1021/acs.jpcc.0c01151.
[86] Y. Yu, S. Yang, M. Dou, Z. Zhang, F. Wang, Photochemically activated atomic ruthenium supported on boron-doped carbon as a robust electrocatalyst for hydrogen evolution, J. Mater. Chem. A. 8 (2020) 16669–16675. Available from: https://doi.org/10.1039/d0ta02107e.
[87] J.N. Hausmann, P.W. Menezes, J.N. Hausmann, P.W. Menezes, Effect of surface-adsorbed and intercalated (Oxy)anions on the oxygen evolution reaction, Angew. Chem. Int. Ed. (2022). Available from: https://doi.org/10.1002/ANIE.202207279.

[88] M. Wu, G. Zhang, W. Wang, H. Yang, D. Rawach, M. Chen, et al., Electronic metal–support interaction modulation of single-atom electrocatalysts for rechargeable zinc–air batteries, Small Methods 6 (2022) 2100947. Available from: https://doi.org/10.1002/SMTD.202100947.
[89] H. Jeong, D. Shin, B.S. Kim, J. Bae, S. Shin, C. Choe, et al., Controlling the oxidation state of Pt single atoms for maximizing catalytic activity, Angew. Chem. Int. Ed. 59 (2020) 20691–20696. Available from: https://doi.org/10.1002/ANIE.202009776.
[90] L. Wang, C. Zhu, M. Xu, C. Zhao, J. Gu, L. Cao, et al., Boosting activity and stability of metal single-atom catalysts via regulation of coordination number and local composition, J. Am. Chem. Soc. 143 (2021) 18854–18858. Available from: https://doi.org/10.1021/JACS.1C09498.
[91] J. Li, Q. Guan, H. Wu, W. Liu, Y. Lin, Z. Sun, et al., Highly active and stable metal single-atom catalysts achieved by strong electronic metal–support interactions, J. Am. Chem. Soc. 141 (2019) 14515–14519. Available from: https://doi.org/10.1021/JACS.9B06482.
[92] X. Wu, B. Feng, W. Li, Y. Niu, Y. Yu, S. Lu, et al., Metal-support interaction boosted electrocatalysis of ultrasmall iridium nanoparticles supported on nitrogen doped graphene for highly efficient water electrolysis in acidic and alkaline media, Nano Energy 62 (2019) 117–126. Available from: https://doi.org/10.1016/j.nanoen.2019.05.034.
[93] K. Liu, G. Wu, G. Wang, Role of local carbon structure surrounding fen4 sites in boosting the catalytic activity for oxygen reduction, J. Phys. Chem. C. 121 (2017) 11319–11324. Available from: https://doi.org/10.1021/acs.jpcc.7b00913.
[94] K. Liu, X. Zhao, G. Ren, T. Yang, Y. Ren, A.F. Lee, et al., Strong metal-support interaction promoted scalable production of thermally stable single-atom catalysts, Nat. Commun. 11 (2020) 1–9. Available from: https://doi.org/10.1038/s41467-020-14984-9. 2020 111.
[95] S. Liang, L.C. Zou, L.J. Zheng, F. Li, X.X. Wang, L.N. Song, et al., Highly stable Co single atom confined in hierarchical carbon molecular sieve as efficient electrocatalysts in metal–air batteries, Adv. Energy Mater. 12 (2022) 2103097. Available from: https://doi.org/10.1002/AENM.202103097.
[96] P. Aggarwal, D. Sarkar, K. Awasthi, P.W. Menezes, Functional role of single-atom catalysts in electrocatalytic hydrogen evolution: current developments and future challenges, Coord. Chem. Rev. 452 (2022) 214289. Available from: https://doi.org/10.1016/j.ccr.2021.214289.
[97] J. Zhang, H. Yang, B. Liu, Coordination engineering of single-atom catalysts for the oxygen reduction reaction: a review, Adv. Energy Mater. 11 (2021) 2002473. Available from: https://doi.org/10.1002/AENM.202002473.
[98] Y. Zhang, L. Guo, L. Tao, Y. Lu, S. Wang, Defect-based single-atom electrocatalysts, Small Methods 3 (2019) 1800406. Available from: https://doi.org/10.1002/SMTD.201800406.
[99] X. Wang, H. Zhu, C. Yang, J. Lu, L. Zheng, H.P. Liang, Mesoporous carbon promoting the efficiency and stability of single atomic electrocatalysts for oxygen reduction reaction, Carbon N. Y. 191 (2022) 393–402. Available from: https://doi.org/10.1016/J.CARBON.2022.01.057.
[100] M. Kottwitz, Y. Li, H. Wang, A.I. Frenkel, R.G. Nuzzo, Single atom catalysts: a review of characterization methods, Chem.—Methods 1 (2021) 278–294. Available from: https://doi.org/10.1002/CMTD.202100020.
[101] H.T. Lien, S.T. Chang, P.T. Chen, D.P. Wong, Y.C. Chang, Y.R. Lu, et al., Probing the active site in single-atom oxygen reduction catalysts via operando x-ray and electrochemical spectroscopy, Nat. Commun. 11 (2020) 1–8. Available from: https://doi.org/10.1038/s41467-020-17975-y. 2020 111.

[102] A.D. Handoko, F. Wei, B.S. Jenndy, et al., Understanding heterogeneous electrocatalytic carbon dioxide reduction through operando techniques, Nat. Catal. 1 (2018) 922−934. Available from: https://doi.org/10.1038/s41929-018-0182-6. 2018 112.
[103] H. Yang, J.N. Hausmann, V. Hlukhyy, T. Braun, K. Laun, I. Zebger, et al., An intermetallic CaFe6Ge6 approach to unprecedented Ca − Fe − O electrocatalyst for efficient alkaline oxygen evolution reaction, ChemCatChem. (n.d.) e202200293. https://doi.org/10.1002/CCTC.202200293.
[104] R. Li, L. Luo, X. Ma, W. Wu, M. Wang, J. Zeng, Single atoms supported on metal oxides for energy catalysis, J. Mater. Chem. A. 10 (2022) 5717−5742. Available from: https://doi.org/10.1039/D1TA08016D.
[105] Z. Zhang, J. Liu, J. Gu, L. Su, L. Cheng, An overview of metal oxide materials as electrocatalysts and supports for polymer electrolyte fuel cells, Energy Environ. Sci. 7 (2014) 2535−2558. Available from: https://doi.org/10.1039/C3EE43886D.
[106] J. Paier, C. Penschke, J. Sauer, Oxygen defects and surface chemistry of ceria: quantum chemical studies compared to experiment, Chem. Rev. 113 (2013) 3949−3985. Available from: https://doi.org/10.1021/cr3004949.
[107] B. Qiao, J. Liu, Y.G. Wang, Q. Lin, X. Liu, A. Wang, et al., Highly efficient catalysis of preferential oxidation of CO in H_2-rich stream by gold single-atom catalysts, ACS Catal. 5 (2015) 6249−6254. Available from: https://doi.org/10.1021/acscatal.5b01114.
[108] Y. Pang, W. Zang, Z. Kou, L. Zhang, G. Xu, J. Lv, et al., Assembling of Bi atoms on TiO2 nanorods boosts photoelectrochemical water splitting of semiconductors, Nanoscale 12 (2020) 4302−4308. Available from: https://doi.org/10.1039/D0NR00004C.
[109] K.L. Zhou, Z. Wang, C.B. Han, X. Ke, C. Wang, Y. Jin, et al., Platinum single-atom catalyst coupled with transition metal/metal oxide heterostructure for accelerating alkaline hydrogen evolution reaction, Nat. Commun. 12 (2021) 1−10. Available from: https://doi.org/10.1038/s41467-021-24079-8. 2021 121.
[110] Y. Yang, Y. Yang, Z. Pei, K.H. Wu, C. Tan, H. Wang, et al., Recent progress of carbon-supported single-atom catalysts for energy conversion and storage, Matter 3 (2020) 1442−1476. Available from: https://doi.org/10.1016/J.MATT.2020. 07.032.
[111] M.B. Gawande, P. Fornasiero, R. Zbořil, Carbon-based single-atom catalysts for advanced applications, ACS Catal. 10 (2020) 2231−2259. Available from: https://doi.org/10.1021/acscatal.9b04217.
[112] C. Rivera-Cárcamo, P. Serp, Single atom catalysts on carbon-based materials, ChemCatChem 10 (2018) 5058−5091. Available from: https://doi.org/10.1002/CCTC.201801174.
[113] K.T. Chan, J.B. Neaton, M.L. Cohen, First-principles study of metal adatom adsorption on graphene, Phys. Rev. B—Condens. Matter Mater. Phys. 77 (2008) 235430. Available from: https://doi.org/10.1103/PHYSREVB.77.235430.
[114] A.V. Krasheninnikov, P.O. Lehtinen, A.S. Foster, P. Pyykkö, R.M. Nieminen, Embedding transition-metal atoms in graphene: structure, bonding, and magnetism, Phys. Rev. Lett. 102 (2009) 126807. Available from: https://doi.org/10.1103/PhysRevLett.102.126807.
[115] J.A. Rodríguez-Manzo, O. Cretu, F. Banhart, Trapping of metal atoms in vacancies of carbon nanotubes and graphene, ACS Nano 4 (2010) 3422−3428. Available from: https://doi.org/10.1021/nn100356q.
[116] Y. Chen, X. Zhang, Z. Zhou, Carbon-based substrates for highly dispersed nanoparticle and even single-atom electrocatalysts, Small Methods 3 (2019) 1900050. Available from: https://doi.org/10.1002/smtd.201900050.
[117] Q. Jia, N. Ramaswamy, H. Hafiz, U. Tylus, K. Strickland, G. Wu, et al., Experimental observation of redox-induced Fe-N switching behavior as a

determinant role for oxygen reduction activity, ACS Nano 9 (2015) 12496–12505. Available from: https://doi.org/10.1021/acsnano.5b05984.

[118] H. Fei, J. Dong, M.J. Arellano-Jiménez, G. Ye, N. Dong Kim, E.L.G. Samuel, et al., Atomic cobalt on nitrogen-doped graphene for hydrogen generation, Nat. Commun. 6 (2015) 1–8. Available from: https://doi.org/10.1038/ncomms9668.

[119] Y. Pan, S. Liu, K. Sun, X. Chen, B. Wang, K. Wu, et al., A bimetallic Zn/Fe polyphthalocyanine-derived single-atom Fe-N_4 catalytic site:a superior trifunctional catalyst for overall water splitting and Zn–air batteries, Angew. Chem.—Int. Ed. 57 (2018) 8614–8618. Available from: https://doi.org/10.1002/anie.201804349.

[120] J.C. Li, M. Cheng, T. Li, L. Ma, X. Ruan, D. Liu, et al., Carbon nanotube-linked hollow carbon nanospheres doped with iron and nitrogen as single-atom catalysts for the oxygen reduction reaction in acidic solutions, J. Mater. Chem. A. 7 (2019) 14478–14482. Available from: https://doi.org/10.1039/C9TA00508K.

[121] F. Yu, X. Xiong, L.Y. Zhou, J.L. Li, J.Y. Liang, S.Q. Hu, et al., Hierarchical nickel/phosphorus/nitrogen/carbon composites templated by one metal–organic framework as highly efficient supercapacitor electrode materials, J. Mater. Chem. A. 7 (2019) 2875–2883. Available from: https://doi.org/10.1039/C8TA11568K.

[122] R. Gusmão, M. Veselý, Z. Sofer, Recent developments on the single atom supported at 2D materials beyond graphene as catalysts, ACS Catal. 10 (2020) 9634–9648. Available from: https://doi.org/10.1021/acscatal.0c02388.

[123] M. Dai, R. Wang, Synthesis and applications of nanostructured hollow transition metal chalcogenides, Small 17 (2021) 2006813. Available from: https://doi.org/10.1002/SMLL.202006813.

[124] J. Zhao, J. Wang, Z. Chen, J. Ju, X. Han, Y. Deng, Metal chalcogenides: an emerging material for electrocatalysis, APL. Mater. 9 (2021) 050902. Available from: https://doi.org/10.1063/5.0049772.

[125] Y. Shi, Y. Zhou, D.R. Yang, W.X. Xu, C. Wang, F. Bin Wang, et al., Energy Level engineering of MoS_2 by transition-metal doping for accelerating hydrogen evolution reaction, J. Am. Chem. Soc. 139 (2017) 15479–15485. Available from: https://doi.org/10.1021/jacs.7b08881.

[126] H. Lu, Y. Zhang, Y. Huang, C. Zhang, T. Liu, Reaction packaging $CoSe_2$ Nanoparticles in N-doped carbon polyhedra with bifunctionality for overall water splitting, ACS Appl. Mater. Interfaces. 11 (2019) 3372–3381. Available from: https://doi.org/10.1021/acsami.8b20184.

[127] Y. Zhao, S. Wei, K. Pan, Z. Dong, B. Zhang, H.H. Wu, et al., Self-supporting transition metal chalcogenides on metal substrates for catalytic water splitting, Chem. Eng. J. 421 (2021) 129645. Available from: https://doi.org/10.1016/J.CEJ.2021.129645.

[128] Y. Zhang, Z. Mu, C. Yang, Z. Xu, S. Zhang, X. Zhang, et al., Rational design of MXene/1T-2H MoS_2-C nanohybrids for high-performance lithium–sulfur batteries, Adv. Funct. Mater. 28 (2018) 1707578. Available from: https://doi.org/10.1002/ADFM.201707578.

[129] J. Teich, R. Dvir, A. Henning, E.R. Hamo, M.J. Moody, T. Jurca, et al., Light and complex 3D MoS_2/graphene heterostructures as efficient catalysts for the hydrogen evolution reaction, Nanoscale 12 (2020) 2715–2725. Available from: https://doi.org/10.1039/c9nr09564k.

[130] H. Li, S. Wang, H. Sawada, G.G.D. Han, T. Samuels, C.S. Allen, et al., Atomic structure and dynamics of single platinum atom interactions with monolayer MoS2, ACS Nano 11 (2017) 3392–3403. Available from: https://doi.org/10.1021/acsnano.7b00796.

[131] D. Liang, Y.W. Zhang, P. Lu, Z.G. Yu, Strain and defect engineered monolayer Ni-MoS_2 for pH-universal hydrogen evolution catalysis, Nanoscale 11 (2019) 18329–18337. Available from: https://doi.org/10.1039/c9nr06541e.

[132] Z. Luo, Y. Ouyang, H. Zhang, M. Xiao, J. Ge, Z. Jiang, et al., Chemically activating MoS_2 via spontaneous atomic palladium interfacial doping towards efficient hydrogen evolution, Nat. Commun. 9 (2018) 2120. Available from: https://doi.org/10.1038/s41467-018-04501-4.
[133] J. Deng, H. Li, J. Xiao, Y. Tu, D. Deng, H. Yang, et al., Triggering the electrocatalytic hydrogen evolution activity of the inert two-dimensional MoS_2 surface via single-atom metal doping, Energy Environ. Sci. 8 (2015) 1594–1601. Available from: https://doi.org/10.1039/c5ee00751h.
[134] Y. Guan, Y. Feng, J. Wan, X. Yang, L. Fang, X. Gu, et al., Ganoderma-Like MoS_2/NiS_2 with single platinum atoms doping as an efficient and stable hydrogen evolution reaction catalyst, Small 14 (2018). Available from: https://doi.org/10.1002/smll.201800697.
[135] J. Zhang, X. Xu, L. Yang, D. Cheng, D. Cao, Single-atom Ru doping induced phase transition of MoS_2 and S vacancy for hydrogen evolution reaction, Small Methods 3 (2019) 1–9. Available from: https://doi.org/10.1002/smtd.201900653.
[136] H. Wang, J.M. Lee, Recent advances in structural engineering of MXene electrocatalysts, J. Mater. Chem. A. 8 (2020) 10604–10624. Available from: https://doi.org/10.1039/d0ta03271a.
[137] M. Pandey, K.S. Thygesen, Two-dimensional MXenes as catalysts for electrochemical hydrogen evolution: a computational screening study, J. Phys. Chem. C. 121 (2017) 13593–13598. Available from: https://doi.org/10.1021/acs.jpcc.7b05270.
[138] M. Naguib, V.N. Mochalin, M.W. Barsoum, Y. Gogotsi, 25th anniversary article: MXenes: a new family of two-dimensional materials, Adv. Mater. 26 (2014) 992–1005. Available from: https://doi.org/10.1002/ADMA.201304138.
[139] J. Peng, X. Chen, W.J. Ong, X. Zhao, N. Li, Surface and heterointerface engineering of 2d mxenes and their nanocomposites: insights into electro- and photocatalysis, Chem 5 (2019) 18–50. Available from: https://doi.org/10.1016/J.CHEMPR.2018.08.037.
[140] L. Zhao, B. Dong, S. Li, L. Zhou, L. Lai, Z. Wang, et al., Interdiffusion reaction-assisted hybridization of two-dimensional metal-organic frameworks and $Ti_3C_2T_x$ nanosheets for electrocatalytic oxygen evolution, ACS Nano 11 (2017) 5800–5807. Available from: https://doi.org/10.1021/acsnano.7b01409.
[141] H. Wang, R. Peng, Z.D. Hood, M. Naguib, S.P. Adhikari, Z. Wu, Titania Composites with 2D transition metal carbides as photocatalysts for hydrogen production under visible-light irradiation, ChemSusChem 9 (2016) 1490–1497. Available from: https://doi.org/10.1002/CSSC.201600165.
[142] J. Zhang, Y. Zhao, X. Guo, C. Chen, C.L. Dong, R.S. Liu, et al., Single platinum atoms immobilized on an MXene as an efficient catalyst for the hydrogen evolution reaction, Nat. Catal. 1 (2018) 985–992. Available from: https://doi.org/10.1038/s41929-018-0195-1.
[143] C. Copéret, A. Comas-Vives, M.P. Conley, D.P. Estes, A. Fedorov, V. Mougel, et al., Surface organometallic and coordination chemistry toward single-site heterogeneous catalysts: strategies, methods, structures, and activities, Chem. Rev. 116 (2016) 323–421. Available from: https://doi.org/10.1021/ACS.CHEMREV.5B00373.
[144] C. Copéret, M. Chabanas, R. Petroff Saint-Arroman, J.M. Basset, Homogeneous and heterogeneous catalysis: bridging the gap through surface organometallic chemistry, Angew. Chem. Int. Ed. 42 (2003) 156–181. Available from: https://doi.org/10.1002/ANIE.200390072.
[145] M.K. Samantaray, E. Pump, A. Bendjeriou-Sedjerari, V. D'Elia, J.D.A. Pelletier, M. Guidotti, et al., Surface organometallic chemistry in heterogeneous catalysis, Chem. Soc. Rev. 47 (2018) 8403–8437. Available from: https://doi.org/10.1039/C8CS00356D.

[146] R.J. Witzke, A. Chapovetsky, M.P. Conley, D.M. Kaphan, M. Delferro, Nontraditional catalyst supports in surface organometallic chemistry, ACS Catal. 10 (2020) 11822–11840. Available from: https://doi.org/10.1021/ACSCATAL.0C03350.
[147] F. Blanc, C. Copéret, A. Lesage, L. Emsley, High resolution solid state NMR spectroscopy in surface organometallic chemistry: access to molecular understanding of active sites of well-defined heterogeneous catalysts, Chem. Soc. Rev. 37 (2008) 518–526. Available from: https://doi.org/10.1039/B612793M.
[148] P.W. Menezes, C. Panda, C. Walter, M. Schwarze, M. Driess, A cobalt-based amorphous bifunctional electrocatalysts for water-splitting evolved from a single-source lazulite cobalt phosphate, Adv. Funct. Mater. 29 (2019) 1808632. Available from: https://doi.org/10.1002/adfm.201808632.
[149] J. Li, Z. Yang, Y. Li, G. Zhang, Advances in single-atom catalysts: design, synthesis and environmental applications, J. Hazard. Mater. 429 (2022) 128285. Available from: https://doi.org/10.1016/J.JHAZMAT.2022.128285.
[150] Q. Zhang, J. Guan, Applications of single-atom catalysts, Nano Res. 15 (2021) 38–70. Available from: https://doi.org/10.1007/S12274-021-3479-8. 2021 151.
[151] J.C. Li, Z. Wei, D. Liu, D. Du, Y. Lin, M. Shao, Dispersive single-atom metals anchored on functionalized nanocarbons for electrochemical reactions, Top. Curr. Chem. 377 (2019) 1–22. Available from: https://doi.org/10.1007/S41061-018-0229-9/TABLES/1.
[152] H. Li, H. Zhu, Z. Zhuang, S. Lu, F. Duan, M. Du, Single-atom catalysts for electrochemical clean energy conversion: recent progress and perspectives, Sustain. Energy Fuels. 4 (2020) 996–1011. Available from: https://doi.org/10.1039/C9SE01004A.
[153] F. Li, G.F. Han, H.J. Noh, S.J. Kim, Y. Lu, H.Y. Jeong, et al., Boosting oxygen reduction catalysis with abundant copper single atom active sites, Energy Environ. Sci. 11 (2018) 2263–2269. Available from: https://doi.org/10.1039/C8EE01169A.

CHAPTER 2

Synthesis techniques for single-atom catalysts

Carsten Walter[1] and Prashanth W. Menezes[1,2]
[1]Materials Chemistry Group for Thin Film Catalysis—CatLab, Helmholtz-Zentrum Berlin für Materialien und Energie, Berlin, Germany
[2]Department of Chemistry, Technische Universität Berlin, Berlin, Germany

2.1 Introduction

Environmental and climatic challenges, as well as diminishing capabilities to meet the global energy demand, are inevitably connected to the continuous increase of world population and industrial progress rooted in the dependence on fossil fuels and irresponsible mining from natural resources [1]. Producing green energy, deriving higher-valued chemicals from previously unused fast renewing sustainable recourses to conserve or even abandon traditional fossil-based resources, and fully recycling and reusing products/by-products from production chains are major challenges in future catalytic applications [2]. Electrochemical methods to drive catalytic reactions such as the CO_2 reduction reaction (CO_2RR) [3], hydrogen evolution reaction (HER) [4], hydrogen oxidation reaction (HOR) [5], oxygen evolution reaction (OER) [6], oxygen reduction reaction (ORR) [7], water electrolysis [8], and production of high value-added chemicals [9] using renewable energy from solar, wind, and tidal power plants are regarded as an environmental-friendly, sustainable, and even cost-efficient strategy. At the current state of research, costly noble metal-based catalysts (e.g., Au, Pt, Pd, Rh, Ru, Ir) still dominate a widespread of these catalytic reactions as benchmarking systems mostly preventing large-scale commercial applications [10,11]. Therefore, great efforts have been made to design and synthesize novel cost-effective, robust electrocatalysts with high catalytic activity and long-term stability, with comparable catalytic properties to that of the noble metals [10]. Many researchers have spent the last decade developing new nanostructured materials with definite sizes, shapes, and compositions based on various earth-abundant and nonprecious materials to enhance stability, activity, and selectivity. Most of the investigated catalysts, such as alloys [12], intermetallics [13,14], chalcogenides [15], pnictides [16], borophosphates [17], phosphides

Single Atom Catalysts
DOI: https://doi.org/10.1016/B978-0-323-95237-8.00006-9

[18], metal-free-based electrocatalysts [19], etc., have shown promising results but still lack high performances required for industrial application. In this regard, single-atom catalysts (SACs) and single-alloy atom catalysts (SAACs) dispersed on various supports have attracted significant attention [20]. Combining the advantages of both homogeneous and heterogeneous catalysts, SACs expose a high number of highly dispersed catalytic active single-atom sites on the surface with nearly 100% metal atom utilization efficiency and are therefore highly advantageous compared to nanoparticulate systems [21]. Since the catalytic reaction occurs at the surface or interfaces of catalysts, only the surface atoms can act as active sites for the reaction and most of the atoms in nanostructured materials do not participate during the reaction [22]. Because of their high activity, selectivity, and stability in catalytic reactions such as CO oxidation [23], CO_2RR [24], and water-gas shift reaction [25], SACs have gained enormous interest and are considered as potential candidates for electrocatalytic application. Recent results have shown that the SACs were also able to catalyze the reactions of HER [26], OER [27], ORR [28], and water electrolysis [29] with good performances [30–32]. To increase the activity further and to improve the efficiency, stability, and selectivity of SACs, several strategies such as modification of the electronic structure, composition, shape, and size on an atomic level have been applied [20]. Lately, apart from SACs, SAACs, a class of single-site catalysts where isolated dispersed metal atoms are hosted on a surface layer of metal have been introduced in the field. They usually combine single atoms of a catalytically active metal alloyed into the surface of less reactive/inactive support. Opposite to the more empiric approach of traditional catalyst synthesis, SAACs emerged from studies of model systems in the field of surface science by scanning probe studies of single crystal samples and to gain knowledge that has been transferred into the catalyst synthesis, characterization, and probing [33]. In this chapter, we focus on various synthetic strategies to access SACs and SAACs for manifold energy applications. The synthetic approaches have been divided into wet chemical approaches, the most common approach to SACs, followed by more energy-intensive methods such as atomic layer deposition (ALD), hydrothermal, and solid-state methods. Since it is impossible to cover all methods in great detail, the reader is encouraged to read the relevant publications for further information.

2.2 Synthetic strategies of single-atom catalysts

Over the last few years, enormous efforts have been dedicated toward the synthesis of SACs for various catalytic applications. However, their stability, recyclability, and scaleup remain significant challenges. Therefore, the rational synthesis of SACs continues to be a task to prevent aggregation of dispersed metal atoms and maintain the structure during catalysis. The variety of synthetic approaches for SACs and SAACs include mixing and activation either wet chemically or in the solid state, ALD, high-temperature vapor transport, galvanic replacement, photochemical reduction, pyrolysis, and hydro(solvo)thermal methods [34]. To address the regulation of the coordination environment, which is crucial for the high catalytic performance of SACs, metal–organic frameworks (MOFs) have been extensively investigated as support materials and as carbon support templates for the synthesis of SACs due to their, highly structured and functional flexibilities. This is because MOFs comprise of an ultrahigh surface area, with a high degree of structural diversity, and compositional variability, which allows the incorporation of diverse catalysts for desired catalytic reactions [35]. In the following sections, we systematically describe various synthetic techniques that have been applied in the field of energy conversion and storage applications.

2.2.1 Wet chemical approaches

The most applied approach to SACs is the mixing-and-activation method. Wet impregnation, (co-)incipient wetness impregnation and (co-)precipitation accomplish the mixing step, and activation, which is usually achieved by thermal treatment under inert or reactive gas. These are well-established and simple techniques for the preparation of SACs and even SAACs [34,36,37]. Here, the precursors are impregnated into a porous framework or onto the surface of the support material [38]. The product is then dried and usually further treated through calcination and/or reduction to obtain the desired catalyst. Optimal synthesis parameters, e.g., the support material, nature of precursor, precursor concentration, temperature, surfactants, the degree of precursor solubility in solvents, and decomposition features are important to adjust, since the size of the active metal (oxide) particles, and their accessibility and distribution over the support are strongly influenced by these parameters [38–40].

2.2.1.1 Metal oxide support

The support materials for SACs are often chosen due to their physical properties. Metal oxides can be easily surface reduced to form cations of lower valence, and often show strong metal–support interaction (SMSI), which is considered a vital factor for the improvement of catalytic activity and chemisorption properties of the supported metal catalyst [41]. Studies have shown that the SMSI is connected with synthetically induced electronic defects in structures of ceria (CeO_2), iron oxide (FeO_x), and titania (TiO_2) [42–44]. On the other hand, the SMSI is not present in alumina (Al_2O_3) and silica (SiO_2), due to the lack of electronic defects in these hard-to-reduce materials. They are most widely used as catalyst support due to their thermal stability and tuneable reactivity [42,45]. Very recently Zhang and coworkers proposed that single metal atoms can be stabilized through a strong covalent metal–support interaction (CMSI) that is not associated with surface defects. Although usually smaller metal crystallites disperse as adatoms, migrating toward larger neighboring particles via Ostwald Ripening, to lower the total surface energy, in their case the opposite was observed. They concluded that isolated Pt atoms are rather stabilized by strong covalent interactions with iron and oxygen atoms on the surface than by surface defects. This effect enables the disintegration of high concentrations of Pt (NPs) to form thermally stable single atoms, even on low-surface-area materials [46,47]. Based on these findings that metal oxide support can promote the dispersion of noble metals and efficiently anchor and stabilize single-site atoms as well as effectively assist in electrochemical conversion, the group of Shao prepared the Ce-modified Fe-N-doped carbon nanowire (Ce/Fe-NCNW). Their ceria-assisted strategy favored the formation of atomically dispersed Fe active sites in an N-doped carbon matrix where ceria spatially confines, traps, and stabilizes the Fe atoms, resulting in a high iron loading of 4.6 wt.% in the final Fe-N-C material for efficient ORR, see Fig. 2.1A for the depiction of the ORR process.

The typical procedure for Ce/Fe-NCNW involved dispersing cetyltrimethylammonium bromide (CTAB) in a 1 M HCl solution under an ice bath and adding ammonium peroxydisulfate (APS) resulting in the formation of a precipitate. To the formed mixture, pyrrole monomer was added dropwise and the mixture was aged to allow polymerization, resulting in polypyrrole nanowires. The polypyrrole nanowires were dispersed in a solution of $Ce(NO_3)_3$ and $Fe(NO_3)_3$, forming a new mixture that was aged for 24 h to allow the Ce^{3+} and Fe^{3+} cations to be adsorbed by the wires. The doped polypyrrole nanowires were filtered, dried, and finally pyrolyzed at 900°C under Ar for 30 min and under NH_3 for 30 min to obtain Ce/Fe-NCNW, as depicted in Fig. 2.1B and C. The Ce/Fe-NCNW catalyst was magnificently active for ORR and showed great performance as a cathode catalyst in anion exchange membrane fuel cells

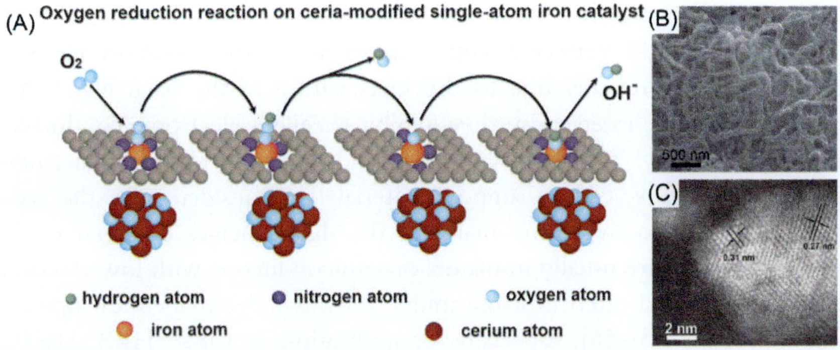

Figure 2.1 (A) Schematic representation of ORR by Ce/Fe-NCNW. (B) Scanning electron microscopy (SEM) image of the Ce-modified Fe-N-doped carbon nanowire (Ce/Fe-NCNW) (C) Z-contrast image of the Ce/Fe-NCNW depicting the isolated metal atoms around the ceria nanoparticles. *Modified with permission from J.-C. Li, S. Maurya, Y.S. Kim, T. Li, L. Wang, Q. Shi, et al., Stabilizing single-atom iron electrocatalysts for oxygen reduction via ceria confining and trapping, ACS Catal. 10 (2020) 2452–2458. https://doi.org/10.1021/acscatal.9b04621; copyright 2020, American Chemical Society.*

[48]. Taking advantage of the base active sites of layered double hydroxides (LDHs) that provide special anchoring sites for the supported noble metal atoms like Au, Sun et al. prepared monatomic Ru sites anchored on cobalt iron LDHs (CoFe-LDHs) nanosheets for the OER. CoFe-LDHs nanosheets prepared by a coprecipitation were slowly added into a pH = 12 adjusted ruthenium chloride ($RuCl_3 \cdot H_2O$) solution. The mixture was stirred at room temperature for 12 h and finally centrifuged, washed with ethanol, and dried at 60°C to give the Ru/CoFe-LDHs with 0.45 wt.% Ru loading. Higher loadings resulted in aggregation and decreased performance in electrocatalysis [49]. Learning from the challenges to prepare SACs without aggregation of the single atoms, which usually afford low loading of the SAC on the support, the group of Gu modified the wet chemical approach to attain the Ir-NiO/CC with a loading of ~18 wt.% of Ir on the nickel oxide (NiO) matrix. They immersed NiO decorated on carbon cloth (NiO/CC) in a solution of chloroiridic acid ($H_2IrCl_6 \cdot 6H_2O$) in ethanol. The dried sample was transferred into a muffle furnace and calcined at 350°C for 2 h, and after washing with water and drying, the Ir-NiO/CC was obtained. The loading of Ir single atoms was adjusted by changing the concentration of $H_2IrCl_6 \cdot$ in ethanol yielding the $Ir_{2\ wt.\%}$-NiO, $Ir_{11\ wt.\%}$-NiO, and $Ir_{18\ wt.\%}$-NiO. They successfully applied their SAC for the electrocatalytic OER with superior activity when compared to the IrO_2 counterpart [50].

2.2.1.2 Carbon support

Though the SMSI between metal centers and carbon support has not been reported, probably due to the inert nature of the basal plane that inhibits electronic interaction, they exhibit excellent electronic conductivity, high porosity, a high surface area, and relatively chemical inertness [51,52]. Therefore, carbon support materials are considered for the synthesis of SACs to overcome many of the disadvantages of metal oxide supports, which are usually insulators or semiconductors with low electron conductivities and often unstable under corrosive electrochemical operating conditions [53–55]. Especially for applications in OER, HER, ORR, zinc-air batteries, and fuel cells SACs on carbon support with various morphologies such as nanowires [54], sheets [55], layered structures [56] nanotubes [57] have been recently reported due to their excellent stability performances and physical properties [58–65].

The CoN_x/C electrocatalyst for efficient HER in alkaline media was prepared by mixing a solution of cobalt/*o*-phenylenediamine composites (Co-*o*PD) in a 1.0 M HNO_3 with a silica colloid solution (HS40, ~12 nm, 40 wt.%), as depicted schematically in Fig. 2.2A. After aging and drying, the powder was pyrolyzed in the N_2 atmosphere followed by etching out the silica template with 2.0 M NaOH. For the final product, the sample was heated in the N_2 atmosphere forming a CoNPs/CoN_x/C composite that was again etched in 0.5 M H_2SO_4 to remove the cobalt-containing particles and heated again in the N_2 atmosphere, leading to homogeneously atomically dispersed Co single atoms, see Fig. 2.2B [66]. Meanwhile, the group of Xie prepared a Pt-CN catalyst for photocatalytic hydrogen evolution. They prepared the catalyst by adding the H_2PtCl_6 to an aqueous dispersion of g-C_3N_4, after aging, washing, and annealing at 125°C, the final product was obtained as Pt-CN. By expanding the aging time, different loadings of Pt can be achieved, but when the loading exceeded 0.16 wt.%, sub-nanometer clusters of Pt were formed [68]. Similarly, the group of Lu presented a low-cost approach to SAC electrocatalyst for OER and ORR by using wood as a template to synthesize the SAC-FeN-WPC-T. In the described procedure, they impregnated dried wood with an aqueous solution of $FeCl_3$ (see Fig. 2.2C for the schematic of the synthesis). The dried $FeCl_3$-wood was then calcined at various temperatures in an Ar-NH_3 (10% NH_3) atmosphere. For the final product, the sample was etched in a 1 M HCl solution to remove inactive Fe species achieving a final Fe loading of 0.8 wt.% on the carbonized

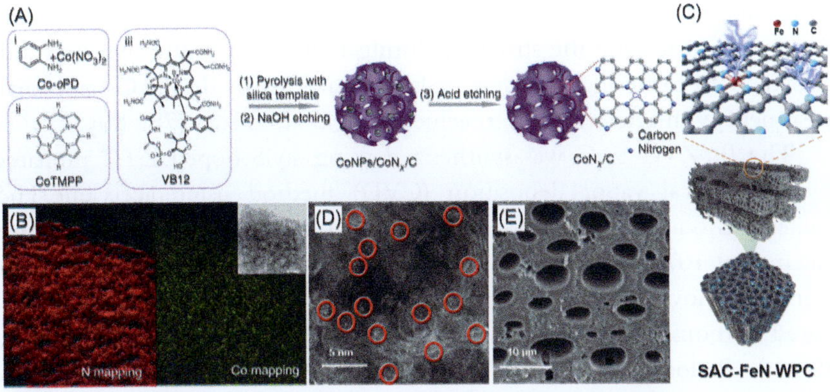

Figure 2.2 (A) Schematic representation of the synthesis of CoN_x/C. (B) Energy-filtered transmission electron microscopy (EFTEM) of CoN_x/C with the elemental mapping and homogeneous distribution of nitrogen (*red*) and cobalt (*green*) at the atomic scale with TEM image as inset. (C) Scheme for the synthetic protocol of SAC-FeN-WPC. (D) High-angle annular dark-field scanning transmission electron microscopy (HAADF-STEM) image of SAC-FeN-WPC with the red circle marking the iron atoms (E) SEM image of SAC-FeN-WPC. *(B) Reprinted with permission from H.-W. Liang, S. Brüller, R. Dong, J. Zhang, X. Feng, K. Müllen, Molecular metal-Nx centres in porous carbon for electrocatalytic hydrogen evolution, Nat. Commun. 6 (2015) 7992. https://doi.org/10.1038/ncomms8992 copyright 2015, Nature Publishing Group (E) Modified with permission from L. Zhong, C. Jiang, M. Zheng, X. Peng, T. Liu, S. Xi, et al., Wood carbon based single-atom catalyst for rechargeable Zn—air batteries, ACS Energy Lett. 6 (2021) 3624—3633. https://doi.org/10.1021/acsenergylett.1c01678; copyright 2021, American Chemical Society.*

wood, with atomically dispersed Fe atoms, see Fig. 2.2D inside a porous carbonized structure, see Fig. 2.2E [67].

Apart from conventional activated carbons, carbon templates such as MOFs and zeolite-templated carbons (ZTCs) possess larger surface areas ($>2000\ m^2\ g^{-1}$) and higher micropore volumes ($>1.0\ cm^3\ g^{-1}$), with superior electrochemical performances and gas adsorption properties and have gained enormous interest as support materials for SACs [69,70]. These materials combine adjusted pore metrics, and predesigned binding sites and possess ligands that can serve to spatially separate the metal nodes to inhibit metal atom migration and hence metal aggregation [69—71]. Therefore, MOFs and ZTCs represent a benign choice as templates and precursors to create porous carbon supports without additional carbon sources [69,70,72]. For several catalyst materials based on Fe, Co, Ni, Cu, Co-Zn, Ni-Zn, Cu-Zn, etc., ZIF and MOF templates were pyrolyzed at high temperatures to derive SACs for various applications [73—77].

Bimetallic zinc-containing MOFs have been found to form N-doped carbon with Zn leaving the structure during high-temperature pyrolysis, but also tend to form agglomerates of larger particles in the SACs and need treatment with acid leaching to achieve single-atom sites [78–80].

The Pt/ZTC SAC was synthesized using an S-doped ZTC prepared by a chemical vapor deposition (CVD) method using acetylene/H_2S mixed gas on NaX zeolite at 823K. After CVD, the sample was further calcined at 1073K in H_2S/He and treated with an HCl/HF aqueous solution to remove the zeolite template. The Pt was supported on the prepared carbons by dispersing the carbon in water containing H_2PtCl_6 (5 wt.% Pt loading) followed by removing the water and reducing the powder in the H_2 atmosphere. The Pt/ZTC selectively catalyzes a two-electron ORR pathway generating H_2O_2 and is considered a promising catalyst for H_2O_2 production [53]. The group of Yao synthesized A-Ni-C with single Ni atom species well dispersed in hollow onion-like carbon nanoshells derived from a MOF precursor for the electrochemical HER, as shown in the synthesis scheme of Fig. 2.3A. The Ni-MOF was synthesized by mixing Ni(L-asp)$(H_2O)_2 \cdot H_2O$ and 4,4′-bipyridyl in water and methanol transferred into an autoclave and heated at 150°C for 48 h. The Ni-MOF was carbonized by heating the sample in the N_2 atmosphere to yield the Ni@C. Further etching with HCl, and activation by electrochemically applied constant potential, and cycling led to single dispersed Ni atoms in A-Ni-C [81].

Iron-based catalysts have emerged as a promising alternative to noble metal catalysts in ORR. To enhance the catalytic activity by increasing the number of exposed active sites, SACs give the maximum atom efficiency and expose the most active sites in a catalyst. To prevent migration and agglomeration during the catalytic process, preparing SACs from suitable templates such as MOFs or ZIFs is a common strategy [82–84].

The electrochemical active Fe/N-GPC for ORR was prepared by using a MOF (MIL-101-NH_2) template with a high surface area and large pore size to accommodate dicyandiamide (DCD) as the nitrogen source with a high nitrogen content and iron chloride ($FeCl_3$) for the active centers. In a typical procedure, the MIL-101-NH_2 suspended in *n*-hexane was mixed with a solution of dimethylformamide (DMF) with DCD and $FeCl_3$ followed by aging and drying in a vacuum to yield the FeD@MIL-101-NH_2, as illustrated in Fig. 2.3B. Further, the FeD@MIL-101-NH_2 was calcined in an Ar atmosphere at various temperatures and unstable components were removed afterward by immersing the samples

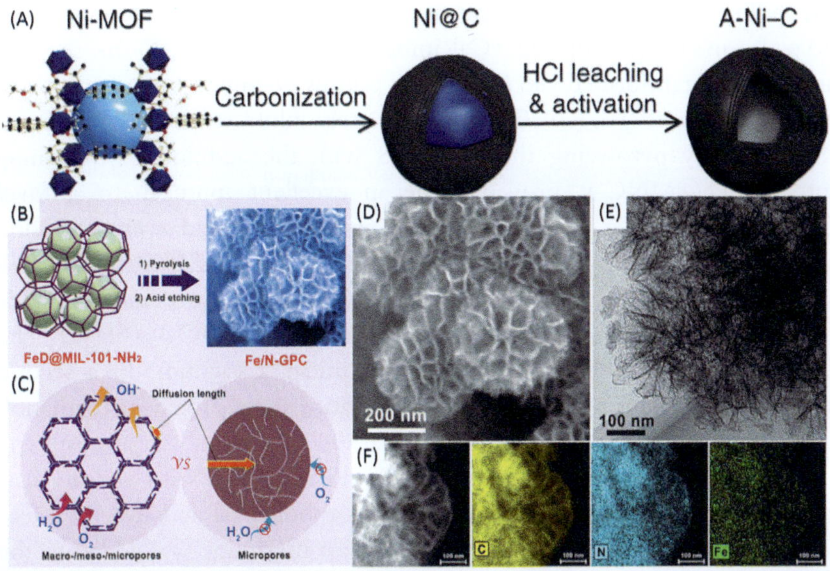

Figure 2.3 (A) Scheme for the synthesis and activation of the A-Ni-C catalysts (C: *black*, H: *white*, O: *red*; N: *dark blue*; Ni: *light blue*). (B) Scheme for the synthesis of the Fe/N-GPC and (C) the hierarchically porous structure advantageous to promote the ORR. (D) SEM, (E) TEM, and (F) HAADF-STEM images of the Fe/N-GPC ORR catalyst with elemental mapping (F) showing the elemental distribution. *(A) Reprinted with permission from L. Fan, P.F. Liu, X. Yan, L. Gu, Z.Z. Yang, H.G. Yang, et al., Atomically isolated nickel species anchored on graphitized carbon for efficient hydrogen evolution electrocatalysis, Nat. Commun. 7 (2016) 10667. https://doi.org/10.1038/ncomms10667, copyright 2016, Nature Publishing Group; (F) Reprinted with permission from Q.-L. Zhu, W. Xia, L.-R. Zheng, R. Zou, Z. Liu, Q. Xu, Atomically dispersed Fe/N-doped hierarchical carbon architectures derived from a metal–organic framework composite for extremely efficient electrocatalysis, ACS Energy Lett. 2 (2017) 504–511. https://doi.org/10.1021/acsenergylett.6b00686. copyright 2017, American Chemical Society.*

in the HF (20 wt.%) solution. After final washing and drying in a vacuum, the electrocatalyst Fe/N-GPC was obtained. The mainly atomically dispersed sample also showed fractions of aggregated Fe_3C particles that most likely were protected by the graphitic layers on the particle surface during the acid treatment [70]. From a porphyrinic MOF (PCN-222) containing one-dimensional mesochannels, the Fe_{SA}-N-C was derived as an ORR catalyst in alkaline and acidic media. By varying the ratio of the precursors Fe-TCPP (TCPP = tetrakis (4-carboxyphenyl)porphyrin) and H_2-TCPP isostructural MOFs, denoted as Fe_x-PCN-222 with x depicting the molar percentage were synthesized. The optimal conditions to assemble Fe-TCPP into 3D networks while preventing molecular stacking and expanding the distance of adjacent Fe-TCPP ligands was found to be 20 wt.% in Fe_{20}-PCN-222. Calcining the Fe_{20}-PCN-222 in the N_2 atmosphere transformed the MOF into porous N-doped carbon with single Fe atom sites (Fe_{SA}-N-C) and completely inhibited particle formation [85]. Directly pyrolyzing the Fe-ZIF-8 with the addition of melamine into the Fe@MNC-x, proved to be an excellent approach to a novel ORR electrocatalyst with high activity and stability, and even surpassing the performance of 20 wt.% commercial Pt/C. The Fe-ZIF-8 was synthesized by dissolving 2-methylimidazole and $Fe(acac)_3$ in methanol and adding a second solution of methanol with dissolved $Zn(NO_3)_2 \cdot 6H_2O$. The obtained product was washed and dried under a vacuum. In a second step, Fe-ZIF-8 and melamine in various ratios were mixed in methanol, dried, and heated in a tube furnace at 900°C in an Ar atmosphere. Depending on the mixture of Fe-ZIF-8 and melamine with various mass ratios, the resulting Fe@NMC-x of the morphological and structural features in the product were adjusted. When no melamine was added to the Fe-ZIF-8 during pyrolysis, resulting in Fe@NMC-0, the structural properties of the Fe-ZIF-8 were maintained, with an intact dodecahedral shape, smooth surfaces, and solid interior. Continuously increasing the melamine ratio led to distinct interior cavities and rougher surfaces, and more interconnected pores as well as a thinner carbon layer until interconnected porous structures with few carbon nanotubes (CNTs) assembling on its irregular polyhedron surface were formed. In excess of melamine, the original Fe-ZIF-8 dodecahedron was completely disposed of, and instead, 3D-connected CNTs were observed [86]. The single-site copper electrocatalyst (Cu-SA/NC(meso)) for ORR was prepared with a ZIF-8 template in several consecutive synthesis steps (see Fig. 2.4A). At first, the ZIF-8 was derived by dissolving 2-methylimidazole in a mixture of dimethylformamide (DMF), methanol (MeOH), and ethanol (EtOH) (3:1:1), and slowly adding a solution of $Zn(NO_3)_2 \cdot 6H_2O$ in DMF and MeOH (1.5:1). The formed precipitate was washed and dried in vacuum to give the ZIF-8. The ZIF-8 and MeIM were then dispersed in H_2O and EtOH (1.5:1) and cetyltrimethylammonium chloride solution (CTAC) was added followed by adding tetraethyl orthosilicate (TEOS). After aging, washing, and drying in a vacuum, the product ZIF-8@mesoSiO_2 was obtained.

The mesoporous nitrogen-doped carbon NC(meso) was prepared by calcining ZIF-8@mesoSiO_2 in an Ar atmosphere and etching the product with aqueous HF (10 wt.%) to remove the SiO_2. In the final step, the

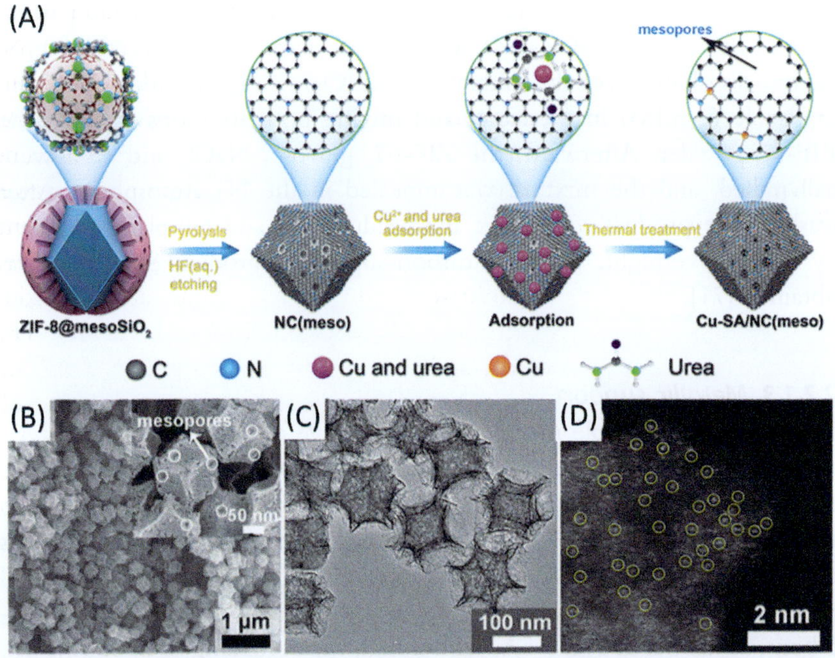

Figure 2.4 (A) Schematic representation for the formation of the Cu-SA/NC(meso) catalysts. (B) SEM image of Cu-SA/NC(meso) and (C) TEM image of the porous structure. (D) AC-HAADF-STEM images of Cu-SA/NC(meso) with atomically dispersed Cu atoms. *Modified with permission from T. Sun, Y. Li, T. Cui, L. Xu, Y.-G. Wang, W. Chen, et al., Engineering of coordination environment and multiscale structure in single-site copper catalyst for superior electrocatalytic oxygen reduction, Nano Lett. 20 (2020) 6206–6214. https://doi.org/10.1021/acs.nanolett.0c02677; copyright 2020, American Chemical Society.*

Cu-SA/NC(meso)-x was synthesized by dissolving $Cu(NO_3)_2 \cdot 3H_2O$ and urea in water and adding NC(meso) to the solution. The product was washed, dried in a vacuum, and transferred into a tube furnace for calcination in an Ar atmosphere, resulting in a porous structure with atomically dispersed Cu atoms, as depicted in Fig. 2.4B–D. Similarly, a nickel SAC Ni-SA/NC(meso) was derived by using $Ni(NO_3)_2 \cdot 6H_2O$ [87]. Lately, the cobalt-based SAC (SGZ-800) was prepared from zeolitic imidazolate frameworks-67 (ZIF-67) by controlling the proportions of ZIF-67 and glucose precursors in a molten salt medium. The Co atoms in the ZIF-67 were transformed by calcination into $Co{-}N_3$ coupling vacancies ($Co{-}N_3$-V) providing unique composition and structure properties that were further employed in triiodine reduction reaction (TRR) in dye-sensitized solar cells (DSSCs) with excellent photovoltaic performance. To attain the SGZ-800, first, the ZIF-67 was synthesized by separately dissolving cobalt nitrate ($Co(NO_3)_2 \cdot 6H_2O$) and 2-methyl imidazole (2-mIM) in methanol and mixing both solutions to yield the ZIF-67 powder. Afterward, the ZIF-67, glucose, NaCl, and KCl were ball milled, and the mixture was annealed in the N_2 atmosphere. After washing the product to remove the residual salts and template, followed by drying overnight, the final carbon-supported product SGZ-800 was obtained [71].

2.2.1.3 Metallic support

Metals, especially noble metals, are an attractive choice as support materials since atomically dispersed noble metals on the surface of other noble metal substrates exhibit enhanced catalytic properties mainly connected to the synergistic effects of the isolated atoms and metal substrates. In this context, a novel Ru SAC was reported where a one-pot wet chemical procedure was used to synthesize single Ru atoms on ultrathin Pd nanoribbons, as shown in Fig. 2.5A, with a high metal loading of about 5.9 wt. %. The nanoribbons were composed of uniformly dispersed Ru atoms on the Pd matrix (Fig. 2.5B and C). The ultrathin Pd/Ru nanoribbons showed higher catalytic performance in the selective hydrogenation of allyl benzyl ether than that of commercial Ru/C and Pd/C catalysts, without any hydrogenolysis of the O-benzyl group [88].

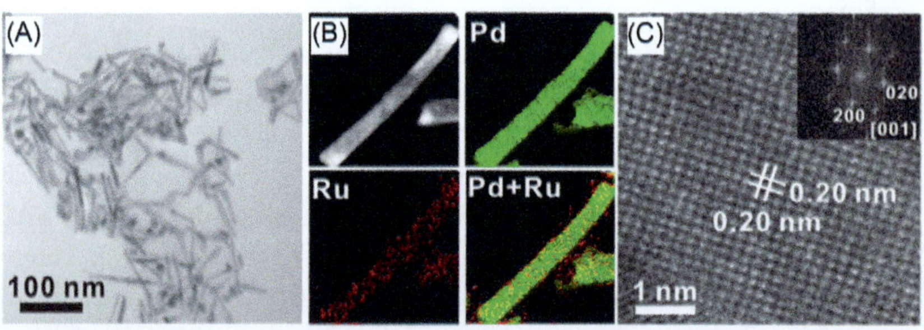

Figure 2.5 (A) TEM and (B) HAADF-STEM image of a Pd/Ru nanoribbon and the corresponding STEM-EDX elemental mapping. (C) HAADF-STEM image of a single nanoribbon with the corresponding FFT pattern (inset). *Modified with permission from J. Ge, D. He, W. Chen, H. Ju, H. Zhang, T. Chao, et al., Atomically dispersed Ru on ultrathin Pd nanoribbons, J. Am. Chem. Soc. 138 (2016) 13850–13853. https://doi.org/10.1021/jacs.6b09246 copyright 2016, American Chemical Society.*

Metallic TiN support was used by Soon and coworkers to stabilize single Pt atoms on N-vacancy sites in the material used in ORR to selectively produce H_2O_2 and application in fuel cells [89,90]. Further, it was revealed that the support may not be innocent in the electrochemical methanol oxidation since the TiN surface may possess a bi-functional effect by providing hydroxy species. Single-atom Pt on acid-treated TiN NPs was prepared by dissolving H_2PtCl_6 $6H_2O$ in ethanol and mixing it with TiN NPs. The vacuum-dried powder was reduced in H_2/N_2 flow and washed again in ethanol and dried. By adjusting the Pt loading with the concentration of the precursor either Pt NPs or atomically dispersed Pt was observed on the surface. When the amount of Pt was as low as 0.35 wt.%, mainly atomically dispersed fractions were observed with only minor content of agglomerated Pt NPs [89]. The interaction between Pt and α-MoC was identified as a key element in the catalytic activity of Pt/MoC. The ability of α-MoC to induce water dissociation and the synergistic activation of methanol by Pt and α-MoC was responsible for the exceptional activity for H_2 production from methanol. The Pt/MoC was synthesized by mixing the Mo_2C support with an aqueous solution of H_2PtCl_6. The mixture was aged, dried, and activated in a CH_4/H_2 (15/85) atmosphere at 300°C for 1 h and additionally at 590°C for 2 h [91].

The wet impregnation method has also been applied to attain SAACs supported on various substrates. The method is conveniently simple, and a desired composition and amount of metal precursor solution is used to fill the pores of the support. But, it generally yields NPs with large average sizes and broad size distributions, and, in the case of multiple metals, inhomogeneous alloying is observed [92]. With modified protocols, however, various ultrasmall alloy nanoparticles (NPs), e.g., Ni-Co, Cu-Co, Cu-Ni, Pt-Co, Pt-Ni, Pt-Cu, Pt-Pd, Co-Pd, Ni-Pd, Pd-Cu, Pd-Zn, and Cu-Pd were accessible using this method [92,93].

2.2.2 Atomic layer deposition

The ALD method is an important technique for thin-film fabrication for various applications. ALD allows atomic-level controlled homogeneous deposition using sequential, self-limiting surface reactions that facilitate the growth of conformal thin films with an accurate thickness on large areas. The film thickness can be controlled by the number of deposition cycles allowing multilayer structures uniformly deposited to form homometallic films or composites with various morphologies and high reproducibility.

Suitable precursors for ALD are volatile and thermally stable, which can be gases, liquids, or solids. To ensure moderate deposition rates and short saturation stages, the precursors chemisorb and react aggressively at the surface or react fast directly with the surface groups [94,95]. The challenge in this method remains to precisely create and tune the anchor sites on the substrate and ensure optimal loading with enough active sites to ensure efficient catalysis but also avoid agglomeration [96].

Learning from flaws of common methods to avoid Pt single-site agglomeration such as solution synthesis, electrodeposition, and physical sputtering, the group of Botton combined the ALD approach and N-doped graphene as support to create single-atom Pt sites on the surface using the obtained material as the electrode in proton-exchange membrane fuel cells (PEMFC). They could prove that Pt was primarily located at the edges of the graphene layers and the Pt was stabilized, and immobilized by the defects within the graphene lattice and the N-dopants (Fig. 2.6A–D).

The Pt atoms were deposited on the N-doped graphene using $MeCpPtMe_3$ as precursors and N_2 as the purge gas. The N-doped graphene was placed inside a steel container with a perforated Al lid and together transferred inside on the heated stage of the reactor. The wall and stage temperatures of the reactor were kept at 150°C and 250°C, and each ALD cycle consisted of a 1 s $MeCpPtMe_3$ pulse, followed by a 20 s N_2 purge, and a 5 s O_2 pulse. The concentration of Pt atoms on the surface was controlled by the number of cycles and up to 150 cycles, no

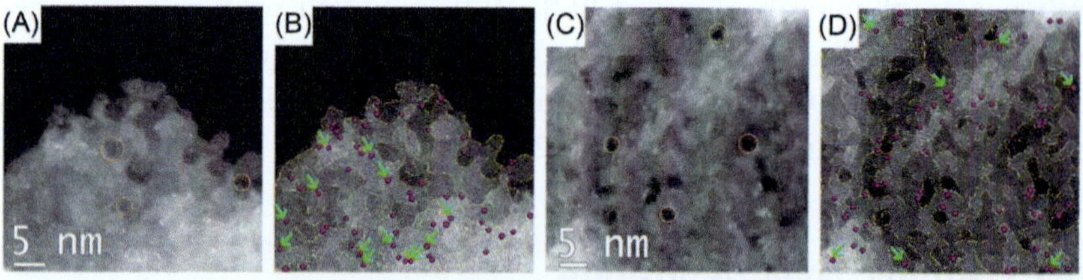

Figure 2.6 Raw (A) and (C) and processed (B) and (D) HAADF images of ALDPt/NGNs after 100, and 150 ALD cycles on N-doped graphene. The orange markings indicate the specific areas with fewer layers in the graphene lattice. The yellow lines in the processed images outline the edges of the graphene nanosheets, and the Pt atoms and clusters are indicated with pink dots. *Modified with permission from S. Stambula, N. Gauquelin, M. Bugnet, S. Gorantla, S. Turner, S. Sun, et al., Chemical structure of nitrogen-doped graphene with single platinum atoms and atomic clusters as a platform for the PEMFC electrode, J. Phys. Chem. C. 118 (2014) 3890–3900. https://doi.org/10.1021/jp408979h; copyright 2014, American Chemical Society.*

formation of Pt NPs was observed [97]. Based on their experience to stabilize Pt species on edge locations of N-doped graphene, they slightly modified the previous synthetic protocol to deposit Pt on N-doped graphene nanosheets (NGNs) (ALDPt/NGNs) with loading of Pt as high as 7.6 wt.% and successfully applied the new SAC for efficient HER [98].

2.2.3 Hydro(solvo)thermal approach

Hydrothermal or solvothermal synthesis is a relatively old but versatile and comparably simple technique to access numerous amorphous and crystalline structures. The chemical reaction occurs in aqueous or nonaqueous solvents at elevated temperatures above the boiling point of the reaction medium and high pressure. Though the reaction proceeds in a "black box" used in an explorative way by controlling the reaction conditions (precursor composition, pH, solvent, heating rate, temperature, pressure, concentration, etc.), it can result in pure phases with distinct sizes and morphology [99]. Despite the simplicity of this method, the synthesis of SAAC and SACs through hydro(solvo)thermal has been highly challenging and hence only a few reports exist.

Bao et al. introduced single Pt atoms into the in-plane domain MoS_2 nanosheets to access the Pt SAC Pt–MoS_2 that was successfully applied for the electrochemical HER. In a hydrothermal approach, they dissolved ammonium molybdate $(NH_4)_6Mo_7O_{24} \cdot 4H_2O$ and H_2PtCl_6 and transferred the solution together with CS_2 into an autoclave. After heating at 400°C for 4 h, the product was treated with saturated NaOH washed and dried at 100°C. Their study showed that the doped Pt atoms tune the adsorption behavior of H atoms on their neighboring S atoms and increase the HER activity [26]. Making use of the boosting effect by doping atoms onto the surface of catalysts, Ma and coworkers reported enhanced OER activity in $Ni(OH)_2$ when doped with W^{6+}. The improved OER activity of the SAC was ascribed to the d^0 characteristic of W^{6+} that stabilized the O radical and also mediated the O–O coupling. The w-$Ni(OH)_2$-e with atomically dispersed tungsten (W) on Ni sites in the layered structure was prepared via an alcohothermal method. Two separate solutions of WCl_6 in ethanol and $NiCl_2$ in ethanol were prepared and mixed slowly under stirring. The newly formed solution was then transferred into an autoclave and held at 150°C for 12 h in an oven. The final product, w-$Ni(OH)_2$, was obtained by treating the collected sample from the autoclave with ultrasound for 24 h in ethanol [100].

2.2.4 Metal-organic frameworks and covalent organic frameworks

MOFs have not only been applied as a template for the SACs preparation but have also been applied directly as SACs due to their unique properties. They provide a high variety of tuneable properties to replace common support materials such as metal oxides as support materials for SACs [101]. MOFs can be precisely designed for catalysis, with the catalytic function implemented at the organic or inorganic component. The pore system can either serve as the anchor for the catalytic species or as the confined space where a chemical reaction is taking place, acting as a passive medium for dispersing the catalytic species [101]. The MOF can also participate in the catalytic reaction, by stabilizing transition states, orienting molecules, or introducing additional active sites [101]. Various MOF-supported catalysts have been reported for single-atom catalyzed reactions such as the hydrosilylation of ketones and alkenes or the hydrogenation of alkenes or photocatalytic HER [102,103].

Bao et al. investigated the surface functionalization of a ZIF Zn(2-methylimidizole)$_2$ (ZIF-8) by ammonium ferric citrate (AFC) as an iron precursor without compromising the underlying coordination framework. They showed that the iron precursor exclusively coordinates to the Zn nodes at the surface of ZIF-8 NPs and after pyrolysis, isolated Fe-N sites in a carbon matrix form on the surface of the ZIF with superior catalytic performance toward electrochemical ORR and CO_2RR. The SAC was prepared by dissolving AFC in water and adding ZIF-8 to the solution under stirring for 24 h as shown in Fig. 2.7. The precipitate was centrifuged, washed several times, and dried at 60°C overnight. After that, the precipitate was immersed again in distilled water for six days, yielding the AFC©ZIF-8 with an iron loading of 0.68 wt.%. The obtained AFC©ZIF-8 was heated to temperatures between 800°C and 1000°C and kept at the chosen temperature for 4 h in an Ar atmosphere. The black pyrolysis products were refluxed in 0.5 M $HClO_4$ at 80°C for 10 h to remove impurities, and after washing and drying, the final product C-AFC©ZIF-8 was produced. To vary the concentration of iron in C-AFC©ZIF-8, the precursor concentration was lowered or raised [104].

At the same time, the group of Wang activated a ZIF-67 for OER by O_2 plasma treatment to form on-site atomic-scale CoO_x in the MOFs while preserving the porous structure. The plasma treatment partially broke the Co-N coordination bonds and the uncoordinated Co species

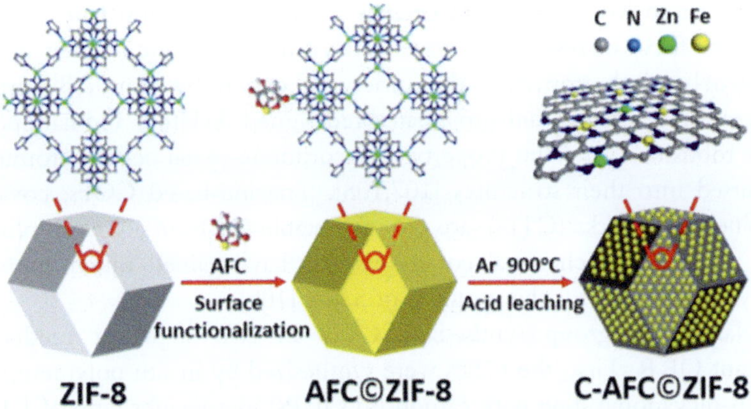

Figure 2.7 Schematic depiction of the synthesis of C-AFC©ZIF-8. *Reprinted with permission from Y. Ye, F. Cai, H. Li, H. Wu, G. Wang, Y. Li, et al., Surface functionalization of ZIF-8 with ammonium ferric citrate toward high exposure of Fe-N active sites for efficient oxygen and carbon dioxide electroreduction, Nano Energy 38 (2017) 281–289. https://doi.org/10.1016/j.nanoen.2017.05.042 copyright 2017, Elsevier.*

reacted with O_2 to form locally a CoO_x species that acted as the active sites to catalyze the OER. The ZIF-67 was synthesized by creating a solution of 2-methylimidazole and cobalt nitrate in methanol. Afterward, the dimethyl imidazole solution was added to the cobalt nitrate solution, and the mixed solution was aged at room temperature for 24 h. The precipitate of the solution was washed and dried to give the ZIF-67. For the O_2 plasma treatment, a radio frequency (RF) power of 200 W was applied, with controlled pressure at 120 Pa, and a treatment time of 1 h. While the RF power up to 300 W did not affect the structural integrity of the MOF, the treatment time when exceeding 2 h resulted in a complete pyrolyzed MOF [105]. The same year, they advanced their plasma approach to activate the ZIF-67 by creating coordinately unsaturated metal sites (CUMSs) in the ZIF. The CUMSs-ZIF-67 was prepared by N_2 dielectric barrier discharge plasma irradiation with a reactor input voltage of 100 V and a current of 3.0 A. They proposed that the activation for OER resulted from the removal of ligands in the ZIF to create the CUMS that provided more active sites for the adsorption of reactants and acted as catalytic centers for OER. In an otherwise untreated ZIF, the Co ions would be coordinated by four strong imidazole ligands without available sites for adsorption and therefore show less activity in electrocatalysis [106].

Similar to MOFs, covalent organic frameworks (COFs) have recently been the focus of research as novel catalyst templates due to their unique physicochemical properties, the variety of design by atomically precise integration of organic units to create predesigned skeletons and nanopores, their robustness, and the property to incorporate metal centers atomically dispersed into their structures [107,108]. Triazine-based COFs, covalent triazine frameworks (CTFs), are the most stable COFs in aqueous solution and under electrochemical conditions and have gained much attention lately as templates for the synthesis of SACs [109].

Nakanishi's group synthesized CTFs in two steps for methanol-tolerant ORR. First, the CTFs were synthesized by in situ polymerization of 2,6-dicyanopyridine with carbon NPs (CPs) in a weight ratio of 1:1 in-molten $ZnCl_2$. Afterward, the CTF/CP was impregnated with a $K_2[PtCl_4]$ solution at 60°C and washed to obtain the Pt-CTF/CP [110]. They later modified the synthetic approach to achieve a higher Pt loading (2.8 wt.%) to use this SAC also for the HOR and ORR [111].

Soon after, Cao, Huang, and coworkers used the rich N-sites in porous CTFs, to anchor and stabilize single Fe sites. The highly stable atomically dispersed Fe-N_4 species on porous porphyrinic triazine-based frameworks (FeSAs/PTFs) showed very good ORR activity, long-term durability, and good methanol tolerance in both alkaline and acidic media (see Fig. 2.8A for the illustrated synthesis). At first, the 5,10,15,20-tetrakis(4-cyanophenyl)porphyrin (TPPCN) was synthesized by mixing 4-formyl benzonitrile and pyrrole in propionic acid and heating the mixture to reflux overnight. The reaction mixture was washed and dried in a vacuum to yield the TPPCN. The TPPCN was then impregnated by a

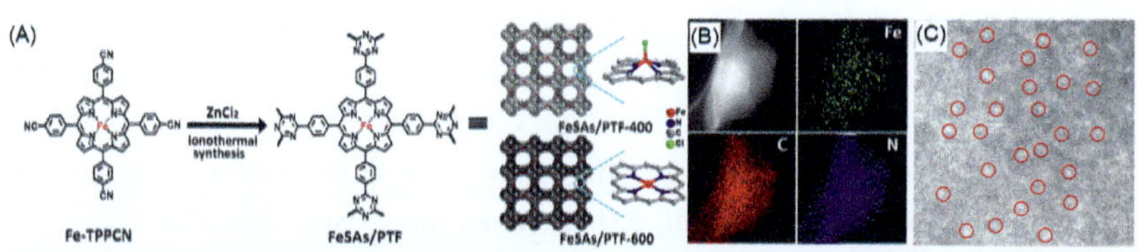

Figure 2.8 (A) Schematic representation of the formation of FeSAs/PTF (B) EDS mapping from TEM showing the homogeneous distribution of Fe, C, and N in FeSAs/PTF. (C) HAADF-STEM image of FeSAs/PTF-600 with red circles marking the single Fe atoms. *Reprinted with permission from J.-D. Yi, R. Xu, Q. Wu, T. Zhang, K.-T. Zang, J. Luo, et al., Atomically dispersed iron–nitrogen active sites within porphyrinic triazine-based frameworks for oxygen reduction reaction in both alkaline and acidic media, ACS Energy Lett. 3 (2018) 883–889. https://doi.org/10.1021/acsenergylett.8b00245; copyright 2018, American Chemical Society.*

solution of $FeCl_2 \cdot 4H_2O$ in DMF under reflux for 6 h. The precipitate was washed and dissolved in $CHCl_3$, followed by washing with 1 M HCl and water. The resulting organic layer was dried over anhydrous magnesium sulfate and evaporated generating the Fe-TPPCN. In a final step, Fe-TPPCN and $ZnCl_2$ (10 mmol) were transferred into an evacuated quartz ampule, sealed, and annealed at different temperatures. The resulting sample was ground and then stirred with water as well as 0.1 M HCl to remove the residual salt and Fe particles. For further purification, the powder was washed with water and methanol and dried in a vacuum to produce FeSAs/PTFs, giving atomically dispersed single Fe atoms on the carbon matrix as shown in Fig. 2.8B and C. They observed the annealing temperature influenced the Fe loading in the final product giving the lowest Fe loading of 2.6 wt.% at 600°C and the highest of 8.3 wt.% at 400°C. Further, they revealed the content of graphitic carbon that forms during partial pyrolysis increased with a higher annealing temperature [112].

2.2.5 Solid-state methods

Common and established techniques to synthesize materials are the thermal annealing method in a protective atmosphere at relatively high temperatures, mechanochemical (grinding) reactions, or molten salt synthesis to produce thermodynamically stable products [113–115]. The solid-state methods have their limitations, and very often these techniques are not suitable to attain NPs. Also, the multiple reaction parameters (precursor, atmosphere, temperature, pressure, milling frequency, milling time, size, and the number of milling balls) complicate a precise control of the reaction pathway and often empiric efforts are required to attain the desired compounds [99,113,114]. Typical procedures involve mixing metal salts or oxides by ball milling and annealing either in an inert or in a reducing atmosphere. The samples are then usually etched or leached in acid or base to remove unwanted species and obtain the final product [116,117].

A series of MN_4/GN composites with M = Mn, Fe, Co, Ni, or Cu have been prepared by Bao et al. using the ball milling technique and applied as highly efficient counter electrode materials for DSSCs. For pure atomically dispersed CoN_4/GN that showed the highest efficiency, a typical procedure included ball milling graphite flakes to graphene nanosheets (GN) at 450 rev. min^{-1} for 20 h in a glove box. Then, the GN was mixed with cobalt (II) phthalocyanine and the process was repeated to obtain the CoN_4/GN without further treatment. The single sites were confirmed by X-ray absorption near edge structure and density functional theory (DFT) calculations [118]. Meanwhile, single-atom Co sites anchored on interconnected B, N-doped CNTs (B, N, Co/C-nanotubes) were achieved via facile molten-salt-assisted pyrolysis of B/N/Co precursors. To prepare B, N, Co/C-nanotubes, sodium chloride (NaCl), glucose ($C_6H_{12}O_6$), boric acid (H_3BO_3), guanidine carbonate ($C_2H_{10}N_6$), and cobalt chloride ($CoCl_2$) were mixed by grinding the powders in a mortar. The mixture was annealed at 750°C in a tube furnace in an Ar atmosphere, and the products were etched with 0.5 M H_2SO_4 yielding B, N, Co/C-nanotubes. The material was additionally activated at 900°C and applied successfully as a catalyst for electrochemical ORR [119]. Lately, the group of Zhang also used a solid-state approach to synthesize Ru-based SACs. The Fe-substituted $MgAl_2O_4$ spinel ($MgAl_{1.2}Fe_{0.8}O_4$, MAFO) was physically mixed with commercial RuO_2 powder and calcined at 900°C in air for 5 h, to attain the Ru_1/MAFO-900 (Ru loadings 2 wt.%). Detailed in situ and postanalysis, as well as DFT calculations, revealed that the dispersion process does not arise from a gas atom trapping mechanism, but from a strong CMSI between FeO_x sites in the MAFO support and the RuO_2 aggregates, as depicted in Fig. 2.9A and B. It was proposed that the CMSI promotes an anti-Ostwald ripening of Ru atoms/RuO_2 subunits and allows the segregation to form atomically dispersed Ru sites on the support surface, as depicted in Fig. 2.9C–E [120]. Based on the findings of Zhang's group, Sun et al. in 2022 prepared $Pt_1/\alpha\text{-}Fe_2O_3$ and $Pt_1/TiO_2\text{-}\alpha\text{-}Fe_2O_3$ SACs similarly. They considered

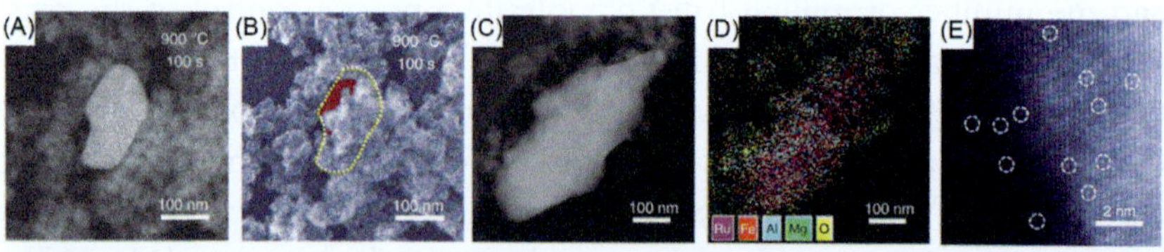

Figure 2.9 (A) AC-HAADF-STEM images with (B) the corresponding secondary electron (SE) images of a RuO_2 + MAFO physical mixture after calcination at 900°C (100 s) in an O_2 atmosphere. The yellow dashed lines in the SE images silhouette the RuO_2 aggregate, and red spots mark the exposed RuO_2 surfaces. (C) HAADF-STEM image of Ru/MAFO-500 with (D) EDX mapping of the sample and (E) AC-HAADF-STEM images of Ru_1/MAFO-900. *Modified with permission from K. Liu, X. Zhao, G. Ren, T. Yang, Y. Ren, A.F. Lee, et al., Strong metal-support interaction promoted scalable production of thermally stable single-atom catalysts, Nat. Commun. 11 (2020) 1263. https://doi.org/10.1038/s41467-020-14984-9 copyright 2020, Nature Publishing Group.*

that due to the chemical similarity of Pt and Ru metals, the observed CMSI between RuO_2 and FeO_x would also apply to PtO_2 for the fabrication of Pt SACs from a physical mixture of PtO_2 with α-Fe_2O_3 and other Fe-containing supports. The Pt_1/α-Fe_2O_3-C700 with a theoretical Pt loading of 0.3 wt.% was attained from the physical mixture of commercial PtO_2 and mesoporous α-Fe_2O_3 calcinated at 700°C in air showing no signs of Pt NPs. The Pt_1/TiO_2-α-Fe_2O_3-C700 with a theoretical Pt loading of 4 wt.% followed the same protocol by mixing PtO_2 and the $TiO_2-\alpha$-Fe_2O_3 complex and calcining the powder at 700°C in air. Since no NPs were found in both samples when heated at the optimal temperature (calcination below 700°C resulted in Pt NPs), they also concluded the anti-Ostwald ripening process promoted by a strong CMSI between Pt and Fe, which indeed stabilized single Pt atoms on the surface [121].

2.2.6 Other methods

Another method to access SACs is the arc-discharge technique. It is a very energy-intensive method to prepare CNTs where a direct-current arc voltage is applied between two graphite electrodes in an inert atmosphere. When the anode contains a metal or metal catalyst, metal-containing carbon is generated in the form of soot [122]. With a modified arc-discharge approach, Nb-in-C was synthesized to attain uniformly dispersed Nb atoms incorporated into an onion-like graphitic shell matrix as efficient catalysts for the electrocatalytic ORR in alkaline media. The synthetic procedure involved a water-cooled niobium rod as an anode, and a carbon rod as a cathode placed into a chamber and a power source to initiate an arc between them. After the evacuation of the chamber, methane (CH_4) was used as a gaseous carbon source. The niobium rod was evaporated by arc discharge, which resulted in a simultaneous decomposition of CH_4. The arc-evaporated niobium and decomposed carbon then formed the Nb-in-C that was deposited on the cooled walls of the chamber [123]. In addition, electrodeposition has been applied as a versatile process of coating a thin metal layer onto the surface of a support material based on the principle of electrolysis to attain SACs. To evaluate the impact of Au doping on the OER activity of transition metal-based electrocatalysts and understand the activity origin at the atomic level, Zhang et al. prepared ^{s}Au supported on NiFe LDH (^{s}Au/NiFe LDH) by electrodeposition as a model system to investigate the nature of active sites by theoretical calculations and supporting experimental data. First, they prepared a NiFe

LDH grown on a nickel mesh via a hydrothermal approach that was applied as a working electrode in the electrodeposition procedure. The electrolyte was a solution of NaCl and $HAuCl_4$. And the Au was deposited onto the NiFe LDH by cycling the potential between −0.6 V and −0.2 V vs. SCE for five cycles to obtain a loading of 0.4 wt.% [124]. Using this method as a general strategy for SACs, Zeng and his group synthesized more than 30 different materials from cathodic or anodic deposition just by varying metal precursors and supports. Electrochemical depositions of single-atom M (M = Ir, Ru, Rh, Pd, Ag, Pt, Au, Fe, Co, Ni, Zn, V, Cr, Mn, and Cu) on supports were conducted using a linear sweep voltammetry method in a standard three-electrode system with a carbon rod used as the counter electrode and an Ag/AgCl electrode as the reference electrode. The SACs were deposited on the glassy carbon working electrodes that were loaded by Nafion with $Co(OH)_2$, MnO_2, MoS_2, or $Co_{0.8}Fe_{0.2}Se_2$ nanosheet supports mixed with active carbon. A $Co_{0.8}Fe_{0.2}Se_2$@NF was prepared separately by directly growing the $Co_{0.8}Fe_{0.2}Se_2$ nanosheets onto the nickel foam (NF) and also used as the working electrode. The electrolyte contained 1 M KOH and $IrCl_4 \cdot xH_2O$, $RuCl_3$, $Na_3RhCl_6 \cdot 12H_2O$, Na_2PdCl_4, $AgNO_3$, H_2PtCl_4, $HAuCl_4 \cdot xH_2O$, $FeCl_3$, $CoCl_2 \cdot 6H_2O$, $NiCl_2$, $Zn(NO_3)_2 \cdot 6H_2O$, VCl_3, $CrCl_3$, $MnCl_2 \cdot 4H_2O$, or $CuCl_2$ as metal precursors for deposition onto the working electrodes. To obtain the final electrodes, the electrochemical depositions were carried out and repeated various times from 0.10 V to −0.40 V for cathodic and from 1.10 V to 1.80 V for anodic deposition with a sweeping rate of 5 mV s^{-1} and washed with water. Especially the cathodically and anodically deposited Ir single atoms on $Co_{0.8}Fe_{0.2}Se_2$@NF showed remarkable performance for overall water splitting, a record-low voltage of 1.39 V was required at 10 mA cm^{-2} in alkaline electrolyte [125]. Likewise, the group of Yan used a similar approach to prepare single-atom Pt immobilized NiO/Ni heterostructure (Pt_{SA}-NiO/Ni) for highly efficient electrochemical HER in an alkaline solution [126]. Lately, Li and coworkers prepared polyoxometalates (POMs) as Pt-based SAC models, in which Pt−O can become a new active site for HER, with a cooperative working mechanism between Pt and O atoms. POMs are nanoscale metal oxide clusters with well-defined crystal structures, abundant and adjustable composition, and unique redox properties. They synthesized four different single-atom Pt-containing POMs, $Na_5[H_3PtW_6O_{24}]$ ($\{PtW_6\}$), $K_4[H_4PtMo_6O_{24}]$ ($\{PtMo_6\}$), $Na_5[H_2PtV_9O_{28}]$ ($\{PtV_9\}$), and $[(C_3H_9)_4N]_4[SiPtW_{11}O_{40}H_2]$ ($\{PtSiW_{11}\}$), as electrode materials for electrochemical HER. The most active species

$\{PtW_6\}/C$, was prepared by forming the POM $Na_5[H_3PtW_6O_{24}]$ first. For this compound, solutions of $Na_2Pt(OH)_6$ and Na_2WO_4 were mixed in a molar ratio of Pt: W = 1:6 and with adjusted pH of 6.2 by nitric acid. After concentration and cooling, pale-yellow crystals of $Na_5[H_3PtW_6O_{24}]$ were obtained. Then, the $Na_5[H_3PtW_6O_{24}]$ and black carbon were homogeneously dispersed in water and stirred for 2 h at ambient temperature. After the addition of Nafion to the solution and further stirring for 2 h, the electrocatalysts were collected by centrifugation and drying [127].

2.3 Conclusion and future outlook

SACs have flourished into a highly active research field, especially in the important category of heterogeneous catalysis. In the last years, significant efforts and devotion have been given to their synthesis as catalyst materials for energy application. We have focused this chapter on various synthetic methods to stabilize the single-atom sites of the catalyst on different support materials. With adequate synthetic techniques and the right choice or design of support, excellent catalytic performances and stability can be achieved. However, many challenges remain such as large-scale preparation methods, stability of single-sites and support during catalysis to prevent agglomeration to NPs, high surface area support, increased loading for a higher number of active sites, etc., have to be addressed before SACs can be considered for industrial application. The migration and agglomeration of single-atom sites to nanoparticulate clusters due to their high surface energy can be prevented by the sophisticated choice of supports. Metal oxide, carbon-based, MOF, COF, or zeolite supports have proven to be suitable to stabilize single sites, but suitable protocols for changing coordination spheres during synthesis and differences in coordination numbers resulting from different defects or vacancies must be developed in an empirical approach. Since the interaction between the single metal atom and the substrate influences the catalytic performance, more focus should be given to revealing the activity and structure relationship. Stronger efforts have to be made, to get more combined experimental/analytical data with theoretical calculations to gain a deeper understanding for educated predictions of structural stabilities of supports, coordinated metal atoms, and coordination spheres. With DFT and state-of-the-art characterization methods such as in situ XPS/XAFS, in situ IR/Raman, etc., it will be possible to specifically predesign catalysts on an atomic level for specific reactions and provide a substantial foundation for optimization. So far, common methods applied for the synthesis require high temperatures to treat the support materials or involve other energy-intensive techniques for the deposition of the single atoms. In addition to that, often costly precursors are utilized making these methods neither sustainable nor green. Nevertheless, precise control over the coordination environment is needed to adjust the intrinsic properties of the active sites on the surface. Slight changes in coordination impact the electronic structure and thus their catalytic properties. Many synthetic approaches (e.g., sequential reduction, photochemical reduction, galvanic replacement, laser-ablation, physical layer deposition, or strong electrostatic adsorption) have been proven to be vital for the production of SACS but have not been applied for the synthesis of energy materials. Also, single-source precursors could play an important role in giving high control during the synthesis of SACs that is needed to precisely modify the local structure of single active sites. Single-source precursors allow for designing SACs by varying the coordination and structure on the atomic scale, which is highly desirable. In addition to that, SAACs have been rarely reported and should be strongly considered due to their synergistic effect among metal atoms where adsorption and activation processes are divided between two or more metal centers. Controlled synthesis of atomically dispersed multiatom catalysts could increase selectivity and activity as well as drastically lower the kinetic barrier in the catalytic reaction and give way to new applications. For the industrial application of SACs or SAACs, however, these earlier-explained challenges have to be tackled first. So far SACs are only fundamentally explored and although their catalytical performances with high selectivity, activity, and stability are promising, challenges remain. Migration and agglomeration of SACs, especially in harsh conditions, are not fully understood and while most SACs tend to form NPs, some undergo segregation, and formed NPs dissolve to single-site atoms. In this regard, the loading of the catalyst is substantially connected to the activity, while a higher concentration of active sites on the surface is needed to efficiently drive the desired catalytic reaction, on the other hand, lowering the active site can lead to catalytic activities that are too low for practical application. To date, SACs have not been industrialized, and a tremendous amount of research in this area is expected in the next years not only in the field of energy but also for various other catalytic reactions.

References

[1] S. Chu, A. Majumdar, Opportunities and challenges for a sustainable energy future, Nature 488 (2012) 294–303. Available from: https://doi.org/10.1038/nature11475.

[2] P. Anastas, N. Eghbali, Green chemistry: principles and practice, Chem. Soc. Rev. 39 (2010) 301–312. Available from: https://doi.org/10.1039/B918763B.

[3] D. Ren, Y. Deng, A.D. Handoko, C.S. Chen, S. Malkhandi, B.S. Yeo, Selective electrochemical reduction of carbon dioxide to ethylene and ethanol on Copper(I) oxide catalysts, ACS Catal. 5 (2015) 2814–2821. Available from: https://doi.org/10.1021/cs502128q.

[4] C. Panda, P.W. Menezes, S. Yao, J. Schmidt, C. Walter, J.N. Hausmann, et al., Boosting electrocatalytic hydrogen evolution activity with a NiPt3@NiS heteronanostructure evolved from a molecular nickel-platinum precursor, J. Am. Chem. Soc. 141 (2019) 13306–13310. Available from: https://doi.org/10.1021/jacs.9b06530.

[5] Z. Zhuang, S.A. Giles, J. Zheng, G.R. Jenness, S. Caratzoulas, D.G. Vlachos, et al., Nickel supported on nitrogen-doped carbon nanotubes as hydrogen oxidation reaction catalyst in alkaline electrolyte, Nat. Commun. 7 (2016) 10141. Available from: https://doi.org/10.1038/ncomms10141.

[6] C. Walter, S. Kalra, R. Beltrán-Suito, M. Schwarze, P.W. Menezes, M. Driess, Manganese sulfide enables the formation of a highly active β-MnOOH electrocatalyst for effective alkaline water oxidation, Mater. Today Chem. 24 (2022) 100905. Available from: https://doi.org/10.1016/j.mtchem.2022.100905.

[7] Z. Chen, J. Wu, Z. Chen, H. Yang, K. Zou, X. Zhao, et al., Entropy enhanced perovskite oxide ceramic for efficient electrochemical reduction of oxygen to hydrogen peroxide, Angew. Chem. Int. Ed. 61 (2022) e202200086. Available from: https://doi.org/10.1002/anie.202200086.

[8] B. Chakraborty, R. Beltrán-Suito, V. Hlukhyy, J. Schmidt, P.W. Menezes, M. Driess, Crystalline copper selenide as a reliable non-noble electro(pre)catalyst for overall water splitting, ChemSusChem 13 (2020) 3222–3229. Available from: https://doi.org/10.1002/cssc.202000445.

[9] P.W. Menezes, C. Walter, B. Chakraborty, J.N. Hausmann, I. Zaharieva, A. Frick, et al., Combination of highly efficient electrocatalytic water oxidation with selective oxygenation of organic substrates using manganese borophosphates, Adv. Mater. 33 (2021) e2004098. Available from: https://doi.org/10.1002/adma.202004098.

[10] M.S. Faber, S. Jin, Earth-abundant inorganic electrocatalysts and their nanostructures for energy conversion applications, Energy Environ. Sci. 7 (2014) 3519–3542. Available from: https://doi.org/10.1039/C4EE01760A.

[11] C. Walter, P.W. Menezes, M. Driess, Perspective on intermetallics towards efficient electrocatalytic water-splitting, Chem. Sci. 12 (2021) 8603–8631. Available from: https://doi.org/10.1039/D1SC01901E.

[12] J. Guan, Y. Liu, Y. Fang, X. Du, Y. Fu, L. Wang, et al., Co-Ni alloy nanoparticles supported by carbon nanofibers for hydrogen evolution reaction, J. Alloy. Compd. 868 (2021) 159172. Available from: https://doi.org/10.1016/j.jallcom.2021.159172.

[13] P.W. Menezes, C. Walter, J.N. Hausmann, R. Beltrán-Suito, C. Schlesiger, S. Praetz, et al., Boosting water oxidation through in situ electroconversion of manganese gallide: an intermetallic precursor approach, Angew. Chem. Int. Ed. 58 (2019) 16569–16574. Available from: https://doi.org/10.1002/anie.201909904.

[14] H. Yang, J.N. Hausmann, V. Hlukhyy, T. Braun, K. Laun, I. Zebger, et al., An intermetallic CaFe 6 Ge 6 approach to unprecedented Ca − Fe − O electrocatalyst for efficient alkaline oxygen evolution reaction, ChemCatChem 14 (2022). Available from: https://doi.org/10.1002/cctc.202200293.

[15] J.N. Hausmann, P.W. Menezes, Why should transition metal chalcogenides be investigated as water splitting precatalysts even though they transform into (oxyhydr) oxides? Curr. Opin. Electrochem. 34 (2022) 100991. Available from: https://doi.org/10.1016/j.coelec.2022.100991.

[16] C. Walter, P.W. Menezes, S. Orthmann, J. Schuch, P. Connor, B. Kaiser, et al., A Molecular Approach to manganese nitride acting as a high performance

electrocatalyst in the oxygen evolution reaction, Angew. Chem. Int. Ed. 57 (2018) 698–702. Available from: https://doi.org/10.1002/anie.201710460.
[17] P.W. Menezes, A. Indra, I. Zaharieva, C. Walter, S. Loos, S. Hoffmann, et al., Helical cobalt borophosphates to master durable overall water-splitting, Energy Environ. Sci. 12 (2019) 988–999. Available from: https://doi.org/10.1039/C8EE01669K.
[18] A.T. Landers, M. Fields, D.A. Torelli, J. Xiao, T.R. Hellstern, S.A. Francis, et al., The predominance of hydrogen evolution on transition metal sulfides and phosphides under CO 2 reduction conditions: an experimental and theoretical study, ACS Energy Lett. 3 (2018) 1450–1457. Available from: https://doi.org/10.1021/acsenergylett.8b00237.
[19] L. Dai, Y. Xue, L. Qu, H.-J. Choi, J.-B. Baek, Metal-free catalysts for oxygen reduction reaction, Chem. Rev. 115 (2015) 4823–4892. Available from: https://doi.org/10.1021/cr5003563.
[20] A. Wang, J. Li, T. Zhang, Heterogeneous single-atom catalysis, Nat. Rev. Chem. 2 (2018) 65–81. Available from: https://doi.org/10.1038/s41570-018-0010-1.
[21] J. Li, M.F. Stephanopoulos, Y. Xia, Introduction: heterogeneous single-atom catalysis, Chem. Rev. 120 (2020) 11699–11702. Available from: https://doi.org/10.1021/acs.chemrev.0c01097.
[22] C. Panda, P.W. Menezes, M. Driess, Nano-sized inorganic energy-materials by the low-temperature molecular precursor approach, Angew. Chem. Int. Ed. 57 (2018) 11130–11139. Available from: https://doi.org/10.1002/anie.201803673.
[23] T. Kropp, Z. Lu, Z. Li, Y.-H.C. Chin, M. Mavrikakis, Anionic single-atom catalysts for CO oxidation: support-independent activity at low temperatures, ACS Catal. 9 (2019) 1595–1604. Available from: https://doi.org/10.1021/acscatal.8b03298.
[24] G. Gao, Y. Jiao, E.R. Waclawik, A. Du, Single atom (Pd/Pt) supported on graphitic carbon nitride as an efficient photocatalyst for visible-light reduction of carbon dioxide, J. Am. Chem. Soc. 138 (2016) 6292–6297. Available from: https://doi.org/10.1021/jacs.6b02692.
[25] J.-X. Liang, J. Lin, J. Liu, X. Wang, T. Zhang, J. Li, Dual metal active sites in an ir1 /feox single-atom catalyst: a redox mechanism for the water-gas shift reaction, Angew. Chem. Int. Ed. 59 (2020) 12868–12875. Available from: https://doi.org/10.1002/anie.201914867.
[26] J. Deng, H. Li, J. Xiao, Y. Tu, D. Deng, H. Yang, et al., Triggering the electrocatalytic hydrogen evolution activity of the inert two-dimensional MoS2 surface via single-atom metal doping, Energy Environ. Sci. 8 (2015) 1594–1601. Available from: https://doi.org/10.1039/C5EE00751H.
[27] J. Liu, M. Jiao, L. Lu, H.M. Barkholtz, Y. Li, Y. Wang, et al., High performance platinum single atom electrocatalyst for oxygen reduction reaction, Nat. Commun. 8 (2017) 15938. Available from: https://doi.org/10.1038/ncomms15938.
[28] P. Chen, T. Zhou, L. Xing, K. Xu, Y. Tong, H. Xie, et al., Atomically dispersed iron-nitrogen species as electrocatalysts for bifunctional oxygen evolution and reduction reactions, Angew. Chem. Int. Ed. 56 (2017) 610–614. Available from: https://doi.org/10.1002/anie.201610119.
[29] C. Zhu, Q. Shi, S. Feng, D. Du, Y. Lin, Single-atom catalysts for electrochemical water splitting, ACS Energy Lett. 3 (2018) 1713–1721. Available from: https://doi.org/10.1021/acsenergylett.8b00640.
[30] P. Aggarwal, D. Sarkar, K. Awasthi, P.W. Menezes, Functional role of single-atom catalysts in electrocatalytic hydrogen evolution: current developments and future challenges, Coord. Chem. Rev. 452 (2022) 214289. Available from: https://doi.org/10.1016/j.ccr.2021.214289.

[31] X. Li, L. Liu, X. Ren, J. Gao, Y. Huang, B. Liu, Microenvironment modulation of single-atom catalysts and their roles in electrochemical energy conversion, Sci. Adv. 6 (2020). Available from: https://doi.org/10.1126/sciadv.abb6833.

[32] W. Liu, H. Zhang, C. Li, X. Wang, J. Liu, X. Zhang, Non-noble metal single-atom catalysts prepared by wet chemical method and their applications in electrochemical water splitting, J. Energy Chem. 47 (2020) 333–345. Available from: https://doi.org/10.1016/j.jechem.2020.02.020.

[33] R.T. Hannagan, G. Giannakakis, M. Flytzani-Stephanopoulos, E.C.H. Sykes, Single-atom alloy catalysis, Chem. Rev. 120 (2020) 12044–12088. Available from: https://doi.org/10.1021/acs.chemrev.0c00078.

[34] F.D. Speck, J.H. Kim, G. Bae, S.H. Joo, K.J.J. Mayrhofer, C.H. Choi, et al., Single-Atom catalysts: a perspective toward application in electrochemical energy conversion, JACS Au 1 (2021) 1086–1100. Available from: https://doi.org/10.1021/jacsau.1c00121.

[35] J. He, N. Li, Z.-G. Li, M. Zhong, Z.-X. Fu, M. Liu, et al., Strategic defect engineering of metal–organic frameworks for optimizing the fabrication of single-atom catalysts, Adv. Funct. Mater. 31 (2021) 2103597. Available from: https://doi.org/10.1002/adfm.202103597.

[36] H.V. Thang, G. Pacchioni, L. DeRita, P. Christopher, Nature of stable single atom Pt catalysts dispersed on anatase TiO2, J. Catal. 367 (2018) 104–114. Available from: https://doi.org/10.1016/j.jcat.2018.08.025.

[37] J. Shan, J. Liu, M. Li, S. Lustig, S. Lee, M. Flytzani-Stephanopoulos, NiCu single atom alloys catalyze the C H bond activation in the selective non-oxidative ethanol dehydrogenation reaction, Appl. Catal. B: Environ. 226 (2018) 534–543. Available from: https://doi.org/10.1016/j.apcatb.2017.12.059.

[38] Z. Liu, B. Liu, D. Ding, M. Liu, F. Chen, C. Xia, Fabrication and modification of solid oxide fuel cell anodes via wet impregnation/infiltration technique, J. Power Sources 237 (2013) 243–259. Available from: https://doi.org/10.1016/j.jpowsour.2013.03.025.

[39] E. Gaigneaux, D.E. De Vos, P.A. Jacobs, J.A. Martens, P. Ruiz, G. Poncelet, P. Grange, Scientific Bases for the Preparation of Heterogeneous Catalysts, first ed., Elsevier, 2006.

[40] F. He, K. Li, C. Yin, Y. Wang, H. Tang, Z. Wu, Single Pd atoms supported by graphitic carbon nitride, a potential oxygen reduction reaction catalyst from theoretical perspective, Carbon 114 (2017) 619–627. Available from: https://doi.org/10.1016/j.carbon.2016.12.061.

[41] S.J. Tauster, S.C. Fung, R.T. Baker, J.A. Horsley, Strong interactions in supported-metal catalysts, Sci. (N. York, N.Y.) 211 (1981) 1121–1125. Available from: https://doi.org/10.1126/science.211.4487.1121.

[42] J.H. Kwak, J. Hu, D. Mei, C.-W. Yi, D.H. Kim, C.H.F. Peden, et al., Coordinatively unsaturated Al3 + centers as binding sites for active catalyst phases of platinum on gamma-Al2O3, Sci. (N. York, N.Y.) 325 (2009) 1670–1673. Available from: https://doi.org/10.1126/science.1176745.

[43] P. Kast, M. Friedrich, D. Teschner, F. Girgsdies, T. Lunkenbein, R. Naumann d'Alnoncourt, et al., CO oxidation as a test reaction for strong metal–support interaction in nanostructured Pd/FeO powder catalysts, Appl. Catal. A: Gen. 502 (2015) 8–17. Available from: https://doi.org/10.1016/j.apcata.2015.04.010.

[44] Q. Fu, H. Saltsburg, M. Flytzani-Stephanopoulos, Active nonmetallic Au and Pt species on ceria-based water-gas shift catalysts, Sci. (N. York, N.Y.) 301 (2003) 935–938. Available from: https://doi.org/10.1126/science.1085721.

[45] F. Yang, H. Zhao, W. Wang, L. Wang, L. Zhang, T. Liu, et al., Atomic origins of the strong metal-support interaction in silica supported catalysts, Chem. Sci. 12 (2021) 12651–12660. Available from: https://doi.org/10.1039/D1SC03480D.
[46] R. Lang, W. Xi, J.-C. Liu, Y.-T. Cui, T. Li, A.F. Lee, et al., Non defect-stabilized thermally stable single-atom catalyst, Nat. Commun. 10 (2019) 234. Available from: https://doi.org/10.1038/s41467-018-08136-3.
[47] G. Vilé, D. Albani, M. Nachtegaal, Z. Chen, D. Dontsova, M. Antonietti, et al., A stable single-site palladium catalyst for hydrogenations, Angew. Chem. Int. Ed. 54 (2015) 11265–11269. Available from: https://doi.org/10.1002/anie.201505073.
[48] J.-C. Li, S. Maurya, Y.S. Kim, T. Li, L. Wang, Q. Shi, et al., Stabilizing single-atom iron electrocatalysts for oxygen reduction via ceria confining and trapping, ACS Catal. 10 (2020) 2452–2458. Available from: https://doi.org/10.1021/acscatal.9b04621.
[49] P. Li, M. Wang, X. Duan, L. Zheng, X. Cheng, Y. Zhang, et al., Boosting oxygen evolution of single-atomic ruthenium through electronic coupling with cobalt-iron layered double hydroxides, Nat. Commun. 10 (2019) 1711. Available from: https://doi.org/10.1038/s41467-019-09666-0.
[50] Q. Wang, X. Huang, Z.L. Zhao, M. Wang, B. Xiang, J. Li, et al., Ultrahigh-loading of Ir single atoms on NiO matrix to dramatically enhance oxygen evolution reaction, J. Am. Chem. Soc. 142 (2020) 7425–7433. Available from: https://doi.org/10.1021/jacs.9b12642.
[51] S.-A. Park, D.-S. Kim, T.-J. Kim, Y.-T. Kim, Strong interaction between Pt and thiolated carbon for electrocatalytic durability enhancement, ACS Catal. 3 (2013) 3067–3074. Available from: https://doi.org/10.1021/cs400649n.
[52] E. Lam, J.H. Luong, Carbon materials as catalyst supports and catalysts in the transformation of biomass to fuels and chemicals, ACS Catal. 4 (2014) 3393–3410. Available from: https://doi.org/10.1021/cs5008393.
[53] C.H. Choi, M. Kim, H.C. Kwon, S.J. Cho, S. Yun, H.-T. Kim, et al., Tuning selectivity of electrochemical reactions by atomically dispersed platinum catalyst, Nat. Commun. 7 (2016) 10922. Available from: https://doi.org/10.1038/ncomms10922.
[54] J.-C. Li, F. Xiao, H. Zhong, T. Li, M. Xu, L. Ma, et al., Secondary-atom-assisted synthesis of single iron atoms anchored on N-doped carbon nanowires for oxygen reduction reaction, ACS Catal. 9 (2019) 5929–5934. Available from: https://doi.org/10.1021/acscatal.9b00869.
[55] X. Zhang, Y. Li, J. Ren, Y. Huang, Synthesis of a Zn/Fe–N–C electrocatalyst towards efficient oxygen reduction reaction via a facile one-pot method, Mater. Res. Express 9 (2022) 25604. Available from: https://doi.org/10.1088/2053-1591/ac569e.
[56] H. Tang, H. Yin, J. Wang, N. Yang, D. Wang, Z. Tang, Molecular architecture of cobalt porphyrin multilayers on reduced graphene oxide sheets for high-performance oxygen reduction reaction, Angew. Chem. Int. Ed. 52 (2013) 5585–5589. Available from: https://doi.org/10.1002/anie.201300711.
[57] Y. Qin, Z. Ou, C. Xu, J. Liu, Q. Lan, R. Jin, et al., Highly accessible single Mn-N3 sites-enriched porous graphene structure via a confined thermal-erosion strategy for catalysis of oxygen reduction, Chem. Eng. J. 440 (2022) 135850. Available from: https://doi.org/10.1016/j.cej.2022.135850.
[58] D. Wang, M. Yuan, J. Xu, Y. Li, K. Shi, H. Yang, et al., Highly active atomically dispersed Co–Nx sites anchored on ultrathin n-doped carbon nanosheets with durability oxygen reduction reaction of zinc–air batteries, ACS Sustain. Chem. Eng. 9 (2021) 16956–16964. Available from: https://doi.org/10.1021/acssuschemeng.1c05352.
[59] L. Li, Y.-J. Chen, H.-R. Xing, N. Li, J.-W. Xia, X.-Y. Qian, et al., Single-atom Fe-N5 catalyst for high-performance zinc-air batteries, Nano Res. (2022). Available from: https://doi.org/10.1007/s12274-022-4424-1.

[60] F. Fan, H. Zhou, R. Yan, C. Yang, H. Zhu, Y. Gao, et al., Anchoring Fe-N-C sites on hierarchically porous carbon sphere and CNT interpenetrated nanostructures as efficient cathodes for zinc-air batteries, ACS Appl. Mater. & interfaces 13 (2021) 41609−41618. Available from: https://doi.org/10.1021/acsami.1c10510.

[61] S.S.A. Shah, T. Najam, M.S. Bashir, L. Peng, M.A. Nazir, M.S. Javed, Single-atom catalysts for next-generation rechargeable batteries and fuel cells, Energy Storage Mater. 45 (2022) 301−322. Available from: https://doi.org/10.1016/j.ensm.2021.11.049.

[62] H.T. Chung, D.A. Cullen, D. Higgins, B.T. Sneed, E.F. Holby, K.L. More, et al., Direct atomic-level insight into the active sites of a high-performance PGM-free ORR catalyst, Sci. (N. York, N.Y.) 357 (2017) 479−484. Available from: https://doi.org/10.1126/science.aan2255.

[63] L. Cao, Q. Luo, J. Chen, L. Wang, Y. Lin, H. Wang, et al., Dynamic oxygen adsorption on single-atomic Ruthenium catalyst with high performance for acidic oxygen evolution reaction, Nat. Commun. 10 (2019) 4849. Available from: https://doi.org/10.1038/s41467-019-12886-z.

[64] J. Ji, Z. Li, C. Hu, Y. Sha, S. Li, X. Gao, et al., Platinum atomic clusters embedded in defects of anatase/graphene for efficient electro-and photocatalytic hydrogen evolution, ACS Appl. Mater. & interfaces 12 (2020) 40204−40212. Available from: https://doi.org/10.1021/acsami.0c06931.

[65] D.A. Bulushev, M. Zacharska, A.S. Lisitsyn, O.Y. Podyacheva, F.S. Hage, Q.M. Ramasse, et al., Single atoms of pt-group metals stabilized by N-doped carbon nanofibers for efficient hydrogen production from formic acid, ACS Catal. 6 (2016) 3442−3451. Available from: https://doi.org/10.1021/acscatal.6b00476.

[66] H.-W. Liang, S. Brüller, R. Dong, J. Zhang, X. Feng, K. Müllen, Molecular metal-Nx centres in porous carbon for electrocatalytic hydrogen evolution, Nat. Commun. 6 (2015) 7992. Available from: https://doi.org/10.1038/ncomms8992.

[67] L. Zhong, C. Jiang, M. Zheng, X. Peng, T. Liu, S. Xi, et al., Wood carbon based single-atom catalyst for rechargeable Zn−air batteries, ACS Energy Lett. 6 (2021) 3624−3633. Available from: https://doi.org/10.1021/acsenergylett.1c01678.

[68] X. Li, W. Bi, L. Zhang, S. Tao, W. Chu, Q. Zhang, et al., Single-atom Pt as Co-catalyst for enhanced photocatalytic H2 evolution, Adv. Mater. (Deerfield Beach, Fla.) 28 (2016) 2427−2431. Available from: https://doi.org/10.1002/adma.201505281.

[69] S. Choi, H. Kim, S. Lee, Y. Wang, C. Ercan, R. Othman, et al., Large-scale synthesis of high-quality zeolite-templated carbons without depositing external carbon layers, Chem. Eng. J. 280 (2015) 597−605. Available from: https://doi.org/10.1016/j.cej.2015.06.055.

[70] Q.-L. Zhu, W. Xia, L.-R. Zheng, R. Zou, Z. Liu, Q. Xu, Atomically dispersed Fe/N-doped hierarchical carbon architectures derived from a metal−organic framework composite for extremely efficient electrocatalysis, ACS Energy Lett. 2 (2017) 504−511. Available from: https://doi.org/10.1021/acsenergylett.6b00686.

[71] H. Jing, Z. Zhao, J. Zhang, C. Zhu, W. Liu, N. Li, et al., Atomic evolution of metal−organic frameworks into Co−N 3 coupling vacancies by cooperative cascade protection strategy for promoting triiodide reduction, J. Phys. Chem. C. 125 (2021) 6147−6156. Available from: https://doi.org/10.1021/acs.jpcc.1c01201.

[72] J. Ding, Y. Tang, S. Zheng, S. Zhang, H. Xue, Q. Kong, et al., The synthesis of MOF derived carbon and its application in water treatment, Nano Res. 15 (2022) 6793−6818. Available from: https://doi.org/10.1007/s12274-022-4327-1.

[73] P. Yin, T. Yao, Y. Wu, L. Zheng, Y. Lin, W. Liu, et al., Single cobalt atoms with precise N-coordination as superior oxygen reduction reaction catalysts, Angew.

Chem. Int. Ed. 55 (2016) 10800–10805. Available from: https://doi.org/10.1002/anie.201604802.

[74] X.X. Wang, D.A. Cullen, Y.-T. Pan, S. Hwang, M. Wang, Z. Feng, et al., Nitrogen-coordinated single cobalt atom catalysts for oxygen reduction in proton exchange membrane fuel cells, Adv. Mater. 30 (2018) 1706758. Available from: https://doi.org/10.1002/adma.201706758.

[75] G.-S. Kang, J.-H. Jang, S.-Y. Son, Y.-K. Lee, D.C. Lee, S.J. Yoo, et al., Pyrrolic N wrapping strategy to maximize the number of single-atomic Fe-Nx sites for oxygen reduction reaction, J. Power Sources 520 (2022) 230904. Available from: https://doi.org/10.1016/j.jpowsour.2021.230904.

[76] J. Li, M. Li, N. An, S. Zhang, Q. Song, Y. Yang, et al., Atomically dispersed Fe atoms anchored on S and N-codoped carbon for efficient electrochemical denitrification, Proc. Natl Acad. Sci. U S A 118 (2021). Available from: https://doi.org/10.1073/pnas.2105628118.

[77] S. Wei, A. Li, J.-C. Liu, Z. Li, W. Chen, Y. Gong, et al., Direct observation of noble metal nanoparticles transforming to thermally stable single atoms, Nat. Nanotechnol. 13 (2018) 856–861. Available from: https://doi.org/10.1038/s41565-018-0197-9.

[78] F. Li, G.-F. Han, H.-J. Noh, S.-J. Kim, Y. Lu, H.Y. Jeong, et al., Boosting oxygen reduction catalysis with abundant copper single atom active sites, Energy Environ. Sci. 11 (2018) 2263–2269. Available from: https://doi.org/10.1039/C8EE01169A.

[79] C. Zhu, Q. Shi, B.Z. Xu, S. Fu, G. Wan, C. Yang, et al., Hierarchically porous M–N–C (M = Co and Fe) single-atom electrocatalysts with robust MNx active moieties enable enhanced ORR performance, Adv. Energy Mater. 8 (2018) 1801956. Available from: https://doi.org/10.1002/aenm.201801956.

[80] J. Gu, C.-S. Hsu, L. Bai, H.M. Chen, X. Hu, Atomically dispersed Fe3 + sites catalyze efficient CO2 electroreduction to CO, Sci. (N. York, N.Y.) 364 (2019) 1091–1094. Available from: https://doi.org/10.1126/science.aaw7515.

[81] L. Fan, P.F. Liu, X. Yan, L. Gu, Z.Z. Yang, H.G. Yang, et al., Atomically isolated nickel species anchored on graphitized carbon for efficient hydrogen evolution electrocatalysis, Nat. Commun. 7 (2016) 10667. Available from: https://doi.org/10.1038/ncomms10667.

[82] Y. Chen, S. Ji, Y. Wang, J. Dong, W. Chen, Z. Li, et al., Isolated single iron atoms anchored on n-doped porous carbon as an efficient electrocatalyst for the oxygen reduction reaction, Angew. Chem. Int. Ed. 56 (2017) 6937–6941. Available from: https://doi.org/10.1002/anie.201702473.

[83] H. Li, X. Chen, J. Chen, K. Shen, Y. Li, Hierarchically porous Fe,N-doped carbon nanorods derived from 1D Fe-doped MOFs as highly efficient oxygen reduction electrocatalysts in both alkaline and acidic media, Nanoscale 13 (2021) 10500–10508. Available from: https://doi.org/10.1039/D1NR01603B.

[84] Z. Zhang, J. Sun, F. Wang, L. Dai, Efficient oxygen reduction reaction (ORR) catalysts based on single iron atoms dispersed on a hierarchically structured porous carbon framework, Angew. Chem. Int. Ed. 57 (2018) 9038–9043. Available from: https://doi.org/10.1002/anie.201804958.

[85] L. Jiao, G. Wan, R. Zhang, H. Zhou, S.-H. Yu, H.-L. Jiang, From metal-organic frameworks to single-atom Fe implanted n-doped porous carbons: efficient oxygen reduction in both alkaline and acidic media, Angew. Chem. Int. Ed. 57 (2018) 8525–8529. Available from: https://doi.org/10.1002/anie.201803262.

[86] X. Chen, N. Wang, K. Shen, Y. Xie, Y. Tan, Y. Li, MOF-derived isolated Fe atoms implanted in N-doped 3D hierarchical carbon as an efficient ORR electrocatalyst in both alkaline and acidic media, ACS Appl. Mater. Interfaces 11 (2019) 25976–25985. Available from: https://doi.org/10.1021/acsami.9b07436.

[87] T. Sun, Y. Li, T. Cui, L. Xu, Y.-G. Wang, W. Chen, et al., Engineering of coordination environment and multiscale structure in single-site copper catalyst for superior electrocatalytic oxygen reduction, Nano Lett. 20 (2020) 6206–6214. Available from: https://doi.org/10.1021/acs.nanolett.0c02677.

[88] J. Ge, D. He, W. Chen, H. Ju, H. Zhang, T. Chao, et al., Atomically dispersed Ru on ultrathin Pd nanoribbons, J. Am. Chem. Soc. 138 (2016) 13850–13853. Available from: https://doi.org/10.1021/jacs.6b09246.

[89] S. Yang, J. Kim, Y.J. Tak, A. Soon, H. Lee, Single-atom catalyst of platinum supported on titanium nitride for selective electrochemical reactions, Angew. Chem. Int. Ed. 55 (2016) 2058–2062. Available from: https://doi.org/10.1002/anie.201509241.

[90] R.-Q. Zhang, T.-H. Lee, B.-D. Yu, C. Stampfl, A. Soon, The role of titanium nitride supports for single-atom platinum-based catalysts in fuel cell technology, Phys. Chem. Chem. Phys. 14 (2012) 16552–16557. Available from: https://doi.org/10.1039/c2cp41392b.

[91] L. Lin, W. Zhou, R. Gao, S. Yao, X. Zhang, W. Xu, et al., Low-temperature hydrogen production from water and methanol using Pt/α-MoC catalysts, Nature 544 (2017) 80–83. Available from: https://doi.org/10.1038/nature21672.

[92] A. Wong, Q. Liu, S. Griffin, A. Nicholls, J.R. Regalbuto, Synthesis of ultrasmall, homogeneously alloyed, bimetallic nanoparticles on silica supports, Sci. (N. York, N.Y.) 358 (2017) 1427–1430. Available from: https://doi.org/10.1126/science.aao6538.

[93] Z. Wang, L. Yang, R. Zhang, L. Li, Z. Cheng, Z. Zhou, Selective hydrogenation of phenylacetylene over bimetallic Pd–Cu/Al2O3 and Pd–Zn/Al2O3 catalysts, Catal. Today 264 (2016) 37–43. Available from: https://doi.org/10.1016/j.cattod.2015.08.018.

[94] S.M. George, Atomic layer deposition: an overview, Chem. Rev. 110 (2010) 111–131. Available from: https://doi.org/10.1021/cr900056b.

[95] M. Leskelä, M. Ritala, Atomic layer deposition (ALD): from precursors to thin film structures, Thin Solid. Films 409 (2002) 138–146. Available from: https://doi.org/10.1016/S0040-6090(02)00117-7.

[96] H. Yan, H. Cheng, H. Yi, Y. Lin, T. Yao, C. Wang, et al., Single-atom Pd_1/graphene catalyst achieved by atomic layer deposition: remarkable performance in selective hydrogenation of 1,3-butadiene, J. Am. Chem. Soc. 137 (2015) 10484–10487. Available from: https://doi.org/10.1021/jacs.5b06485.

[97] S. Stambula, N. Gauquelin, M. Bugnet, S. Gorantla, S. Turner, S. Sun, et al., Chemical structure of nitrogen-doped graphene with single platinum atoms and atomic clusters as a platform for the PEMFC electrode, J. Phys. Chem. C. 118 (2014) 3890–3900. Available from: https://doi.org/10.1021/jp408979h.

[98] N. Cheng, S. Stambula, M.N. Da Wang, et al., Platinum single-atom and cluster catalysis of the hydrogen evolution reaction, Nat. Commun. 7 (2016) 13638. Available from: https://doi.org/10.1038/ncomms13638.

[99] H. Kohlmann, Looking into the black box of solid-state synthesis, Eur. J. Inorg. Chem. 2019 (2019) 4174–4180. Available from: https://doi.org/10.1002/ejic.201900733.

[100] J. Yan, L. Kong, Y. Ji, J. White, Y. Li, J. Zhang, et al., Single atom tungsten doped ultrathin α-Ni(OH)2 for enhanced electrocatalytic water oxidation, Nat. Commun. 10 (2019) 2149. Available from: https://doi.org/10.1038/s41467-019-09845-z.

[101] J. Gascon, A. Corma, F. Kapteijn, F.X. Llabrés i Xamena, Metal Organic Framework catalysis: quo vadis? ACS Catal. 4 (2014) 361–378. Available from: https://doi.org/10.1021/cs400959k.

[102] T. Sawano, Z. Lin, D. Boures, B. An, C. Wang, W. Lin, Metal-organic frameworks stabilize mono(phosphine)-metal complexes for broad-scope catalytic reactions, J. Am. Chem. Soc. 138 (2016) 9783–9786. Available from: https://doi.org/10.1021/jacs.6b06239.
[103] T. Zhou, Y. Du, A. Borgna, J. Hong, Y. Wang, J. Han, et al., Post-synthesis modification of a metal–organic framework to construct a bifunctional photocatalyst for hydrogen production, Energy & Environ. Sci. 6 (2013) 3229. Available from: https://doi.org/10.1039/c3ee41548a.
[104] Y. Ye, F. Cai, H. Li, H. Wu, G. Wang, Y. Li, et al., Surface functionalization of ZIF-8 with ammonium ferric citrate toward high exposure of Fe-N active sites for efficient oxygen and carbon dioxide electroreduction, Nano Energy 38 (2017) 281–289. Available from: https://doi.org/10.1016/j.nanoen.2017.05.042.
[105] S. Dou, C.-L. Dong, Z. Hu, Y.-C. Huang, J. Chen, L. Tao, et al., Atomic-scale CoOx species in metal–organic frameworks for oxygen evolution reaction, Adv. Funct. Mater. 27 (2017) 1702546. Available from: https://doi.org/10.1002/adfm.201702546.
[106] L. Tao, C.-Y. Lin, S. Dou, S. Feng, D. Chen, D. Liu, et al., Creating coordinatively unsaturated metal sites in metal-organic-frameworks as efficient electrocatalysts for the oxygen evolution reaction: insights into the active centers, Nano Energy 41 (2017) 417–425. Available from: https://doi.org/10.1016/j.nanoen.2017.09.055.
[107] X. Feng, X. Ding, D. Jiang, Covalent organic frameworks, Chem. Soc. Rev. 41 (2012) 6010–6022. Available from: https://doi.org/10.1039/C2CS35157A.
[108] S.-Y. Ding, J. Gao, Q. Wang, Y. Zhang, W.-G. Song, C.-Y. Su, et al., Construction of covalent organic framework for catalysis: Pd/COF-LZU1 in Suzuki-Miyaura coupling reaction, J. Am. Chem. Soc. 133 (2011) 19816–19822. Available from: https://doi.org/10.1021/ja206846p.
[109] K. Kamiya, Selective single-atom electrocatalysts: a review with a focus on metal-doped covalent triazine frameworks, Chem. Sci. 11 (2020) 8339–8349. Available from: https://doi.org/10.1039/D0SC03328F.
[110] K. Kamiya, R. Kamai, K. Hashimoto, S. Nakanishi, Platinum-modified covalent triazine frameworks hybridized with carbon nanoparticles as methanol-tolerant oxygen reduction electrocatalysts, Nat. Commun. 5 (2014) 5040. Available from: https://doi.org/10.1038/ncomms6040.
[111] R. Kamai, K. Kamiya, K. Hashimoto, S. Nakanishi, Oxygen-tolerant electrodes with platinum-loaded covalent triazine frameworks for the hydrogen oxidation reaction, Angew. Chem. Int. Ed. 55 (2016) 13184–13188. Available from: https://doi.org/10.1002/anie.201607741.
[112] J.-D. Yi, R. Xu, Q. Wu, T. Zhang, K.-T. Zang, J. Luo, et al., Atomically dispersed iron–nitrogen active sites within porphyrinic triazine-based frameworks for oxygen reduction reaction in both alkaline and acidic media, ACS Energy Lett. 3 (2018) 883–889. Available from: https://doi.org/10.1021/acsenergylett.8b00245.
[113] A. Stein, S.W. Keller, T.E. Mallouk, Turning down the heat: design and mechanism in solid-state synthesis, Science 259 (1993) 1558–1564. Available from: https://doi.org/10.1126/science.259.5101.1558.
[114] A. Stolle, Ball Milling Towards Green Synthesis: Applications, Projects, Challenges, first ed., Royal Society of Chemistry, Cambridge, 2015.
[115] S.K. Gupta, Y. Mao, A review on molten salt synthesis of metal oxide nanomaterials: status, opportunity, and challenge, Prog. Mater. Sci. 117 (2021) 100734. Available from: https://doi.org/10.1016/j.pmatsci.2020.100734.
[116] J.-C. Li, Y. Meng, R. Ma, H. Hu, S. Zhao, Y. Zhu, et al., Ionothermal-transformation strategy to synthesize hierarchically tubular porous single-iron-atom

catalysts for high-performance zinc-air batteries, ACS Appl. Mater. Interfaces 13 (2021) 58576–58584. Available from: https://doi.org/10.1021/acsami.1c16915.

[117] G. Sun, Z.-J. Zhao, R. Mu, S. Zha, L. Li, S. Chen, et al., Breaking the scaling relationship via thermally stable Pt/Cu single atom alloys for catalytic dehydrogenation, Nat. Commun. 9 (2018) 4454. Available from: https://doi.org/10.1038/s41467-018-06967-8.

[118] X. Cui, J. Xiao, Y. Wu, P. Du, R. Si, H. Yang, et al., A graphene composite material with single cobalt active sites: a highly efficient counter electrode for dye-sensitized solar cells, Angew. Chem. Int. Ed. 55 (2016) 6708–6712. Available from: https://doi.org/10.1002/anie.201602097.

[119] R. Zhao, J. Chen, Z. Chen, X. Jiang, G. Fu, Y. Tang, et al., Atomically dispersed CoN 4 /B, N-C nanotubes boost oxygen reduction in rechargeable Zn–air batteries, ACS Appl. Energy Mater. 3 (2020) 4539–4548. Available from: https://doi.org/10.1021/acsaem.0c00215.

[120] K. Liu, X. Zhao, G. Ren, T. Yang, Y. Ren, A.F. Lee, et al., Strong metal-support interaction promoted scalable production of thermally stable single-atom catalysts, Nat. Commun. 11 (2020) 1263. Available from: https://doi.org/10.1038/s41467-020-14984-9.

[121] W. Jianchen, K. Yong, F. Sun, Mass production of thermally stable Pt single-atom catalysts for the catalytic oxidation of sulfur dioxide, Catal. Sci. & Technol. 12 (2022) 124–134. Available from: https://doi.org/10.1039/D1CY01578H.

[122] J.H. Byeon, J.H. Park, K.Y. Yoon, J. Hwang, Ambient spark generation to synthesize carbon-encapsulated metal nanoparticles in continuous aerosol manner, Nanoscale 1 (2009) 339–343. Available from: https://doi.org/10.1039/b9nr00058e.

[123] X. Zhang, J. Guo, P. Guan, C. Liu, H. Huang, F. Xue, et al., Catalytically active single-atom niobium in graphitic layers, Nat. Commun. 4 (2013) 1924. Available from: https://doi.org/10.1038/ncomms2929.

[124] J. Zhang, J. Liu, L. Xi, Y. Yu, N. Chen, S. Sun, et al., Single-atom Au/NiFe layered double hydroxide electrocatalyst: probing the origin of activity for oxygen evolution reaction, J. Am. Chem. Soc. 140 (2018) 3876–3879. Available from: https://doi.org/10.1021/jacs.8b00752.

[125] Z. Zhang, C. Feng, C. Liu, M. Zuo, L. Qin, X. Yan, et al., Electrochemical deposition as a universal route for fabricating single-atom catalysts, Nat. Commun. 11 (2020) 1215. Available from: https://doi.org/10.1038/s41467-020-14917-6.

[126] K.L. Zhou, Z. Wang, C.B. Han, X. Ke, C. Wang, Y. Jin, et al., Platinum single-atom catalyst coupled with transition metal/metal oxide heterostructure for accelerating alkaline hydrogen evolution reaction, Nat. Commun. 12 (2021) 3783. Available from: https://doi.org/10.1038/s41467-021-24079-8.

[127] F.-Y. Yu, Z.-L. Lang, Y.-J. Zhou, K. Feng, H.-Q. Tan, J. Zhong, et al., Revealing hydrogen evolution performance of single-atom platinum electrocatalyst with polyoxometalate molecular models, ACS Energy Lett. 6 (2021) 4055–4062. Available from: https://doi.org/10.1021/acsenergylett.1c01911.

CHAPTER 3

Characterization techniques for single-atom catalysts

Weiju Hao, Jinli Fan and Chengyu Fu
School of Materials and Chemistry, University of Shanghai for Science and Technology, Shanghai, P.R. China

As one of the most popular catalytic materials in recent years, single-atom catalysts (SACs) show unparalleled catalysis performance in the field of energy [1]. Generally, the large surface-free energy of the active sites increases the agglomeration tendency and the impossibility of independent existence. Therefore, it is necessary to employ the appropriate substrates to make them stable [2,3]. For instance, porous crystalline materials such as molecular sieves, metal-organic frameworks (MOFs), etc., are promising SACs carriers owing to their confinement effect and ultra-high durability [4,5]. Although these highly-active SACs mono-dispersed on porous crystalline materials show extraordinary talents, it is still challenging to unveil the formation mechanism of SACs, their geometric local electronic structures of single atoms [6,7]. Therefore, series characterization instruments play an important role in the field of single atoms. In this chapter, a series of characterization instruments and methods of in situ X-ray diffraction (XRD), X-ray photoelectron spectroscopy (XPS), Fourier transform infrared spectroscopy (FTIR), Raman spectroscopy, along with geometric-phase analysis (GPA), electron energy loss spectroscopy (EELS), X-ray absorption spectroscopy (XAS), and first principle calculations for SACs in the energy fields of electrocatalysis, photocatalysis, and pollutant degradation, etc., are, respectively, discussed in detail (Fig. 3.1) [8].

3.1 X-ray diffraction and in situ X-ray diffraction

The crystal consists of a unit cell with atoms regularly arranged. When a monochromatic X-ray beam is cast incidentally on the crystalline substances in regular arrangement if the regularly arranged distances between atoms share the same magnitude order with the wavelength of the incident X-ray (λ), the scattered X-rays from these different atoms will

Single Atom Catalysts
DOI: https://doi.org/10.1016/B978-0-323-95237-8.00007-0

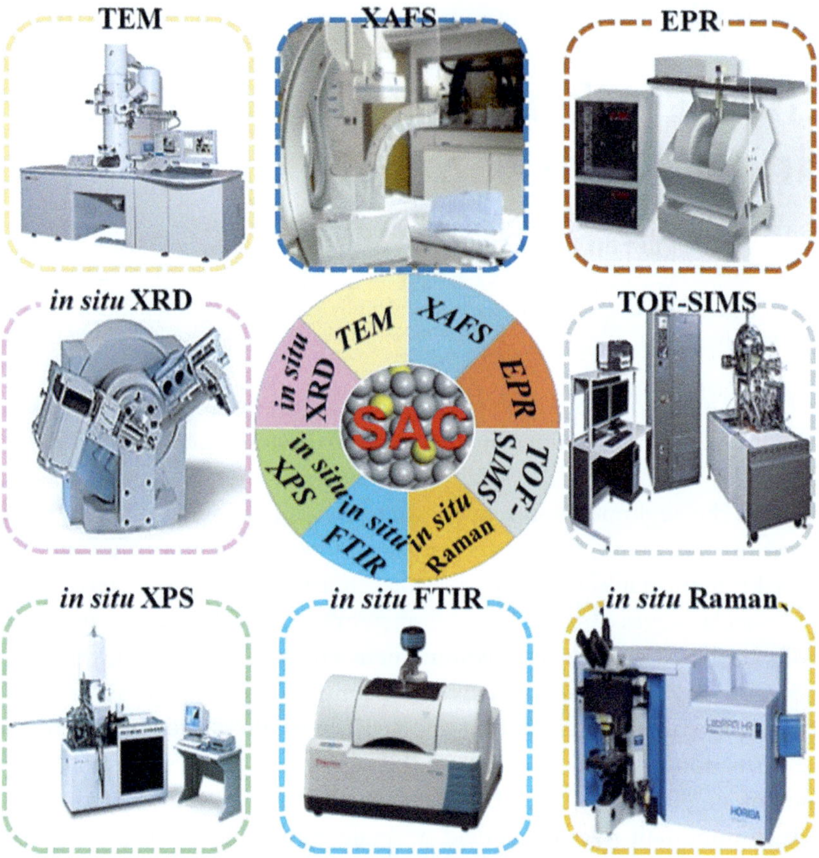

Figure 3.1 Series of characterization instruments and methods.

interfere with each other and produce enhanced XRD in the certain directions. If the Bragg equation $2d\sin\theta = n\lambda$ is agreed, where d is the crystal plane spacing, θ is the angle between the incident X-ray and the corresponding crystal plane, λ is the wavelength of X-ray, and n is the diffraction order, that is to say, the optical path difference is equal to an integer multiple of the wavelength, enhanced XRD will occur at the same phase [9].

Based on neutron diffraction data, Rietveld first proposed a method to modify the structure by the peak shape intensity of powder diffraction stepwise scanning in 1969. Subsequently, in situ XRD technique is introduced to further characterize single atoms. The whole process is to obtain the variation of peak intensity with different temperature points for the certain position of materials, so the more compatible information reflects a structural change in real time, which is helpful to deeply understand the charging/discharging process and phase transformation and extend the knowledge into further development. For example, Huang et al. demonstrated the fabrication of single-atom Ag chains on Hollandite-type manganese oxide (HMO) nanorods. In order to clarify the formation and location of single atom Ag chains, Ag/HMO was measured by in situ XRD at every 50°C interval. The diffraction peaks of both HMO and metal Ag appeared in the XRD patterns of Ag/HMO at 80°C. When the temperature increased to 230°C, the peak intensity in the (111) crystal plane of Ag gradually decreased, and its position moved toward the lower direction, which also illustrated the gradual disappearance of Ag atoms. The crystal structure of Hollandite did not slightly change, and no other diffraction peaks occurred during the whole process, indicating that Ag atoms migrated along the surface and were highly dispersed on the HMO surface [10].

3.2 Inductively coupled plasma atomic emission spectroscopy

For qualitative and quantitative analysis of the elemental composition of materials, inductively coupled plasma (ICP) is the main light source employed in atomic emission spectroscopy (AES), which has the advantages of low detection limit, wide linear range, less ionization and chemical interference, high accuracy and precision. In the 1850s, Kirchhoff and Bunsen made the first spectroscope for spectral analysis and obtained the characteristic spectra of some elements, which laid the foundation of qualitative spectral analysis. In the 1920s, Gerlach proposed an internal standard method to solve the instability problem of light source, which provided feasibility for quantitative analysis. The introduction of ICP light source in the 1960s has greatly promoted the development of emission spectrum analysis. In recent years, with the use of detection devices, the ability of multielement simultaneous analysis has been greatly improved [11]. He et al. proposed a general strategy to increase the yield of SACs *via* introducing structural defects into PCN-222(Co)[Zr_6O_8(TCPP-CO)$_2$(H_2O)$_8$] through defect engineering and then converting them into Co SACs *via* carbonizing and acid etching. In the ICP-AES analysis, the Co content in the DPCN-222(Co)-X sample decreased gradually, but the Zr content remained relatively constant, confirming the existence of structural defects in DPCN-222-X and DPCN-222(Co)-X, which were

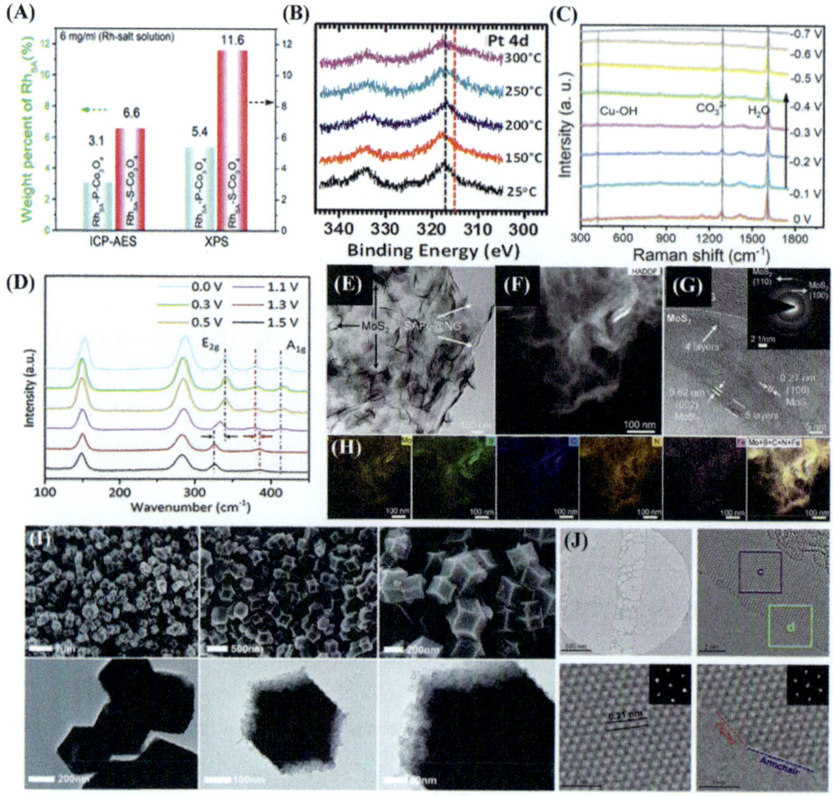

Figure 3.2 (A) Rh_{SA} loading (wt.%) on the surface of S-Co_3O_4 and P-Co_3O_4 determined by ICP-AES (bulk-technique); (B) XRD of Pt 4d of 0.1 atom % Pt/Co_3O_4; (C) 2Bn-ACu@UiO-67; (D) Raman spectra of the $MoReS_3$ sample with different voltages applied; (E) TEM image of MoS_2/SAFe@NG; (F) HAADF image of MoS_2/SAFe@NG; the inset image in (G) is the SAED pattern; (H–I) corresponding EDS mapping images of MoS_2/SAFe@NG and SEM images and TEM images of Fe_1/d-CN; (J) AC-HRTEM image of an MPG ribbon. *(A) A. Kumar, X. Liu, J. Lee, B. Debnath, A.R. Jadhav, X. Shao, et al., Discovering ultrahigh loading of single-metal-atoms via surface tensile-strain for unprecedented urea electrolysis, Energy Environ. Sci. 14 (2021) 6494–6505 (Copyright 2021-RSC) (B) L. Nguyen, S. Zhang, L. Wang, Y. Li, H. Yoshida, A. Patlolla, et al., Reduction of nitric oxide with hydrogen on catalysts of singly dispersed bimetallic sites Pt1Com and Pd1Con, ACS Catal. 6 (2016) 840–850 (Copyright 2016-ACS) (C) S. Chen, W.H. Li, W. Jiang, J. Yang, J. Zhu, L. Wang, et al., MOF encapsulating N-heterocyclic carbene-ligated copper single-atom site catalyst towards efficient methane electrosynthesis, Angew. Chem. Int. Ed. Engl., 61 (2022) e202114450 (Copyright 2021 Wiley-VCH GmbH) (D) Y. Fu, Y. Shan, G. Zhou, L.Y. Long, L.L. Wang, K. Yin, et al., Electric strain in dual metal janus nanosheets induces structural phase transition for efficient hydrogen evolution, Joule, 3 (2019) 2955–2967 (Copyright 2019-Elsevier) (G) B. Chen, T. Wang, S. Zhao, J. Tan, N. Zhao, S.P. Jiang, et al., Efficient reversible conversion between MoS2 and Mo/Na2S enabled by graphene-supported single atom catalysts, Adv. Mater., 33 (2021) e2007090*

the result of intermittent missing joints [12]. Kumar et al. synthesized Rh_{SA}-S-Co_3O_4 bifunctional electrocatalyst *via* liquid nitrogen quenching method, and the loading amount of Rh_{SA}-P-Co_3O_4 was determined to be 3.1 wt.% while that of Rh_{SA}-S-Co_3O_4 was 6.6 wt.% (Fig. 3.2A), whose specific composition was effectively analyzed by ICP-AES [13].

3.3 X-ray photoelectron spectroscopy and in situ X-ray photoelectron spectroscopy

XPS technique provides a powerful platform for the analysis of surface chemical valence. The theoretical basis of XPS method developed in 1905 was Einstein's photoelectric effect that the concept of photons was used to describe the ejection of electrons from a surface when photons were impinged upon it. In 1981, Siegbahn et al. were awarded the Nobel Prize in Physics for their first development of the XPS technique. With a bundle of a certain energy X-ray irradiation solid samples, incoming photons interact with the samples, the photons are absorbed and energy transferred to the atoms of a bound electron shell, the electronic part of the energy used to overcome the binding energy and work function, the remaining energy as its kinetic energy and emitted, become photoelectron, this process is called the photoelectric effect. The formula is as follows:

$$hv = E_k + E_b + E_r \quad (3.1)$$

where hv stands for the energy of X photon (h is Planck's constant, v is the frequency of light); E_k stands for the energy of photoelectrons; E_b stands for the binding energy of electrons; and E_r is the recoil energy of an atom.

X-rays with sufficient energy interact with the sample, and X photons transfer all the energy to bound electrons in atoms or molecules, making electrons of different energy levels ionize with a specific probability. Therefore, XPS technique can detect the intensity distribution of

(Copyright 2021 Wiley-VCH GmbH) (H)–(I) M.Q. Zhao, H.R. Liu, H.W. Zhang, W. Chen, H.Q. Sun, Z.H. Wang, et al. ORR catalyst with single-atom iron sites derived from a double-layer MOF for superior flexible quasi-solid-state rechargeable Zn–air batteries, Energy Environ. Sci. 14 (2021) 6455–6463 (Copyright 2021-RSC). (J) M. Chen, Z.X. Wang, X. Ge, Z. Wang, K. Fujisawa, J. Xia, et al., Controlled fragmentation of single-atom-thick polycrystalline graphene, Matter, 2 (2020) 666–679 (Copyright 2020-Elsevier).

photoionization electrons of different energies. In the analysis of the traditional XPS measurement, it is inevitable to be exposed to atmospheric environment in the process of sample preparation, and the properties of catalysts are certain to be greatly affected by the large errors, which limits the application of practical dynamic photoelectrochemical analysis. For this, in situ XPS technique is proposed to enable us to intuitively obtain the intensity information in equilibrium and nonequilibrium state of materials, which plays an important role in the field of single-atom catalysis [20]. Nguyen et al. obtained Co_3O_4 modified with highly dispersed Pt SACs for the reduction of nitric oxides, and in situ XPS was used to study the cation status of Pt atoms in bimetallic Pt_1Co_m active sites during the catalytic reduction of NO at different temperatures (Fig. 3.2B). In the Pt 4*d* spectrum, when the temperature was lower than 300°C, these separately dispersed cationic Pt atoms did not aggregate into nanoparticles on the surface of Co_3O_4. In addition, the different results between Pt/Co_3O_4 and the control group of Pt/SiO_2 strongly indicated their significantly different coordination environment of Pt, which clearly confirmed that Pt atoms of the separately dispersed bimetallic sites were in a cationic state during the H_2 reduction of NO at up to 300°C [14].

3.4 Fourier transform infrared spectrometer and in situ FTIR

The theoretical principle is that in organic molecules, atoms that make up chemical bonds or functional groups are in a state of constant vibration, which is comparable to the vibration frequency of infrared light. Therefore, the chemical bonds or functional groups in the molecules have forced vibration absorption when infrared light is used to irradiate organic molecules. Different chemical bonds or functional groups have different absorption frequencies and will be in different positions in the infrared spectrum. FTIR spectrum is used to observe vibrations and rotations between different atoms, which obtains the basic molecular structure through transmission [21]. In situ FTIR is divided into two kinds of transmission and diffuse reflection. The former is referred to the use of high vacuum probe molecule adsorption and adsorption of small molecules in the sample surface activation process, while the latter can carry out atmospheric adsorption, pressure adsorption, and vacuum adsorption. In situ FTIR is more commonly used for the study of the mechanism of single atoms catalysis. Yan et al. designed a single-atom Pt catalyst supported on hydroxyapatite (HAP) by a simple hydrothermal strategy [22]. The Pt_1/HAP catalyst showed high selectivity and excellent stability in the selective oxidation of C_2-C_4 polyols to hydroxyl acids. In situ FTIR showed that the oxidation of glycerol to glyceric acid is mainly through the activation of the primary hydroxyl group on Pt_1/HAP. The Pt-O-P bond in PO_4^{3-}(Pt_1-OPO_4^{3-}) with strong electron absorption not only realized the activation of C-H bond but also made the transition state of the adsorbed hydroxy acid to C-C cleavage unstable, leading to a sharp increase in the selectivity of hydroxy acid. In addition, PO_4^{3-} species were strongly coordinated with Pt single atoms, ensuring the excellent stability of Pt_1/HAP for more than 160 h efficient catalysis.

3.5 Raman and in situ Raman

Raman spectroscopy is a kind of scattering spectrum based on Raman scattering effect. The information of molecular vibration and rotation can be obtained and applied to the study of molecular structure by analyzing the scattering spectrum with different frequencies from the incident light. The Raman scattering is referred to the inelastic scattering caused by molecular vibration, optical phonon excitation in solid, and then the interaction with laser light. The surface coarsening method is employed to enhance the Raman intensity. Taking advantage of in situ Raman spectroscopy scattering the incident light with a great change in frequency, the monochromatic incident light excited on the surface of the electrode potential modulated by the electrode, by measuring the relationship between the intensity of the scattered Raman spectral signal and the change of polarization properties and the change of electrode potential or current intensity [23].

Chen et al. designed MOF encapsulating N-heterocyclic carbene-ligated Cu SACs catalyst (2Bn-Cu@UiO-67) for methane electrosynthesis. In situ Raman spectra showed a peak belonging to Cu-OH at 409.2 cm^{-1}, which did not weaken in intensity even when different potential conditions were applied, providing direct evidence for the stable modification of Cu species (Fig. 3.2C). The peaks at 1250 and 1600 cm^{-1} can be attributed to the signal peak of CO_3^{2-} and chemisorbed H_2O. Compared with 2Bn-Cu@UiO-67, a new peak attributed to Cu_2O was observed at 630 cm^{-1}, which provided further evidence that Cu species in 2Bn-ACu@UiO-67 were not linked by NHCs. Therefore, in situ Raman spectroscopy at various potentials confirmed that NHCs ligands contribute to the stable existence of Cu monoatomic catalysts during the CO_2 reduction [15]. Fu et al. designed Janus $MoReS_3$ nanosheets with asymmetric out-of-plane T' phase, which was sensitive to applied voltage and easy to transition to T'' phase, which was further verified by in situ Raman spectrum. Typical peaks at 416.3 and 341.2 cm^{-1} were corresponded to the out-of-plane A_{1g} and the in-plane E_{2g} patterns (Fig. 3.2D). The peak of A_{1g} mode remained constant at 404.3 cm^{-1} under low applied voltage, but dropped sharply at 1.1 V and disappeared at 1.3 V. In comparison, the peak of E_{2g} mode rapidly decreased from 341.8 cm^{-1} (0.0 V) to 334.8 cm^{-1} (1.1 V) and anchored at 327.4 cm^{-1} (1.3 V), which strongly proved that the structural transition from T' to T'' phase was efficiently achieved at low voltage [16].

3.6 Transmission electron microscopy and high-angle annular dark-field scanning transmission electron microscopy

To fully understand the special catalytic performance of SACs, it is necessary to directly observe the single dispersed catalytic sites and reaction pathways on the support bases. The morphology and metal dispersion of the prepared nano-catalysts is usually determined by scanning electron microscopy (SEM) and transmission electron microscopy (TEM). Conventional high-resolution transmission electron microscopy (HRTEM) has been served as a powerful tool to identify structural features and basic morphology in the nanoscale. However, it is far from direct evidence that metal atoms are mono-dispersed anchoring on the substrate surface, and the aberration of the magnetic lens becomes the main reason for the application limitation [24]. For this, Nellist and Pennycook first observed a single isolated metal atom on a carrier from STEM by focusing an electron probe on a sample to detect electrons scattered to a ring dark-field detector in 1996, which could generate images in STEM with elemental contrast Z (the atomic number difference), which was defined as high-angle annular dark field-scanning transmission electron microscopy (HAADF-STEM), whose resolution is dependent on the element contrast Z. HAADF-STEM technique is considered to be the most effective visualization method for single atoms anchored on a carrier composed of light elements [25].

Zhang et al. synthesized Pt single atoms supported on iron oxide (Pt_1/FeO_x), and observed that Pt single atom by HAADF image and then proposed the concept of "SAC" [26]. Chen et al. screened the SAM'@NG catalyst to enhance the electrochemical performance of MoS_2 through theoretical simulation, and the interface of SAFe@NG to MoS_2 showed the largest binding energy, which illustrated strong stability. TEM showed that MoS_2 nanosheets were uniformly anchored on the surface of SAFe@NG, forming a typical sandwich structure with SAFe@NG as the intermediate layer (Fig. 3.2E). The MoS_2 distribution on SAFe@NG was further investigated by EDS element mapping (Fig. 3.2F, H), and all elements were in good agreement with each other, indicating that the MoS_2 nanosheet was completely fixed on the SAFe@NG substrate. Two typical lattice spacing of 0.62 and 0.27 nm were implied from the HRTEM images (Fig. 3.2G), which was attributed to the lattice plane of (002) and (100) for MoS_2. This result was in good agreement with selected electron diffraction (SAED) patterns [17,27].

3.7 Aberration-corrected high-resolution transmission electron microscopy

In recent years, great progress has been made in the field of experimental and theoretical analysis of the atomic structure of materials. Especially, the aberration correction is realized on high-resolution transmission electron microscope, which becomes a milestone in the development of electron microscopy. This enables us not only to observe in the subangstrom scale but also to measure the atomic positions of local structures such as material surfaces, interfaces, and catalyst particles with micromicron precision, which is comparable to the measurement accuracy of atomic positions in macroscopic single crystal by XRD technique. Pan et al. developed a scalable two-step annealing method for preparing ultra-high-density SACs (UHD-SACS) libraries to demonstrate the effectiveness of nitrogen-doped carbon (NC), polymeric carbon nitride (PCN), and metal oxides substrates impregnated with various metals combined with two-step carbonization for the synthesis of ultra-high density monoatomic catalysts. Experimental and theoretical analyses showed that controlled fractional carbonization can gradually remove weakly interacting equilibrium ion species (e.g., Cl ions) to stabilize isolated metal atoms (e.g., Ni ions), thereby avoiding agglomeration caused by complete decomposition [28]. Zhao et al. developed Fe_1/D-CN catalyst pH universal electrolyte that showed excellent ORR electrocatalytic activity. Isolated single Fe sites were directly confirmed by Aberration-corrected high-resolution transmission electron microscopy (AC-HRTEM) analysis, indicated by a number of individual bright spots circled with orange markers (Fig. 3.2I). The Fe content in Fe_1/D-CN catalyst was 2.04%, which was used by ICP mass spectrometry (ICP-Ms) [18]. Chen et al. designed an effective method to control the fragmentation of MPG sheets by managing the local strain during the neck shrinkage of thermoplastic polymers. The atomic structure and mass of the MPG band and the effect of grain boundaries on the crack propagation path were further investigated by low pressure AC-HRTEM. Through the AC-HRTEM image around the edge area of two adjacent bands (width ~ 810 nm) formed by the neck shrinkage process, it can be observed that the MPG sheet is straight and brittle fractured after the neck shrinkage process (Fig. 3.2J). The lattice fringes are found to be spaced 0.21 nm apart, which matches well with the (100) plane of carbon atoms arranged hexagonally in graphene, confirming that the MPG bands exhibit highly crystalline qualities [19].

3.8 Time-of-flight secondary ion mass spectrometry

Modern time-of-flight secondary ion mass spectrometers have minimum measurable concentrations on the order of PPM (parts per million) for almost all elements and even reach PPB (parts per billion). The second feature is high longitudinal resolution. The latest-generation Time-of-flight secondary ion mass spectrometry (TOF-SIMS) have a resolution of two to three atomic layers. Meanwhile, with the improvement of technology, the analysis area becomes smaller and smaller [29]. Hence, TOF-SIMS has an irreplaceable position in the analysis of the material composition, doping, and impurity contamination. The principle is that when the surface of the sample is bombarded by a high-energy focused primary ion, the primary ion is injected into the sample to be analyzed, and the kinetic energy is transferred to the solid atoms. By measuring the time-of-flight of different secondary ions and measuring their mass/charge-to-mass ratio, the surface and interior element distribution characteristics of the bombarded samples were analyzed. TOF-SIMS analyze almost all elements, including hydrogen, can give isotopic information, and analyze the composition and molecular composition of compounds. It has good sensitivity, high mass resolution, wide range of measurable molecular weight, and can also perform microregional composition imaging and depth profile analysis. Yang et al. reported the synthesis of

Au single atoms on amino-rich graphite carbon nitride by a mild and environmentally friendly reduction of urea. The composition of U-ACN was revealed by TOF-SIMS. The mass-to-charge ratio peaks located at 1, 26, and 64 were the characteristic peaks of H^- CN^- and $C_2N_3^-$ ionic groups, respectively. Two peaks were corresponded to Au fragment ions, $C_2N_2Au^-$ and $AuN_2O_4^-$, indicating that Au atoms interacted with other atoms in the CN matrix [30]. Chen et al. prepared multifunctional sulfones for oxidation via heterogeneous single-atom intercalation catalysts. The catalyst consists of vertically grown MoS_2 nanosheets on carbon paper, ensuring a high surface area for edge exposure sites for catalysis. By using the sample destruction technique of MoS_2 TOF-SIMS, it was proved that the Co encapsulation was a single crystal, and even after prolonged sputtering, the uniform Co^+ signal can be clearly seen in the horizontal and vertical directions. The uniform distribution of single Co between MoS_2 layers was confirmed by atom-resolution scanning transmission electron microscopy (STEM) images [31].

3.9 X-ray absorption spectroscopy

The appearance of X-ray absorption spectrum can be traced back to the first discovery of X-ray by Rontgen in 1895, the first measurement of absorption edge by Maurice de Broglie in 1913, and the discovery of fine structure by Fricke and Hertz in 1920. Louis de Broglie eventually took advantage of the atomic photoelectric effect and won the 1929 Nobel Prize in physics for measuring the XAS spectrum by a laboratory light source. The principle is illustrated as follows: When the incident X-ray energy is lower than the binding energy of electrons in elemental orbitals (such as *s* orbital), these electrons cannot be in the highest unoccupied or vacuum state, and such low-intensity interaction between X-ray and electrons leads to a flat region. However, when the X-ray is in the energy range of 1–200 keV that qualifies the sample for the excited state, it will emit fluorescence, photoelectrons, Auger electrons, and inelastic scattering electrons during the relaxation back to the ground state. Commonly, due to absorption and scattering, the X-ray intensity is attenuated, and the absorption intensity is much greater than the scattering. Scattering is the basis of diffraction, and the smooth back bottom of spectral line is from scattering. Absorption is related to elements and valence and is used for composition and valence analysis of materials. As an advanced characterization method with ultra-high resolution (approximately 0.01 Å), XAS based on synchrotron radiation sources enables researchers to analyze metal SACs at the atomic level, which greatly promotes the development of such catalytic materials [27]. On this basis, X-ray absorption fine structure (XAFS) spectrogram is developed to obtain the electronic and geometric local structure information of the system by studying these oscillatory structures. The measurement range of XAFS covers single crystal, noncrystal, and mixtures. It mainly provides information about the electronic structure and geometric structure of atomic clusters in a small range. The XAFS data analysis steps mainly include the following four steps: [32] (1) conduct wavelet-transformed EXAFS (WT-EXAFS) to roughly determine the types of coordination elements around atoms; (2) extract the atomic spacing and coordination number information by Fourier-transformed EXAFS (FT-EXAFS); (3) analyze the 3D configuration of atoms through XANES, establish a clear model according to the existing conditions and experience, and perform theoretical calculation to iterate the atomic structure most in line with the actual and test data; (4) design the most appropriate DFT model to predict the change of atomic structure, reaction activation energy, reaction trend, and other information.

According to the difference in action mechanism and peak shape, XAFS are usually divided into two regions: (1) extended X-ray absorption fine structures (EXAFS), the region above the absorption edge of 50−1000 eV, is mainly characterized by small amplitudes resembling sinusoidal fluctuations; (2) X-ray absorption near edge structures (XANES) or near-edge X-ray absorption fine structures (NEXAFS), the region from the front to 50 eV above the absorption edge, the region from the front to 50 eV above the absorption edge [33]. The former EXAFS results from the multiple scattering of the excited photoelectrons by the surrounding atoms, which reflects not only the geometric configuration of atoms in the environment around absorbing atoms but also the structure of electronic states at low energy sites near the Fermi level of condensed matter, so it becomes a useful tool for studying condensed matter. The latter XANES is mostly used to distinguish the near-edge structure of Z atomic coefficient, which is usually divided into three parts (XANES = pre-edge + edge + XANES), according to the signal mechanisms, and each part can be separately analyzed in practice.

The differences between XANES and EXAFS are as follows: (1) XANES sends shock waves, absorbs clear signals, and is easy to measure; (2) the collection time of XANES spectrum is short and is suitable for time-resolved experiments; (3) XANES is more sensitive to chemical information such as valence, unoccupied electron orbitals, and charge transfer; (4) XANES is weakly dependent on temperature and can be used for high temperature in situ chemistry experiments; (5) XANES has the "fingerprint effect," that is, the feature identifies the local structure around the absorbing atoms, which can quickly identify elements, but it is not suitable for the multimixture system.

According to the basic formula of EXAFS, improving the resolution of EXFS data signal in K-space is helpful to distinguish the contribution of different elements, and a series of calculation and transformation of EXAFS formula is carried out. Obtaining Fourier transform EXAFS spectrum (FT-EXAFS) and wavelet transform EXAFS spectrum (WT-EXAFS) are used to illustrate the local structure of matter and meet the requirements of extensive applications in the future. Jiang et al. investigated Sb single atoms anchored on nitrogen-doped carbon nanosheets (denoted as Sb SA/NC) toward carbon dioxide reduction reaction [30]. Firstly, the electronic structures of N and C species in Sb SA/NC were determined by K-edge soft XAS spectrum. It was found that the four peaks in the C K-edge spectrum at 285.4, 287, 288.6, and 292.4 eV were attributed to the antibonding orbitals of $\pi^*C=C$, π^*C-N and σ^*C-C, respectively. The Sb-N-C bond was determined by peaks between 286 and 289 eV (Fig. 3.3A). Moreover, the XANES spectra of Sb K-edge showed that Sb SA/NC catalyst lay between the absorption edges of standard Sb Foil and Sb_2O_3, indicating that the valence state of Sb was between 0 and +3. XANES spectral fitting result showed that the average oxidation state of Sb in Sb SA/NC was +2.1 (Fig. 3.3B). The FT-EXAFS indicated that an obvious peak near 1.36 Å was attributed to the Sb-N bond (Fig. 3.3C). Compared with the standard Sb Foil, no peak from the Sb-Sb bond was detected near 2.18 Å in the Sb SA/NC sample, indicating that Sb atoms were monodispersed on the carbon matrix and N atoms were served as the main binding sites to form the coordination of the Sb-N bond. The coordination configuration of Sb SA/NC was quantitatively analyzed by least-squares EXAFS fittings (Fig. 3.3D). The result showed that the coordination number of Sb center was 4, and the average Sb-N bond length was 1.98 Å, indicating the local structure of $Sb\text{-}N_4$. The WT-EXAFS was further employed to determine the dispersion position of Sb single atoms in the Sb SA/NC sample (Fig. 3.3E). Compared with the WT contour plots of standard Sb Foil and Sb_2O_3, the strong WT signal at 3.5 $Å^{-1}$ was originated from the Sb-N bond.

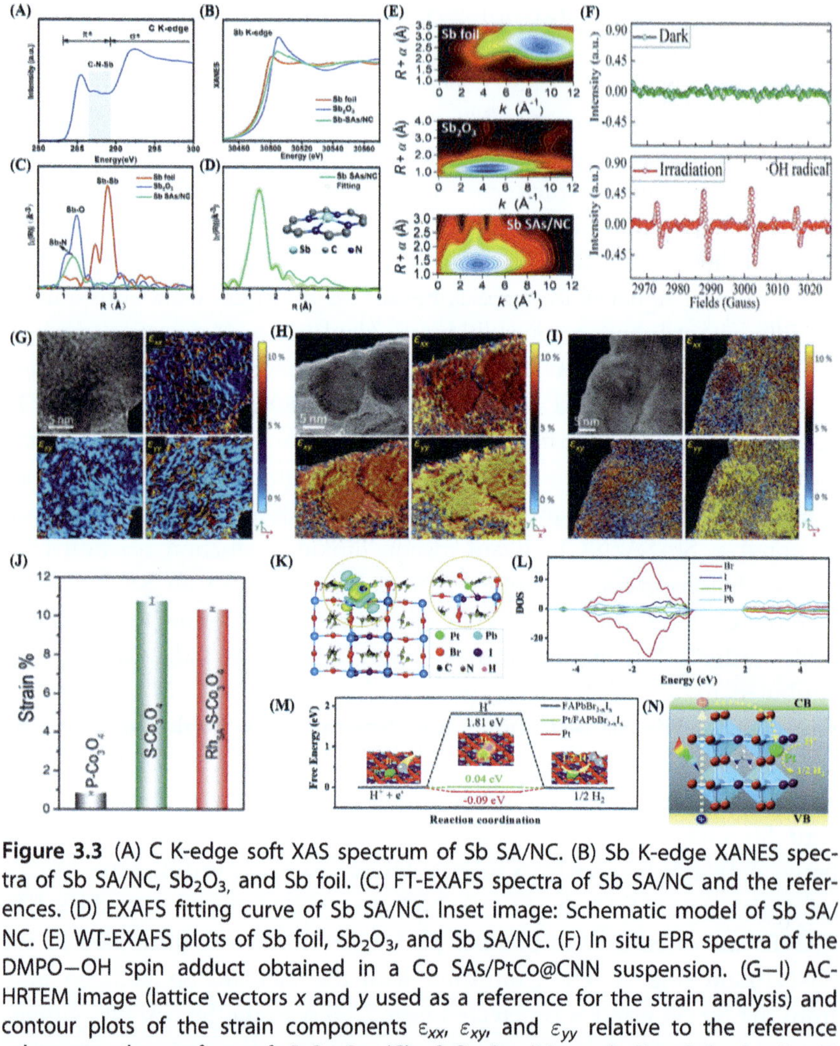

Figure 3.3 (A) C K-edge soft XAS spectrum of Sb SA/NC. (B) Sb K-edge XANES spectra of Sb SA/NC, Sb_2O_3, and Sb foil. (C) FT-EXAFS spectra of Sb SA/NC and the references. (D) EXAFS fitting curve of Sb SA/NC. Inset image: Schematic model of Sb SA/NC. (E) WT-EXAFS plots of Sb foil, Sb_2O_3, and Sb SA/NC. (F) In situ EPR spectra of the DMPO—OH spin adduct obtained in a Co SAs/PtCo@CNN suspension. (G—I) AC-HRTEM image (lattice vectors x and y used as a reference for the strain analysis) and contour plots of the strain components ε_{xx}, ε_{xy}, and ε_{yy} relative to the reference values on the surface of P-Co_3O_4 (G), S-Co_3O_4 (H), and Rh_{SA}-S-Co_3O_4 (I). (J) Comparison of tensile strain values on the surface of P-Co_3O_4, S-Co_3O_4, and Rh_{SA}-S-Co_3O_4 calculated from their respective AC-HRTEM images. Theoretical calculation and investigations: (K) Differential charge densities of Pt/$FAPbBr_{3-x}I_x$, and the inset shows the corresponding local structure. The isosurface value is set to be 0.001 e $Å^{-3}$, and the positive and negative charges are shown in yellow and cyan, respectively. (L) Projected density of states (PDOS) of Pt/$FAPbBr_{3-x}I_x$. (M) Calculated free energy of H* adsorption for Pt, $FAPbBr_{3-x}I_x$ and Pt/$FAPbBr_{3-x}I_x$, and the inset shows the corresponding schematic mechanism. (N) Schematic illustration of the photocatalytic process of Pt/$FAPbBr_{3-x}I_x$. *(E) Y. Yang, F. Li, J. Chen, J. Fan, Q. Xiang, Single Au atoms anchored on amino-group-enriched graphitic carbon nitride for*

3.10 Electron paramagnetic resonance

Electron Paramagnetic Resonance (EPR) is a spectral technique for detecting and studying paramagnetic materials containing unpaired electrons. The research objects include atoms and molecules with odd number of electrons, ions with unfilled inner electron shells, free radicals generated by radiation, semiconductors, metals, etc. [36]. Through the EPR spectrum, we know the states of unpaired electrons in molecules, atoms and ions, and their surroundings, and thus the structure and chemical bonds of matter can be obtained. As is known to us all, the electron is a kind of elementary particle with a certain mass and a negative charge, which can be carried out in two kinds of motion: One is in orbit around the nucleus, and the other is the spin of an axis through its center. Due to the movement of electrons, a moment of force or torque is created, which in turn produces an electric current and a magnetic moment. In the external constant magnetic field H, the magnetic moment of the electron acts like a small magnetic rod or needle. Since the spin quantum number of the electron is 1/2, there are only two orientations of the electron in the external magnetic field: (1) It corresponds to a low energy level and the energy is $-1/2g\beta H$ when it is parallel to H; (2) when antiparallel to H, corresponding to the higher level, the energy is $+1/2g\beta H$, the energy difference between the two levels is $g\beta h$. If, in the direction perpendicular to H, an electromagnetic wave with a frequency of ν is added to satisfy the condition of $h\nu = g\beta H$ (h is Planck's constant, $h = 6.62620 \times 10{-}27$ g s, g is the spectral splitting factor, $g = 2.00232$, β is the natural unit of the electron magnetic moment, which is called Bohr magneton, $\beta = 9.2710 \times 10^{-21}$ g $\cdot$ Gauss^{-1}), electrons in lower energy levels absorb the energy of the electromagnetic wave and transition to higher energy levels, producing EPR. Yao et al. prepared the CoSAs/PtCo@CNN

photocatalytic CO2 Reduction, ChemSusChem, 13 (2020) 1979–1985 (Copyright 2020 Wiley-VCH GmbH) (F) Z.L. Jiang, T. Wang, J.J. Pei, H.S. Shang, D.N. Zhou, H.J. Li, et al., Discovery of main group single Sb-N4 active sites for CO2 electroreduction to formate with high efficiency, Energy Environ. Sci., 13 (2020) 2856–2863 (Copyright 2020-RSC) (J) A. Kumar, X. Liu, J. Lee, B. Debnath, A.R. Jadhav, X. Shao, et al., Discovering ultrahigh loading of single-metal-atoms via surface tensile-strain for unprecedented urea electrolysis, Energy Environ. Sci., 14 (2021) 6494–6505 (Copyright 2021-RSC) (N) Y.Q. Wu, Q. Wu, Q.Q. Zhang, Z.Z. Lou, K.F. Liu, Y.D. Ma, et al., An organometal halide perovskite supported Pt single-atom photocatalyst for H2 evolution, Energy Environ. Sci., 15 (2022) 1271–1281 (Copyright 2022-RSC).

catalyst for photocatalytic overall water splitting (POWS) and observed the characteristic signal of •OH in the suspension by in situ EPR spectroscopy (Fig. 3.3F) [34]. The results indicated that the POWS involves the oxidation of a single electron to produce •OH when employing the CoSAs/PtCo@CNN catalyst. However, the •OH signal could not be detected in the control groups of Pt/PtCo@CNN, Pt/PtCo@CN, CoSAs@CNN, and PtCo@CNN samples. Therefore, a large number of •OH groups were generated under aqueous environment, mainly due to the synergistic interaction between Co single atoms and PtCo alloy at the CNN interfaces, indicating the spillover phenomenon of protons or hydroxyl groups between the CoSAs and PtCo alloys.

3.11 Electron energy loss spectroscopy

EELS possesses the incident electron beam to generate inelastic scattering in the sample. The electron loss energy *DE* directly reflects the scattering mechanism, chemical composition, and thickness of the sample, so it can analyze the element composition, chemical bonds, and electronic structure of the thin sample microregion [37]. Promoted by the electron inelastic collision theory, it has high sensitivity to the surface and adsorbed molecules including the adsorbed hydrogen. More importantly, it can identify the atomic and molecular structure and chemical properties of the adsorbed surface. The incident electron backscattering phenomenon occurs when the incident electron beam irradiates the specimen surface. The electrons backscattered back to the surface are composed of two parts: one part without energy loss, called elastic scattered electrons; the other part has energy loss, which is called inelastic scattered electrons, in accordance with the law of conservation of energy and momentum. Among the inelastic scattered electrons, there are some Auger electrons with a certain characteristic energy, whose characteristic energy is only related to the energy of the material. If the number of these Auger electrons is detected on the sample according to the energy distribution, the composition of each element of the material can be calibrated, which is called Auger electron spectroscopy analysis technology. If its characteristic energy is not only related to the elements of the material but also to the energy of the incident electron, it is called characteristic energy loss electron spectroscopy. If the number of energy loss electrons is detected on the sample according to the energy distribution, a series of spectral peaks can be obtained, which is called electron energy loss spectrum analysis technology and has a wide range of applications in the field of single atom catalysis [38]. David et al. prepared atomically dispersed Ni-PACN electrocatalysts for CO_2 reduction and conducted atom-resolution EELS spectra by placing STEM probes directly at the Ni atom [39]. The results showed that for the intensities of N k-edge, Ni L3-edge and Ni L2-edge were 400, 850, and 866 eV, respectively. The Ni L-edge and N k-edge were observed to be almost completely reduced, proving that N and Ni were colocated at the Ni single atom sites and coordinate with each other directly. The earlier provided direct microscopic evidence for the coexistence of N and Ni in Ni-N-C materials and motivated further investigations into the exact ligand environment of these NiN_x species.

3.12 Geometric-phase analysis

The GPA method was first proposed by Professor M.J. Hytch in the Journal Ultramicroscopy in 1998. The basic principle of the geometric phase method is to obtain the magnitude of the displacement field and strain field in related regions by high-resolution image analysis. The specific operation method is to perform Fourier transform on HRTEM, select two nonlinear vectors, and then perform inverse Fourier transform, that is, to get the displacement field and strain field of the microscopic measurement scale with high precision. Though there are two common methods to obtain strain fields, including peak-finding method and GPA method, the latter taking advantage of HRTEM processing to analyze the displacement field and strain field has a higher accuracy and wider application range than the former one, with an accuracy up to 0.003 nm. The GPA method has been widely applied to measure the displacement field and strain field around the precipitated phase of cracks, dislocations, and structural alloys in semiconductor materials. Kumar et al. proposed a surface strain strategy to prepare a high-loading Ru monoatomic catalyst Rh_{SA}-S-Co_3O_4 for urea oxidation reaction on a large scale [13]. They used aberration-corrected high-resolution transmission electron microscope (AC-HRTEM) to verify the surface lattice strain. The spatial distribution of lattice strain components was plotted by GPA. According to AC-HRTEM image, the mean spacing *d* (311) of P-Co_3O_4 was 2.43 Å, closer to the theoretical value of pure Co_3O_4 (2.41 Å), while the mean spacing *d* (311) of S-Co_3O_4 increases to 2.67 Å. Therefore, ε_{xx}, ε_{xy}, and ε_{yy} obtained by GPA analysis are approximately zero (~0.8%) on the surface of P-Co_3O_4, while ε_{yy} gradually increased to a large positive value

(~10.7%) on the surface of S-Co_3O_4, especially the maximum value of the strain component of ε_{yy} was ~10.5% (Fig. 3.3G, H and J). This indicated that there was a tensile strain on the surface of S-Co_3O_4. The S-Co_3O_4 support was immersed in ethanol Rh-salt solution (6 mg mL^{-1}) for 20 min, followed by thermal annealing at 250°C in air for 2 h. The surface tensile strain values were slightly reduced (~10.4%) compared with the S-Co_3O_4 support, which may be due to the reduction of surface tensile strain by the subsequent thermal annealing process, confirming the mono-dispersion of Rh on the S-Co_3O_4 support (Fig. 3.3I, J).

3.13 First-principles calculations

The basic idea of first-principles calculation is to treat a system composed of multiple atoms as a system composed of multiple electrons and nuclei and to deal with the problem as "nonempirical" as possible according to the basic principles of quantum mechanics. It only needs five fundamental constants (m_0, e, h, c, and kB) to calculate the physical properties of the system, such as energy and electronic structure. It can determine the structure and basic properties of known materials and achieve accurate control at the atomic level [40,41]. It is a powerful tool for solving experimental theoretical problems and predicting the structural properties of new materials at the present stage. When the size of catalysts is reduced to clusters or single atoms, their energy level structure and electronic structure will undergo fundamental changes, and their activity, selectivity, and stability are worth investigating [42].

3.13.1 Atomic charge

The electrons are always distributed in infinite three dimensions. In order to conveniently study the electrical properties of the system, the three-dimensional space is divided into a series of atomic spaces. The electrons in each subspace are considered "subordinate" to an atom, and the number of electrons assigned to each atom can be obtained by subtracting it from the nuclear charge to know the atomic charge, which helps to determine either the selected atoms are relatively positively or negatively charged. The unmeasurable atomic charge is largely dependent on the segmentation of real space. Various definitions, such as Mulliken charge, Bader (AIM) charge, Hirshfeld charge, etc., have their own characteristics and are suitable for different occasions, with qualitative and quantitative differences.

3.13.2 Charge density difference

The differential density usually in the form of an iso-surface is the total density minus that of each existing atom in the multiple systems. When the combination of A and B is investigated, we often consider the function of $D(AB)-D(A)-D(B)$, where D stands for the charge density, which visually shows the accumulation and decrease in electron density before and after the combination of A and B.

3.13.3 Density of states

The density of states (DOS) refers to the number of states of the system in the interval of $E + dE$, which reflects the energy level structure of the periodic system and can be used to analyze the covalent action in the periodic system.

3.13.4 Projected density of states

Projected DOS (PDOS) represents the curve of the contribution of a particular segment to DOS. In particular, DOS can be decomposed into PDOS contributed by each atom to investigate the contribution of different atoms and atomic orbitals to the electronic structure.

3.13.5 Density functional theory

Density functional theory (DFT) refers to the method of solving the Kohn-Sham (KS) equation from the electron density and obtaining the energy and other properties of the multielectron system through various approximation theories. It can be traced back to the Thomas-Fermi model developed by Thomas and Fermi in the 1920s. The kinetic energy of an atom is expressed as a function of electron density, and classical expressions of nucleus–electron and electron–electron interactions are added to obtain atomic energy [43]. Wu et al. prepared an organometallic halide perovskite-supported Pt single-atom photocatalyst (denoted as Pt/$FAPbBr_{3-x}I_x$) for hydrohalic acid decomposition via self-adsorption and photo-reduction [35]. Fig. 3.3K shows that the loading of Pt single atoms increased the charge density around it, so the modification of Pt single atoms was favorable for accepting photogenerated electrons and providing a better electronic environment for HER catalysis. PDOS diagram verified a significant orbital hybridization between the introduced Pt single atoms and the coordinated Br/I atom, which was conducive to promoting charge transfer from the perovskite substrate to the Pt monatomic site, thus accelerating the subsequent proton reduction process (Fig. 3.3L). In order to further reveal the reason for the enhancement of HER activity in Pt/$FAPbBr_{3-x}I_x$ photocatalysis, the absolute value of hydrogen adsorption-free energy (ΔG_{H^*}) was studied. As shown in Fig. 3.3M, the ΔG_{H^*} value of $FAPbBr_{3-x}I_x$ was 1.18 eV, which meant that it should overcome huge adsorption energy for hydrogen evolution. However, the low adsorption energy was reduced to $\Delta G_{H^*} = 0.04$ eV after the modification of Pt with high HER activity (Fig. 3.3N). As a result of the enhanced local charge distribution, a strong electron reserve capacity was generated, acting as a proton acceptor to anchor the dissociated proton (H^*) on HER active sites, thus promoting the formation and release of molecular hydrogen.

Acknowledgment

The authors declare no competing financial interest.

References

[1] V. Hasija, S. Patial, P. Raizada, A. Aslam Parwaz Khan, A.M. Asiri, Q. Van Le, V.-H. Nguyen, P. Singh, Covalent organic frameworks promoted single metal atom catalysis: strategies and applications, Coord. Chem. Rev. 452 (2022) 214298.

[2] Y. Wang, J. Mao, X. Meng, L. Yu, D. Deng, X. Bao, Catalysis with two-dimensional materials confining single atoms: concept, design, and applications, Chem. Rev. 119 (2019) 1806–1854.

[3] S. Swain, A. Altaee, M. Saxena, A.K. Samal, A comprehensive study on heterogeneous single atom catalysis: current progress, and challenges, Coord. Chem. Rev. 470 (2022) 214710.

[4] K. Qi, M. Chhowalla, D. Voiry, Single atom is not alone: metal–support interactions in single-atom catalysis, Mater. Today 40 (2020) 173–192.

[5] I.C. Gerber, P. Serp, A theory/experience description of support effects in carbon-supported catalysts, Chem. Rev. 120 (2020) 1250–1349.

[6] Y. Pan, C. Zhang, Z. Liu, C. Chen, Y. Li, Structural regulation with atomic-level precision: from single-atomic site to diatomic and atomic interface catalysis, Matter 2 (2020) 78–110.

[7] Y. Yang, Y. Yang, Z. Pei, K.-H. Wu, C. Tan, H. Wang, et al., Recent progress of carbon-supported single-atom catalysts for energy conversion and storage, Matter 3 (2020) 1442–1476.

[8] F. Zhang, Y. Zhu, Q. Lin, L. Zhang, X. Zhang, H. Wang, Noble-metal single-atoms in thermocatalysis, electrocatalysis, and photocatalysis, Energy. Environ. Sci. 14 (2021) 2954–3009.

[9] L.B. McCusker, Electron diffraction and the hydrogen atom, Science 355 (2017) 136.

[10] Z. Huang, X. Gu, Q. Cao, P. Hu, J. Hao, J. Li, et al., Catalytically active single-atom sites fabricated from silver particles, Angew. Chem. Int. Ed. Engl. 51 (2012) 4198–4203.

[11] A. Marcos, S.J. Hill, A drift correction procedure for ICP-AES systems, Analyst 125 (2000) 1015–1020.
[12] J. He, N. Li, Z.G. Li, M. Zhong, Z.X. Fu, M. Liu, et al., Strategic defect engineering of metal–organic frameworks for optimizing the fabrication of single-atom catalysts, Adv. Func. Mater. 31 (2021) 2103597.
[13] A. Kumar, X. Liu, J. Lee, B. Debnath, A.R. Jadhav, X. Shao, et al., Discovering ultrahigh loading of single-metal-atoms via surface tensile-strain for unprecedented urea electrolysis, Energy Environ. Sci. 14 (2021) 6494–6505.
[14] L. Nguyen, S. Zhang, L. Wang, Y. Li, H. Yoshida, A. Patlolla, et al., Reduction of nitric oxide with hydrogen on catalysts of singly dispersed bimetallic sites Pt_1Co_m and Pd_1Co_n, ACS Catal. 6 (2016) 840–850.
[15] S. Chen, W.H. Li, W. Jiang, J. Yang, J. Zhu, L. Wang, et al., MOF encapsulating N-heterocyclic carbene-ligated copper single-atom site catalyst towards efficient methane electrosynthesis, Angew. Chem. Int. Ed. Engl. 61 (2022) e202114450.
[16] Y. Fu, Y. Shan, G. Zhou, L.Y. Long, L.L. Wang, K. Yin, et al., Electric strain in dual metal janus nanosheets induces structural phase transition for efficient hydrogen evolution, Joule 3 (2019) 2955–2967.
[17] B. Chen, T. Wang, S. Zhao, J. Tan, N. Zhao, S.P. Jiang, et al., Efficient reversible conversion between MoS_2 and Mo/Na_2S enabled by graphene-supported single atom catalysts, Adv. Mater. 33 (2021) e2007090.
[18] M.Q. Zhao, H.R. Liu, H.W. Zhang, W. Chen, H.Q. Sun, Z.H. Wang, et al., ORR catalyst with single-atom iron sites derived from a double-layer MOF for superior flexible quasi-solid-state rechargeable Zn–air batteries, Energy Environ. Sci. 14 (2021) 6455–6463.
[19] M. Chen, Z.X. Wang, X. Ge, Z. Wang, K. Fujisawa, J. Xia, et al., Controlled fragmentation of single-atom-thick polycrystalline graphene, Matter 2 (2020) 666–679.
[20] Q. Wang, H. Wu, Y. Wang, et al., XPS analysis of yolk-shell Sb_2O_3/WO_3 for ultra-fast acetone resistive sensor, J. Hazard. Mater. 412 (2021) 125175.
[21] J.M. Moreira, G.F. Campos, L.M. de Campos Pinto, G.R. Martins, B. Tirloni, C.S. Schwalm, et al., Copper (II) complexes with novel Schiff-based ligands: synthesis, crystal structure, thermal (TGA–DSC/FTIR), spectroscopic (FTIR, UV-Vis) and theoretical studies, J. Therm. Anal. Calorim. 147 (2021) 4087–4098.
[22] H. Yan, M. Zhao, X. Feng, S. Zhao, X. Zhou, S. Li, et al., PO_4^{3-} coordinated robust single-atom platinum catalyst for selective polyol oxidation, Angew. Chem. Int. Ed. Engl. 61 (2022). e202116059.
[23] B.K. Raman Govindhan, Nano Cu interaction with single amino acid tyrosine derived self-assemblies; study through XRD, AFM, confocal Raman microscopy, SERS and DFT methods, J. Phys. Chem. Solids 111 (2017) 123–134.
[24] B. Apicella, A. Ciajolo, A. Tregrossi, J. Abrahamson, R.L. Vander Wal, C. Russo, HRTEM and EELS investigations of flame-formed soot nanostructure, Fuel 225 (2018) 218–224.
[25] M. Nikulshina, P. Blanchard, C. Lancelot, A. Griboval-Constant, M. Marinova, V. Briois, et al., Genesis of active phase in MoW/Al_2O_3 hydrotreating catalysts monitored by HAADF and in situ QEXAFS combined to MCR-ALS analysis, Appl. Catal. B: Environ. 269 (2020) 118766.
[26] B. Qiao, A. Wang, X. Yang, L.F. Allard, Z. Jiang, Y. Cui, et al., Single-atom catalysis of CO oxidation using Pt1/FeOx, Nat. Chem. 3 (2011) 634–641.
[27] P. Zimmermann, S. Peredkov, P.M. Abdala, S. DeBeer, M. Tromp, C. Müller, et al., Modern X-ray spectroscopy: XAS and XES in the laboratory, Coordin. Chem. Rev. 423 (2020) 213466.

[28] X. Hai, S. Xi, S. Mitchell, K. Harrath, H. Xu, D.F. Akl, et al., Scalable two-step annealing method for preparing ultra-high-density single-atom catalyst libraries, Nat. Nanotechnol. 17 (2022) 174–181.

[29] S. Selvakumar, N. Nuns, M. Trentesaux, V.S. Batra, J.M. Giraudon, J.F. Lamonier, Reaction of formaldehyde over birnessite catalyst: a combined XPS and ToF-SIMS study, Appl. Catal. B: Environ. 223 (2018) 192–200.

[30] Y. Yang, F. Li, J. Chen, J. Fan, Q. Xiang, Single Au atoms anchored on amino-group-enriched graphitic carbon nitride for photocatalytic CO_2 reduction, ChemSusChem 13 (2020) 1979–1985.

[31] Z. Chen, C. Liu, J. Liu, J. Li, S. Xi, X. Chi, et al., Cobalt single-atom-intercalated molybdenum disulfide for sulfide oxidation with exceptional chemoselectivity, Adv. Mater. 32 (2020) e1906437.

[32] S. Yu, X. Wang, S. Yang, G. Sheng, A. Alsaedi, T. Hayat, et al., Interaction of radionuclides with natural and manmade materials using XAFS technique, Sci. China Chem. 60 (2016) 170–187.

[33] M.Y. Wang, L. Arnadottir, Z.C.J. Xu, Z.X. Feng, In situ X-ray absorption spectroscopy studies of nanoscale electrocatalysts, Nano-Micro Lett. 11 (2019).

[34] Z.L. Jiang, T. Wang, J.J. Pei, H.S. Shang, D.N. Zhou, H.J. Li, et al., Discovery of main group single Sb-N_4 active sites for CO_2 electroreduction to formate with high efficiency, Energy Environ. Sci. 13 (2020) 2856–2863.

[35] Y.Q. Wu, Q. Wu, Q.Q. Zhang, Z.Z. Lou, K.F. Liu, Y.D. Ma, et al., An organometal halide perovskite supported Pt single-atom photocatalyst for H_2 evolution, Energy Environ. Sci. 15 (2022) 1271–1281.

[36] F. Geng, Q. Yang, C. Li, et al., EPR and EPR imaging study on a $NaCrO_2$ cathode: electronic property and structural degradation with Cr dissolution, J. Phys. Chem. Lett. 12 (2021) 781–786.

[37] O.L. Krivanek, N. Dellby, T.C. Lovejoy, R.F. Egerton, P. Rez, Advances in atomic-resolution and molecular-detection EELS, Microsc. Microanal. 23 (2017) 1028–1029.

[38] T.C. Lovejoy, G.J. Corbin, N. Dellby, N. Johnson, M.V. Hoffman, A. Mittelberger, et al., Progress in ultra-high energy resolution EELS, Microsc. Microanal. 25 (2019) 628–629.

[39] D.M. Koshy, A.T. Landers, D.A. Cullen, A.V. Ievlev, H.M. Meyer III, C. Hahn, et al., Direct characterization of atomically dispersed catalysts: nitrogen-coordinated Ni sites in carbon-based materials for CO_2 electroreduction, Adv. Energy Mater. 10 (2020) 2001836.

[40] F. Ling, W. Xia, L. Li, X. Zhou, X. Luo, Q. Bu, et al., Single transition metal atom bound to the unconventional phase of the MoS_2 monolayer for catalytic oxygen reduction reaction: a first-principles study, ACS Appl. Mater. Interfaces 13 (2021) 17412–17419.

[41] K. Schwarz, R. Sundararaman, The electrochemical interface in first-principles calculations, Surf. Sci. Rep. 75 (2020) 100492.

[42] B. Liu, J. Zhao, Y. Liu, J. Xi, Q. Li, H. Xiang, et al., Application of high-throughput first-principles calculations in ceramic innovation, J. Mater. Sci. Tech. 88 (2021) 143–157.

[43] B.G. Janesko, Replacing hybrid density functional theory: motivation and recent advances, Chem. Soc. Rev. 50 (2021) 8470–8495.

CHAPTER 4

Single-atom catalysts for electrocatalytic oxygen reduction

Yuan Ha
School of Advanced Materials and Nanotechnology, Xidian University, Xi'an, P.R. China

4.1 Introduction

Severe energy crises and ever-deteriorating environmental problems have made the search for sustainable and renewable environmentally-friendly energy storage and conversion technologies [1–5]. Fuel cells and metal-air batteries have been recognized as the most promising candidate to meet this demand for energy storage and conversion owing to their high energy efficiency and renewable electrochemical energy devices [6–9]. Despite there are many merits of fuel cells and metal-air batteries, large-scale commercial applications are limited by the sluggish kinetics of oxygen reduction reaction (ORR) at the cathode electrode [10–14]. Electrocatalysts play a critical role in obtaining the high performance of fuel cells and metal-air batteries [15,16]. Therefore the rational design of efficient and robust catalysts to reduce the reaction energy barrier and enhance the activity is vital but challenging for ORR. Nowadays, conventional noble-metal Pt- and Ru-based catalysts show state-of-the-art catalytic activity and good stability in acidic and alkaline media for ORR, which are the most commonly used as commercial ORR catalysts for fuel cells and metal-air batteries [17]. Unfortunately, their scarcity and high costs hinder the widespread application of these energy devices. Inexpensive and earth-abundant nonnoble metal (Fe, Co, Ni, etc.)-based catalysts have demonstrated their activities similar to commercial Pt/C catalysts, which enable to replace the noble metal-based catalysts; yet their catalytic stability in acidic and alkaline media is still unsatisfactory [18]. In order to cater to the requirement of low-cost, high efficiency, and stability for ORR catalysts, reducing the particle size of the catalyst to the atomic-level structure is one of the most effective ways to dramatically boost the utilization efficiency of the metal sites [19–21]. Thus, single-atom catalysts (SACs) have been regarded as a new family of ORR catalysts due to their maximum atom

Single Atom Catalysts
DOI: https://doi.org/10.1016/B978-0-323-95237-8.00004-5

utilization efficiency (100%), adjustable electronic structures, and excellent catalytic performance [22–24]. Based on a large number of experimental studies and theoretical calculations, noble and nonnoble metal-based SACs with exceptional catalytic activities have demonstrated promising perspectives in sustainable fuel cells and metal-air batteries for ORR in both acidic and alkaline media [25–27]. Herein, recent achievements of single-atom ORR catalysts in fuel cells and metal-air batteries are systematically summarized, focusing on noble and nonnoble metal-based electrocatalysts including single Pt, Ru, Ir, Fe, Co, Ni, and Mn sites, etc. Meanwhile, theoretical studies of ORR mechanisms and the relationships between active sites and structure property are comprehensively dissected. Finally, to facilitate further development, the remaining challenges and future directions of SACs in fuel cells and metal-air batteries are presented.

4.2 The mechanisms for oxygen reduction reactions

Comprehensive understandings of ORR mechanisms on SACs are extremely helpful in developing high-performance SACs in both fuel cells and the discharging metal-air batteries (Fig. 4.1). It is generally accepted that the mechanism of ORR can be divided into two-electron ($2e^-$) and four-electron ($4e^-$) reaction pathways [28], where O_2 molecules are reduced to H_2O (or OH^-) and H_2O_2 (or HOO^-) products [29]. In other words, the $2e^-$ reduction of O_2 firstly generates intermediate products H_2O_2 in acidic media or HO_2^- in basic media, whereas the $4e^-$ reduction is a one-step reduction process of O_2 to directly generate H_2O in acidic media or OH^- in alkaline media [30–33]. The reaction pathways of ORR in different electrolytes are listed in Table 4.1.

For practical application (e.g., fuel cells and metal-air batteries), the H_2O_2 or HO_2^- intermediates generated from the $2e^-$ reduction pathway has been considered one of the competitive strategies for producing H_2O since they reduce energy conversion efficiency and damage the electrodes [34,35]. Thus, the $4e^-$ pathway directly reducing O_2 to H_2O is the ideal ORR reaction process for fuel cells and metal-air batteries. For the detailed procedure of $4e^-$ reduction, several possible transformations between *O, *OH, and *OOH intermediates make the ORR process more complex, which can be described as the following equations [Eqs. (4.1)–(4.12)]:

ORR in acidic media (*represents a surface-free site on the catalyst):

$$O_2 + * \rightarrow *O_2 \quad (4.1)$$

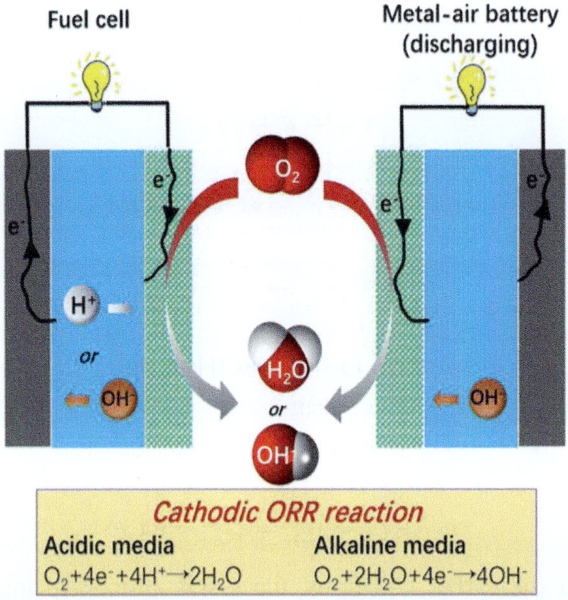

Figure 4.1 Illustrations of cathodic ORR in fuel cells and metal-air batteries (discharging). *Reproduced with permission L. Du, L. Xing, G. Zhang, M. Dubois, S. Sun, Strategies for engineering high-performance PGM-free catalysts toward oxygen reduction and evolution reactions, Small Methods, 4 (6) (2020) 2000016. Copyright 2020, WILEY-VCH.*

Table 4.1 ORR pathways in both acidic and alkaline media.

Pathway	Electrolyte	Reactions
$4e^-$	Acid	$O_2 + 4H^+4e^- \rightarrow 2H_2O$
	Alkaline	$O_2+2H_2O + 4e^- \rightarrow 4OH^-$
$2e^- + 2e^-$	Acid	$O_2+2H^+ +2e^- \rightarrow H_2O_2$; $H_2O_2+2H^+ +2e^- \rightarrow 2H_2O$
	Alkaline	$O_2+H_2O+2e^- \rightarrow HOO^- +OH^-$; $HOO^- +H_2O +2e^- \rightarrow 3OH^-$

$$*O_2+H^+ +e^- \rightarrow *OOH \tag{4.2}$$

$$*OOH+H^+ +e^- \rightarrow *O+H_2O \tag{4.3}$$

$$*O+H^+ +e^- \rightarrow *OH \tag{4.4}$$

$$*OH+H^+ +e^- \rightarrow *+H_2O \tag{4.5}$$

$$\text{Total: } O_2 + 4H^+ + 4e^- \rightarrow 2H_2O \tag{4.6}$$

ORR in alkaline media

$$O_2 + * \rightarrow *O_2 \tag{4.7}$$

$$*O_2 + H_2O + e^- \rightarrow *OOH + OH^- \tag{4.8}$$

$$*OOH + e^- \rightarrow *O + OH^- \tag{4.9}$$

$$*O + H_2O + e^- \rightarrow *OH + OH^- \tag{4.10}$$

$$*OH + e^- \rightarrow * + OH^- \tag{4.11}$$

$$\text{Total: } O_2 + 4H^+ + 4e^- \rightarrow 2H_2O \tag{4.12}$$

Based on the earlier equations, the ORR process mainly undergoes species adsorption, electron transfer, proton transfer, bond breaking and formation, and species desorption [36]. At first, oxygen absorbs on the surface-free site of the catalyst ($^*O_2$, *represents a surface-free site on the catalyst). The $^*O_2$ trapped electrons are transferred from the anode and are gradually reduced to *O, *OH, and *OOH intermediates. The final step is a desorption process of OH^- to H_2O. However, despite the massive efforts made to clarify the detailed kinetics of the whole $4e^-$ pathway in ORR, the rate-determining step is still no definitive conclusion. That is because the reaction pathway of ORR depends on many factors, such as the structure of catalysts, chemical composition, and experimental parameter conditions. Therefore rationally designing a suitable electrocatalyst with highly active and high $4e^-$ reduction process selectivity is the key to practical applications in fuel cells and metal-air batteries (Fig. 4.1) [37].

4.3 Introduction to SACs catalysts for oxygen reduction reactions

According to density functional theory (DFT) calculations, Nørskov et al. have put forward a volcano-shaped relationship between the optimal catalytic activity of the metal and the oxygen binding energy [38–40]. It could be found that the position of noble metals (such as Pt, Pd, Ru, etc.) is closer to the peak of the volcano map while nonnoble metals (such as

Fe, Co, Ni, etc.) come second, which makes them possible to be the most promising SACs catalysts to achieve the high ORR activities. SACs are usually coordinated with four and two nonmetal atoms (mostly N or C atoms) in the same plane to form a well-defined M−N−C structure, which has exhibited promising perspectives in sustainable fuel cells and Zn−air batteries [41]. In general, the M−N−C sites are further anchored on the carbon substrate exhibiting a large specific surface area as well as a high pore volume, promoting electron conductivity [42]. Therefore in this part, we introduce SCAs for ORR, which were classified into two categories based on the metal types: (1) noble metal-based SACs for ORR and (2) nonnoble metal-based SACs for ORR.

4.3.1 Noble metal-based single-atom catalysts for ORR

4.3.1.1 Pt-based single-atom catalysts for ORR

Currently, the most effective catalysts for fuel cells and metal-air batteries are platinum (Pt)-based materials numerous advanced nanomaterials with unique structures have been explored with excellent ORR performances. Platinum (Pt)-based catalysts represent the most widely explored ORR catalyst and the state-of-the-art performance benchmark. Several pioneering works have proved the progress of Pt-based SACs in electrocatalytic ORR. For example, to achieve Pt SACs with high activity and stability, Sun et al. prepared the high density of Pt isolated single atoms on the MOF-derived N-doped carbon (MOF-NC) support by adjusting the ALD exposure time of the Pt precursor (Fig. 4.2A−D). Owing to a rich concentration of nitrogen-doping sites, the MOF-NC support is expected to form a strong covalent bonding with Pt SAs, thereby resulting in a strong Pt SA−support interaction, which contributes to the enhanced stability of Pt SACs. Compared with Pt NPs, the as-prepared Pt SA catalyst shows much-enhanced mass activity (6.5 times) and good stability in electrocatalysis of the ORR. DFT calculations revealed that Pt SAs exhibit versatile electronic structures with different adsorption species during the ORR. With coadsorption of both hydroxyl and oxygen, the unoccupied orbitals of Pt SAs could be modulated with enhanced activity in the ORR [43]. Liu et al. designed atomically dispersed Pt supported on a mesoporous orthorhombic molybdenum carbide (β-Mo_2C) substrate as a model catalyst by taking ORR as a model reaction. They found that the atomically dispersed Pt monomers strongly interacting on Mo_2C support is demonstrated as a model catalyst in the four-electron ORR, and the beneficial interactions between two closely neighboring and yet

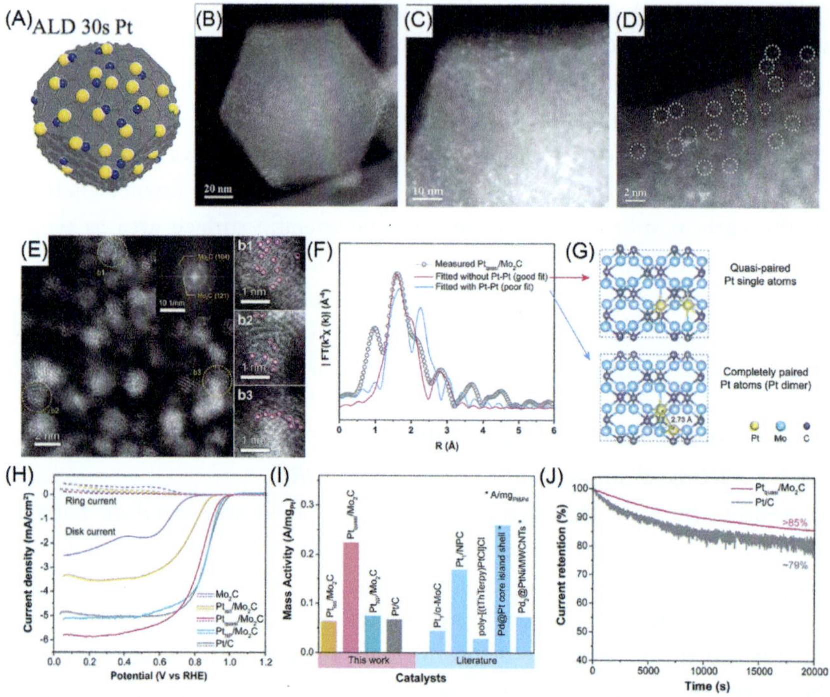

Figure 4.2 (A–D) HADDF-STEM images for samples of Pt SAs-ZIF-NC prepared with 30 s ALD (isolated Pt single atoms marked in the circle). (E) HAADF-STEM image of Pt_{quasi}/Mo_2C, inset showing the FFT pattern of Pt_{quasi}/Mo_2C. (F) Fitted EXAFS results of Pt_{quasi}/Mo_2C, without and with Pt–Pt coordination. (G) Structural models of the cases for quasi-paired Pt single atoms and completely paired Pt atoms (Pt dimers). (H) The ORR polarization curves and (I) the mass activity of Pt_{quasi}/Mo_2C and other samples. (J) Stability of Pt_{quasi}/Mo_2C determined by chronoamperometry test conducted at a potential of 0.85 V. *(A–D) Reproduced with permission Z. Song, Y. N. Zhu, H. Liu, M. N. Banis, L. Zhang, J. Li, et al., Engineering the low coordinated Pt single atom to achieve the superior electrocatalytic performance toward oxygen reduction, Small 16 (43) (2020) 2003096. Copyright 2020, Wiley-VCH. (J) Reproduced with permission L. Zhang, T. Yang, W. Zang, Z. Kou, Y. Ma, M. Waqar, et al., Quasi-paired Pt atomic sites on Mo2C promoting selective four-electron oxygen reduction, Adv. Sci. 8 (18) (2021) 2101344. Copyright 2021, Wiley-VCH.*

noncontiguous Pt single atom sites (named as quasipaired Pt single atoms, Pt_{quasi}/Mo_2C) are shown Fig. 4.2E-G. Compared to isolated Pt single atom sites (Pt_{iso}/Mo_2C), the Pt_{quasi}/Mo_2C shows a higher half-wave potential of 0.83 V, superior mass activity of 0.224 A mg^{-1} Pt, and near-100% selectivity toward four-electron ORR due to the synergistic interaction from the two quasipaired Pt atom sites in modulating the binding mode of reaction intermediates (Fig. 4.2H–J) [44]. Nakanishi et al. reported that a hybrid material (Pt-CTF/CP) consisting of Pt-CTF and conductive carbon nanoparticles (CPs) can be used as a methanol-tolerant oxygen reduction electrocatalyst [45]. The CTF had many nitrogen sites, which behave as anchoring sites for Pt single-atom deposition. Thus, the hybridized material (Pt-CTF/CP) exhibits electrocatalytic activity for ORR in an acidic solution with a high methanol tolerance, which is a very attractive property for direct methanol fuel cells as the methanol crossover effect is one of the issues to be addressed. The nitrogen sites next to single-atom Pt may play an important role in breaking O − O bonds for the ORR with the four-electron pathway. Methanol oxidation also requires Pt ensemble sites, and the single-atom Pt in Pt-CTF/CP could not catalyze methanol oxidation, providing methanol-tolerant behavior.

4.3.1.2 Ru-based single-atom electrocatalysts

Single-atom Ru catalysts were capable of boosting ORR kinetics and lowering the cost, which displayed promising potential in fuel cells and Zn-air batteries (ZABs). Xing and coworkers designed a Ru–N–C single-site catalyst (Ru-SSC) for high-efficiency ORR catalysis by microporous metal-organic frameworks (MOFs) confined strategy [46]. The atomic structure was carefully examined by aberration-corrected high-angle annular dark-field scanning transmission electron microscopy (HAADF-STEM) and X-ray absorption spectroscopy (XAS), revealing isolated Ru atom coordinated with N and O structures. DFT calculations suggest that in situ generated OH ligand tailored the d-orbital electronic structure of Ru and thus optimized the adsorption-desorption behaviors of oxygen-containing intermediates on the Ru site toward boosted ORR kinetics. Thus, the as-developed Ru-SSC exhibits a positive half-wave potential of 0.824 V versus RHE and superb long-term stability, surpassing the state-of-the-art Pt/C catalyst. Luo et al. first reported single atomic Ru sites anchored onto $Ti_3C_2T_x$ MXene nanosheets (Ru-SA/$Ti_3C_2T_x$) as electrocatalysts for catalyzing acidic ORR (Fig. 4.3). The as-prepared Ru-SA/$Ti_3C_2T_x$ exhibits the highest activity with the most positive onset potential (E_{onset}) of 0.92 V versus RHE and an $E_{1/2}$ of 0.80 V versus RHE, which are similar to those of Pt/C ($E_{onset} = 0.92$ V and $E_{1/2} =$ 0.82 V vs RHE, respectively). Theoretical calculations reveal that isolated $Ru-O_2$ sites can effectively optimize the adsorption of reactants/intermediates and lower the energy barriers for the potential determining steps,

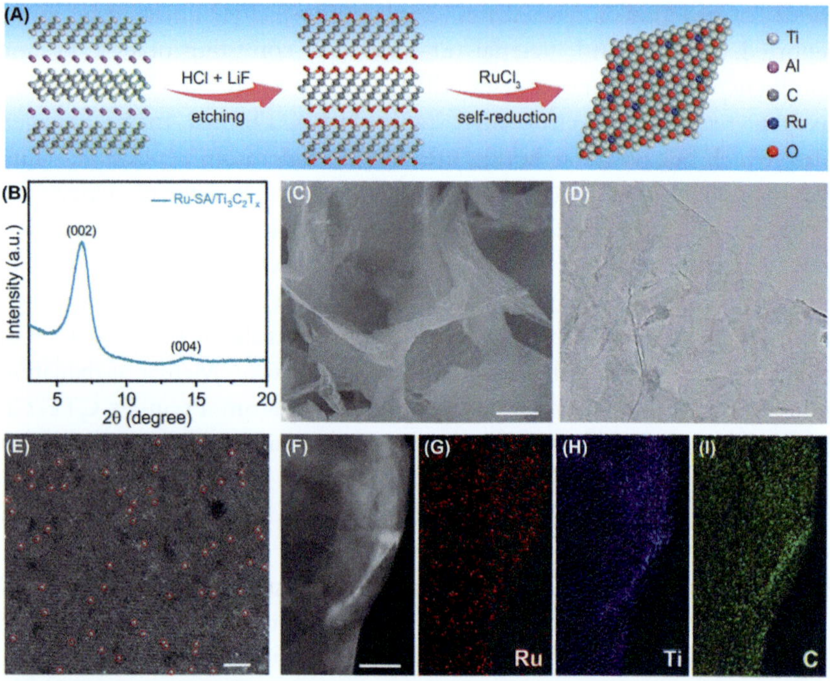

Figure 4.3 (A) Schematic of the synthesis route of Ru-SA/$Ti_3C_2T_x$. (B) XRD pattern. (C) Low-magnification FESEM image. (D) TEM image. (E) Aberration-corrected HAADF-STEM image. (F–I) HAADF-STEM image and the corresponding EDS elemental mapping of Ru-SA/$Ti_3C_2T_x$. *Reproduced with permission X. Peng, S. Zhao, Y. Mi, L. Han, X. Liu, D. Qi, et al., Trifunctional single-atomic Ru sites enable efficient overall water splitting and oxygen reduction in acidic media, Small 16 (33) (2020) 2002888. Copyright 2020, Wiley-VCH.*

thereby accelerating the ORR kinetics [47]. Generally, a nitrogen-doped carbon matrix would be an ideal choice as a support, as it not only can endow the catalyst with high conductivity but also can impart the uniform dispersion of the active sites; hence, the active sites in the electrolyte can be well-protected and the catalyst could maintain high durability in long-term operations. For example, Tour group reported an atomically dispersed Ru on nitrogen-doped graphene (Ru-N/G-750) that performs as an electrocatalyst for the ORR in an acidic medium [48]. The Ru/nitrogen-doped GO catalyst exhibits excellent four-electron ORR activity, offering onset and half-wave potentials of 0.89 and 0.75 V (vs RHE) in 0.1 M $HClO_4$, respectively, together with better durability and tolerance toward methanol and carbon monoxide poisoning than seen in commercial Pt/C catalysts. DFT calculations further indicated that a Ru-oxo-N_4 structure appears to be responsible for the ORR catalytic activity in the acidic medium.

4.3.1.3 Other precious-metal-based single-atom electrocatalysts

Compared with other precious metal specimens, the adsorption-free energy of OH* of the Ir, Rh, and Pd are closest to the peak of the volcano map, implying their high catalytic activity for ORR. For instants, Chen's group developed an Ir-N-C SAC (Ir-SAC) for the ORR by using ZIF-8 as an active site host and then in situ encapsulated Ir acetylacetonate as a guest (Fig. 4.4A–C). The density functional theory (DFT) calculations confirmed that the unique electronic structure of the Ir atom center induced by the Ir-N_4 configuration contributes to optimal adsorption energy that facilitates both the adsorption and desorption of ORR intermediates, thus leading to unprecedented catalytic performance. As expected, the obtained Ir–SAC exhibited a positive half-wave potential of 0.864 V versus RHE and a small Tafel slope of 41 mV dec^{-1} in 0.1 M $HClO_4$ for ORR (Fig. 4.4D). The finding provides an in-depth understanding of the future development of high-performance SACs [49]. Shui et al. reported a sequential coordination method to dope a large number of PM atoms (Ir, Rh, Pt, and Pd) into a zeolite imidazolate framework (ZIF), which are further pyrolyzed into nitrogen-coordinated PM SACs. The PM loadings are as high as 1.2–4.5 wt.%, achieving the highest PM loadings in ZIF-derived SACs to date (Fig. 4.4E). In the acidic half-cell, Ir1-N/C and Rh1-N/C exhibit much higher ORR activities than nanoparticle catalysts Ir/C and Rh/C. DFT calculations reveal that the ORR activity of PM SAC depends on the match between the OH* adsorption on PM and the electronegativity of coordinating anions, and the stronger OH* adsorption is, the higher electronegativity is needed for the coordinating anions. The revealed principle provides guidance for designing future PM SACs for ORR [50]. Wu et al. reported that single Pd atoms supported on graphitic carbon nitride (Pd1/g-C_3N_4) would act as an electrocatalyst for the ORR, leading to the four-electron pathway by the sequential hydrogenation of O^{2*} to $O^* + H_2O^*$. They claimed that Pd1/g-C_3N_4 has a lower energy barrier than Pt, and possibly has higher ORR activity than Pt. The ORR on Pd single atoms supported on graphitic carbon nitride (Pd/g-C_3N_4) has been studied computationally at different Pd coverages. It was found that the ORR activity depends on the coverage of Pd, and the SAEC with a 25% coverage showed the

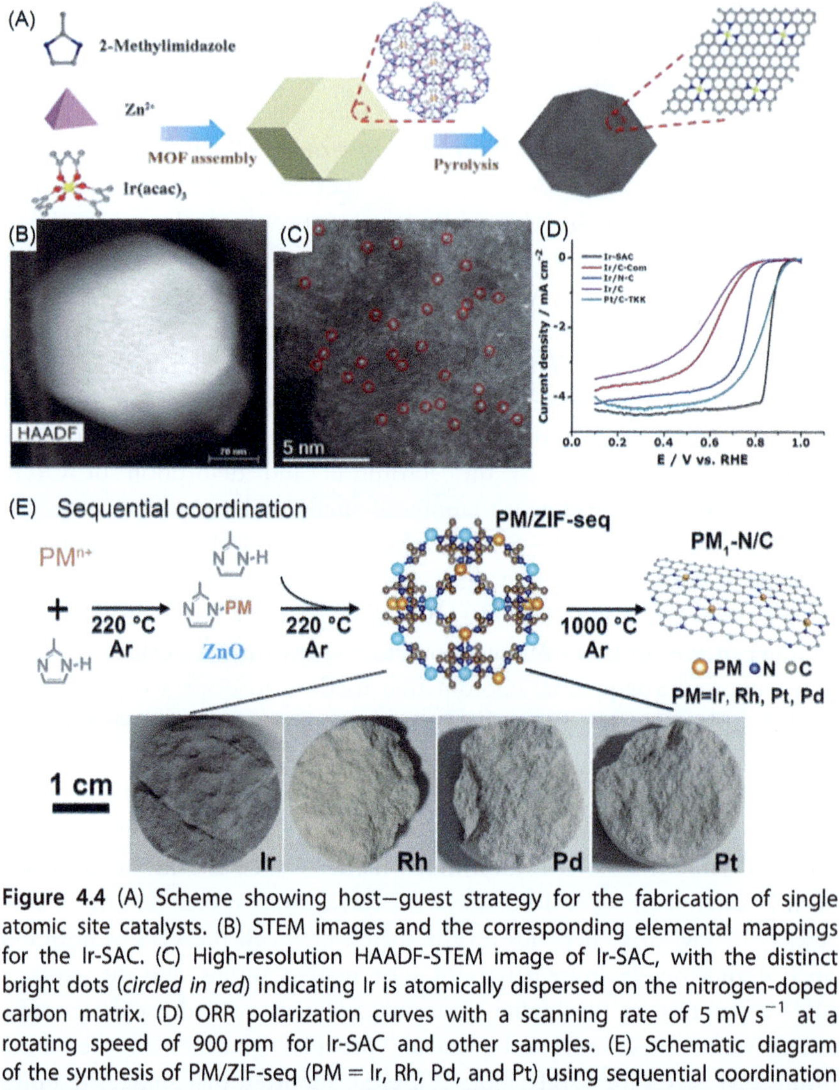

Figure 4.4 (A) Scheme showing host–guest strategy for the fabrication of single atomic site catalysts. (B) STEM images and the corresponding elemental mappings for the Ir-SAC. (C) High-resolution HAADF-STEM image of Ir-SAC, with the distinct bright dots (*circled in red*) indicating Ir is atomically dispersed on the nitrogen-doped carbon matrix. (D) ORR polarization curves with a scanning rate of 5 mV s^{-1} at a rotating speed of 900 rpm for Ir-SAC and other samples. (E) Schematic diagram of the synthesis of PM/ZIF-seq (PM = Ir, Rh, Pd, and Pt) using sequential coordination method, which was converted to PM1-N/C SACs after carbonization. The photo images show the uniform color of the PM/ZIF-seq. *(D)Reproduced with permission M. Xiao, J. Zhu, G. Li, N. Li, S. Li, Z.P. Cano, et al., A single-atom iridium heterogeneous catalyst in oxygen reduction reaction, Angew. Chem. Int. Ed. 58 (28) (2019) 9640 – 9645. Copyright 2019, Wiley-VCH GmbH. (E) Reproduced with permission Q. Liu, Y. Li, L. Zheng, J. Shang, X. Liu, R. Yu, et al., Sequential synthesis and active-site coordination principle of precious metal single-atom catalysts for oxygen reduction reaction and PEM fuel cells, Adv. Energy Mater. 20 (10) (2020) 2000689. Copyright 2020, Wiley-VCH.*

lowest energy barrier of 0.39 eV for ORR among the series, which was even lower than that (0.80 eV) of Pt. Validation of this prediction has not been demonstrated in experiments yet, which might be challenging due to the relatively low conductivity of the C_3N_4 substrate [51].

4.3.2 Nonprecious metal SACs for ORR

4.3.2.1 Fe-based single-atom electrocatalysts

Nonprecious and earth-abundant transition metal (Fe, Co, Ni, etc.)-based electrocatalysts with low cost and high electrocatalytic activity and durability have been extensively investigated in recent years [52]. Especially, a hybrid composite electrocatalyst consisting of transition metal and nitrogen-doped carbonaceous materials (TM-N-C) has shown even higher activity than commercial Pt/C for the ORR in alkaline conditions [53]. Among them, atomically dispersed Fe-N-C materials held an important status in Zn–air batteries and fuel cells, where Fe-N_x moieties can strongly adsorb O_2 to speed up the ORR kinetics [54]. Based on this, Fe-N-C catalysts have been intensively investigated in the past few decades and great progress has been made. For instance, Zhang et al. reported that high purity pyrrole-type FeN_4 site could be obtained by calcining the pyridine-type FeN_4 site under NH_3 atmosphere, while FeN_4 site with higher content of pyridine type was achieved in Ar atmosphere as shown in Fig. 4.5A–D [55]. The results showed that the onset potential and the half-wave potential for HP-FeN_4 are 0.95 and 0.80 V versus RHE, while those for pyridine-type FeN_4 are 0.86 and 0.71 V versus RHE. DFT calculations revealed that the Fe atom in HP-FeN_4 accepted fewer electrons from the pyrrolic-N compared with the case of the Fe atom in pyridine-type FeN_4, making the valent state of Fe in HP-FeN_4 more positive. Then, Feng and coworkers found the Fe–N–C with symmetrically coordinated Fe–N_4 active site could be used as a highly efficient electrocatalyst for ORR with outstanding kinetic current density at 0.8 V ($J_k = 32.04$ mA cm^{-2}) and high $E_{1/2}$ of 0.93 V in the alkaline electrolyte [56]. Hu et al. innovatively explored a hierarchical ordered porous carbon catalyst with atomic Fe–N_4 sites using the Fe-doped single-crystalline ZIF-8 structure by hierarchical ordered porosity as a precursor (Fig. 4.5E–I). The increasingly accessible active sites and faster mass transfer endow the FeN_4/HOPC-c-1000 with an exciting $E_{1/2}$ of 0.80 V versus RHE, approaching that of commercial Pt/C (0.82 V vs RHE) in 0.5 M H_2SO_4, which is far higher than FeN_4/C without hierarchically ordered pores (Fig. 4.5J) [57]. Furthermore, introducing defects in carbon substrate is

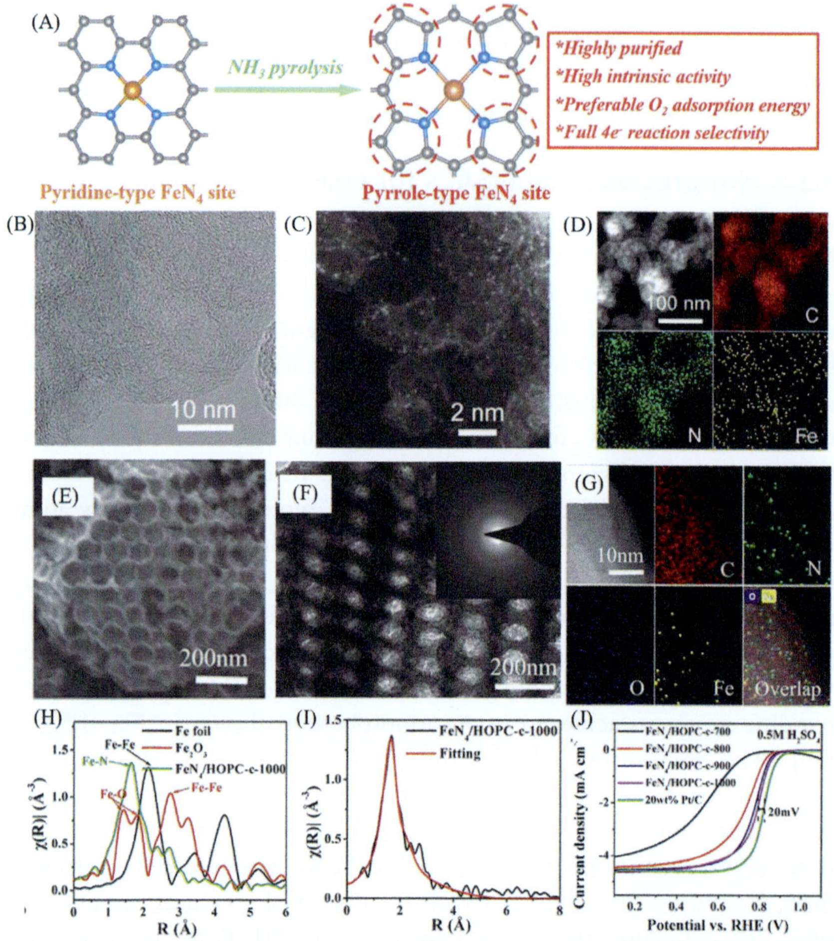

Figure 4.5 (A) Preparation process of high-purity pyrrole-type FeN_4 structure. The balls in grey, blue, and orange represent C, N, and Fe atoms, respectively. (B) HRTEM and (C) HAADF-STEM image of HP-FeN_4 material. (D) EDS elemental mapping profiles of HP-FeN_4. (E) SEM images FeN_4/HOPC-c-1000. (F) TEM image of FeN_4/HOPC-c-1000, with the inset showing the corresponding SAED pattern. (G) EELS mapping of FeN_4/HOPC-c-1000. (H) Fourier transform Fe K-edge EXAFS spectra, and (I) EXAFS fitting results of FeN_4/HOPC-c-1000. (J) ORR polarization curves of FeN4/HOPC-c prepared at different annealing temperatures in O_2-saturated 0.5 M H_2SO_4 solution with a scan rate of 10 mV s^{-1}. *(D) Reproduced with permission N. Zhang, T. Zhou, M. Chen, H. Feng, R. Yuan, W. Yan, et al., High-purity pyrrole-type FeN4 sites as a superior oxygen reduction electrocatalyst, Energy Environ. Sci. 13 (1) (2020) 111–118. Copyright 2020, The Royal Society of Chemistry (J) Reproduced with permission M. Qiao, Y. Wang, Q. Wang, G. Hu, X. Mamat, S. Zhang, et al., Hierarchically ordered porous carbon with atomically dispersed FeN4 for ultraefficient oxygen reduction reaction in proton-exchange membrane fuel cells, Angew. Chem. Int. Ed. 59 (7) (2020) 2688–2694. Copyright 2019, Wiley-VCH.*

also widely adopted to promote the dissociation of O_2 molecules via optimizing the electronic structure. For example, Chen et al. intelligently designed an NH_4Cl-assisted strategy to tailor Fe−N_x/GM carbon with abundant FeN_4 sites on edge [58]. A mass of released gas (NH_3 and HCl) from the decomposition of NH_4Cl salt at the high temperature promotes the formation of pores and nitrogen-doped edges in the carbon skeleton and increases the number of FeN_x active sites, as proved by the XAS, STEM, and nitrogen adsorption-desorption, which results in an enhanced ORR activity ($E_{1/2} = 0.80$ V vs RHE, in 0.5 M H_2SO_4) for Fe−N_x/GM with FeN_4 sites cooperated by adjacent pore defects, transcending the FeN_x/GP ($E_{1/2} = 0.78$ V vs RHE). Despite a high ORR activity with acceptable stability for Fe−N−C-based SACs, Fe with high Fenton reaction activity is still an issue for practical application. Therefore designing other non-Fe-based atomically dispersed M−N−C (M = Co, Ni, Mn, Cu, etc.) catalysts with negligible Fenton reaction activity, promising activity and high stability for ORR is significant.

4.3.2.2 Co-based single-atom electrocatalysts

Among the transition metals, a Co-based single atom electrocatalyst has received increasing interest since it exhibits a suppressed Fenton reactivity in comparison with Fe-based SACs while maintaining a remarkable catalytic activity. Similar to Fe-based SACs, Co-based SACs are usually stabilized as single-atom sites via coordination with substrates, typically in the form of M − N − C. For instance, Yin et al. developed stable Co SAECs by pyrolysis of predesigned Zn/Co bimetallic MOFs [59]. The introduction of Zn can efficiently space Co atoms from agglomerating and generate micropores during pyrolysis. The obtained Co-N_x single sites exhibited a remarkable ORR performance ($E_{1/2} = 0.881$ V vs RHE) and durability in alkaline solution. Recently, a catalyst with atomically dispersed CoN_4 sites has been synthesized through a simple one-step thermal activation of Co-doped ZIF precursors. The best Co SAECs, obtained by the most optimal thermal activation condition, possessed an excellent ORR activity ($E_{solution} = 0.80$ V) and good stability in acid media. However, severe aggregation occurred at high Co loadings, which limited the maximum ORR activity for this type of catalyst. As shown in Fig. 4.6A−D, Han et al. investigated the size effect on the electrocatalytic activity of Co catalysts from nanometer to single-atom scale, demonstrating that cobalt single atoms on N-doped carbon could achieve a higher half-wave potential (0.82 V) and a larger limiting diffusion current density (4.96 mA cm^{-2})

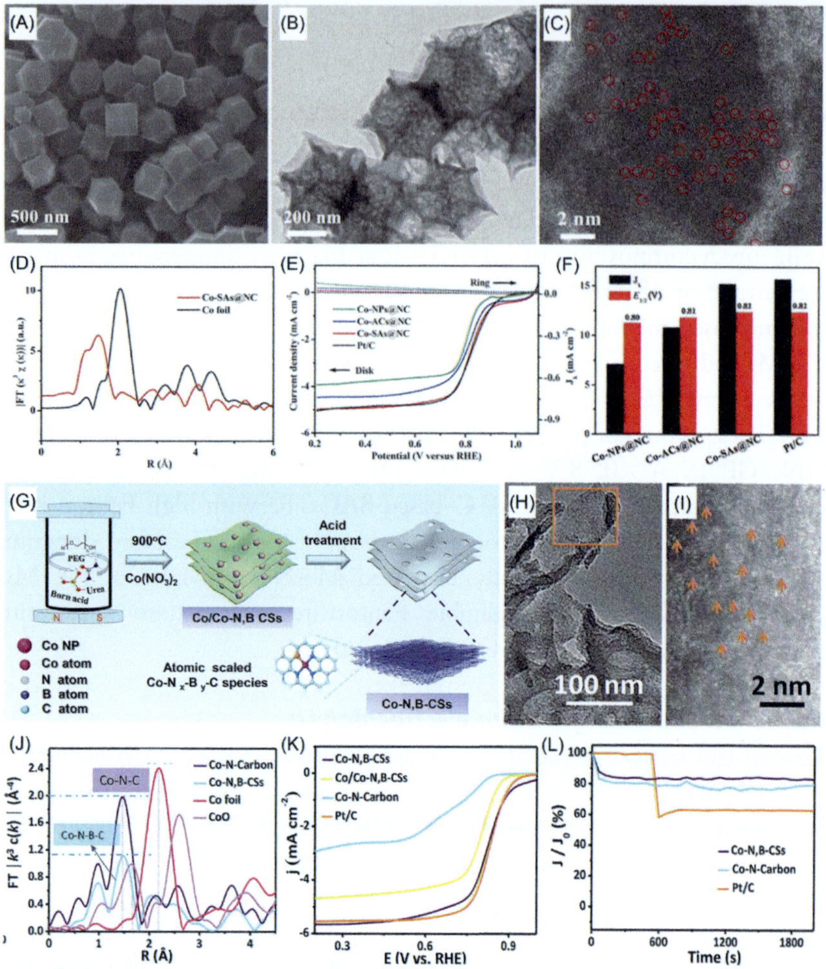

Figure 4.6 (A) SEM, (B) TEM images and (C) HAADF-STEM images of Co-SAs@NC. (D) Fourier transforms of the Co K-edge spectra of Co-SAs@NC and Co foil. (E) ORR polarization curves of the Co-SAs@NC using RRDEs at 1600 rpm. (F) J_k at 0.6 V and $E_{1/2}$ for synthesized Co-NPs@NC, Co-ACs@NC, Co-SAs@NC and Pt/C catalysts. (G) Synthetic procedure of the Co/N−B-doped carbon nanosheet (Co-N,B-CSs). (H) TEM, and (I) HAADF-STEM images of Co-N,B-CSs. (J) Fourier transformed EXAFS spectra of the Co foil and Co-N,B-CSs and Co−N−Carbon. (K) LSV curves of Co-N,B-CSs, Co/Co-N,B-CSs, Co−N−Carbon, and Pt/C catalysts in O_2-saturated 0.1 M KOH solution. (L) Chronoamperometric response for Co-N,B-CSs, Co−N−Carbon, and Pt/C electrode at 0.75 V (vs RHE) after the introduction of 9.7 mL of CH_3OH into 230.3 mL of 0.1 M KOH solution. *(F) Reproduced with permission X. Han, X. Ling, Y. Wang, T. Ma, C. Zhong, W. Hu, et al., Generation of nanoparticle, atomic-cluster, and single-atom cobalt catalysts from zeolitic imidazole frameworks by spatial isolation and their use in zinc-air batteries, Angew. Chem. 131 (16) (2019) 5413 − 5418. Copyright 2019, Wiley-VCH. (L) Reproduced with permission Y. Guo, P. Yuan, J. Zhang, Y. Hu, I.S. Amiinu, X. Wang, et al., Carbon nanosheets containing discrete Co-Nx-By-C active sites for efficient oxygen electrocatalysis and rechargeable Zn-Air batteries, ACS Nano 12 (2) (2018) 1894 − 1901. Copyright 2018, American Chemical Society.*

than atomic Co clusters (0.81 V, 4.44 mA cm^{-2}) and Co nanoparticles counterpart (0.80 V, 3.86 mA cm^{-2}) (Fig. 4.6E–F) [60]. Guo et al. also figured out that B-doped Co–N–C active sites (Co–N_3B) prefer to adsorb oxygen molecules in side-on mode and therefore afford better ORR activity than the conventional Co–N_4 configuration (Fig. 4.6G–J). The introduced B element gives an electron-deficient site that can activate the electron transfer around the Co – N – C sites, strengthen the interaction with oxygenated species, and thus accelerate reaction kinetics in the $4e^-$ processed ORR and OER. Thus, the catalyst showed Pt-like ORR performance with a half-wave potential (E1/2) of 0.83 V versus RHE, a limiting current density of about 5.66 mA cm^{-2}, and higher durability than Pt/C catalysts (Fig. 4.6K–L) [61]. Recently, Wu et al. innovatively developed a highly effective lysozyme (Lys)-assisted MOF approach to prepare single-atom Co implanted pyridinic-N doped porous carbon catalysts (Co-Pyridinic N-C). The Co-pyridinic N-C catalyst exhibited superior ORR activity in both alkaline and acidic media with a half-wave potential of 0.87 and 0.83 V, respectively. X-ray absorption fine structure measurements and aberration-corrected HAADF-STEM as well as DFT calculations suggest that quaternary pyridinic N-coordinated Co sites were favorable for oxygen intermediate absorption/desorption, leading to accelerated reaction kinetics and improved catalytic oxygen reduction activity [62].

4.3.2.3 Other nonprecious metal-based single-atom electrocatalysts

SACs of 4d and 5d nonprecious metals have also been reported for the ORR. For example, Li et al. reported an efficient ORR catalyst that consists of atomically dispersed nitrogen-coordinated single Mn sites on partially graphitic carbon (Mn-N-C). The Mn-N-C catalyst exhibited a half-wave potential of 0.80 V versus RHE and a power density of 0.46 Wcm^{-2}. First-principles calculations further support the MnN_4 sites as the origin of the ORR activity via a $4e^-$ pathway in acidic media [63]. Yang et al. dispersed Cu–N_4 molecules on graphene nanosheets through a covalently integrated manner, in which the electronic structure of the Cu active site could be adjusted by the hetero* orbital symmetry of the simplified O_2 molecule and improve the ORR activity. The obtained Cu–N_4–C catalyst with a well-defined Cu–N_4 coordination center exhibits a surprisingly high ORR activity ($E_{1/2}$ = 0.84 V vs RHE). This work provides an avenue for the bottom-up fabrication of low-priced highly active M–N–C catalysts [64]. Gu and coworkers also reported a new type of nitrogen-triggered Zn SAC (ZnN_x/C) and proved that the

O_2 adsorption on the $Zn-N_4$ structure could occur at a low overpotential of 1.70 V. This work indicates that the ZnN_x/C catalyst is one of the most promising Pt alternatives with best performance/price ratio for ORR process in fuel cells [65]. Besides, a new kind of W SAC anchored on a porous O, N-doped carbon nanosheet (W_1/NO-C) has been prepared by Wang's group, which possesses a tridentate $W_1N_1O_2$ with the coordination of two O atoms and one N atom. The as-prepared W_1/NO-C catalyzes the ORR via a $2e^-$ pathway with high onset potential, high H_2O_2 selectivity in the wide potential range, and excellent operation durability in 0.1 M KOH solution, superior to most of the state-of-the-art H_2O_2 electrocatalysts ever reported [66]. Chen group also reported a high-performance Mg-based electrocatalyst with a super high ORR activity in alkaline and acidic condition (Fig. 4.7A–B). DFT calculations reveal that Mg-N_2C_2-C sites outperformed the conventional $Mg-N_4-C$ sites in terms of intrinsic ORR activity, endowing the

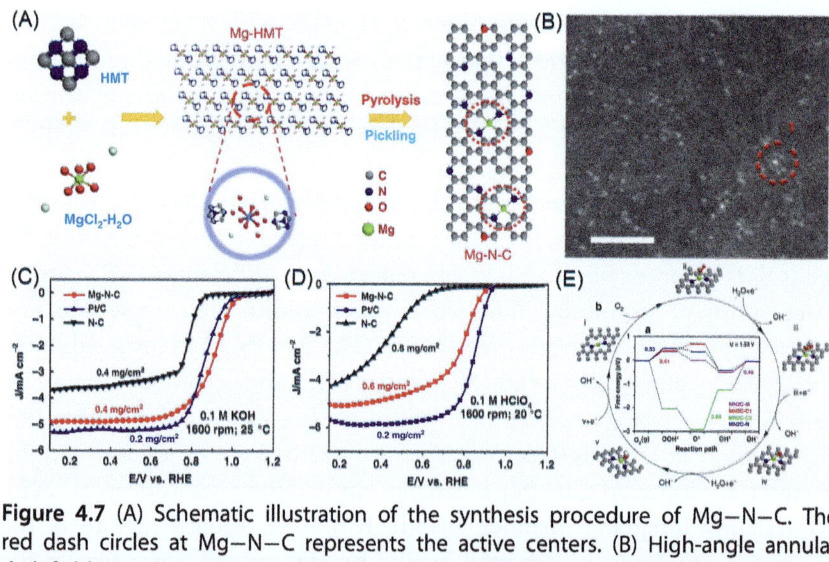

Figure 4.7 (A) Schematic illustration of the synthesis procedure of Mg–N–C. The red dash circles at Mg–N–C represents the active centers. (B) High-angle annular dark-field-scanning transmission electron microscopy (HAADF-STEM) image of Mg–N–C. Scale bar: 2 nm. (C) LSV curves after normalization by glass carbon (GC) electrode in O_2-saturated 0.1 M aqueous KOH and (D) 0.1 M aqueous $HClO_4$ electrolyte solution. (E) The free energy diagram of MN_2C and favorable reaction pathway of MN_2C. *Reproduced with permission S. Liu, Z. Li, C. Wang, W. Tao, M. Huang, M. Zuo, et al., Turning main-group element magnesium into a highly active electrocatalyst for oxygen reduction reaction, Nat. Commun. 11 (1) (2020) 938. Copyright 2020, Springer Nature.*

previously catalytic inactive Mg–N–C SACs with a remarkable ORR performance ($E_{1/2} = 0.91$ V) (Fig. 4.7C–E) [67].

4.4 Application of SACs for the ORR in electrochemical energy devices

4.4.1 SACs catalyst development for fuel cells

Fuel cells are energy conversion and storage technologies that can convert chemical energy in the form of fuels (methanol, ethanol, hydrogen, and oxygen, etc.) into electrical and thermal energy by electrochemical reactions. For most explored fuel cell devices, proton exchange membrane fuel cells (PEMFCs) are the most promising technologies due to they have the advantages of zero carbon emissions, high power density, fast start-up, and low operating temperatures, and show great potential for commercial applications [68]. The electrocatalytic hydrogen oxidation reaction (HOR), and ORR are the key processes for the earlier-mentioned PEMFCs. Nevertheless, cathode ORR kinetics of PEMFCs is very slower than that of the HOR and requires a large overpotential to make the reaction occur, which greatly limits the energy output of fuel cells [69]. Currently, Pt and Pt-based metals are recognized as commercial ORR electrocatalysts in the PEMFCs. Unfortunately, the cost and poor durability of the Pt and Pt-based electrocatalysts severely hamper their large-scale application. Thus, in the past decades, substantial efforts have been made in researching cost-effective catalysts to drive the development of fuel cells. Single-metal-site SACs have been explored to boost the ORR kinetics, which can be used as effective cathode catalysts in fuel cells. For example, Liu et al. reported that a carbon-defect-anchored Pt SAC (Pt1-N/BP) was favorable for $4e^-$ ORR, and can be applied in an acidic H_2/O_2 fuel cell as the cathode catalyst [70]. The Pt1.1/BP_{defect} with a 1.1 wt.% Pt content shows an enhanced ORR performance and excellent durability in the acidic medium. The $Pt_{1.1}/BP_{defect}$-based single-cell delivers a super high platinum utilization of 0.09 g_{Pt} kW^{-1}. The Pt single atoms anchored by four carbon atoms neighboring carbon divacancies (Pt-C_4 site) led to high ORR activity and stability, which was favorably assembled into a single $Pt_{1.1}/BP_{defect}$-based PEMFC, giving a maximum power density of 520 mW cm^{-2} at 80°C, suggesting that the $Pt_{1.1}/BP_{defect}$ as a promising alternative catalyst for acidic fuel cells. Besides, atomic non-Pt-based M-N-C catalysts have currently exhibited promising perspectives in acidic fuel cells due to their low cost, where the isolated metal can be coordinated with adjacent N or C atoms to form various active sites for ORR. For example, Wu et al. synthesized a unique porous Co−N−C nanofiber hierarchical structure through electrospinning cobalt-doped zeolitic imidazolate frameworks into selected polyacrylonitrile and poly(vinylpyrrolidone) polymers (Fig. 4.8A-D), which exposed plentiful $Co-N_4$ active sites and facilitated to electronic conductivity. The membrane electrode assembly achieves a power density of 0.40 W cm^{-2} in a practical H_2/air cell (1.0 bar) and demonstrates significantly enhanced durability under accelerated stability tests [71]. Similarly, Jiang et al. constructed atomically dispersed Fe supported on N-doped graphene (FeSA-G) with a high Fe SAC loading of $\approx 7.7 \pm 1.3$ wt.%, where a possible planar Fe-N_4 configuration was proposed [72]. The as-synthesized FeSA-G exhibited a high onset potential of 0.950 V versus RHE and halfwave potential of 0.804 V versus RHE in acid electrolyte for the ORR, which can be applied in a polymer electrolyte membrane fuel cell with a peak power density of 325 mW cm^{-2}, better than 313 mW cm^{-2} obtained on the cell with a Pt/C cathode. Subsequently, Xu et al. synthesized atomically dispersed Mn−N−C catalysts, which contain more $Mn-N_4$ active sites and high specific surface area (Fig. 4.8E−F). The Mn−N−C catalyst exhibits promising ORR activity along with significantly enhanced stability, achieving a peak power density of 0.39 W cm^{-2} under 1.0 bar H_2-air condition in an MEA, outperforming most PGM-free ORR catalysts (Fig. 4.8G−I) [73]. Interestingly, Sun et al. constructed atomically dispersed bimetallic Zn-Co supported on N-doped carbon (Zn/CoN-C) electrocatalyst for ORR. Such architecture offers the enhanced binding ability of O_2 due to the formation of N-coordinated dual-metal sites and thus facilitates the cleavage of O − O bond. As a result, the Zn/CoN-C catalyst shows outstanding ORR performance in both alkaline and acid conditions with a half-wave potential of 0.861 and 0.796 V, respectively. Meanwhile, the excellent catalytic activity in acid was verified by H_2/O_2 fuel cell tests (peak power density of 705 mW cm^{-2}) [74].

4.4.2 SACs catalyst development for metal-air batteries

Up to now, metal-air batteries have been receiving much attention and achieving great progress. Typically, metal-air batteries are composed of a metal anode [like lithium (Li), aluminum (Al), magnesium (Mg), zinc (Zn), and iron (Fe)], separator, electrolyte, air cathode, and proper

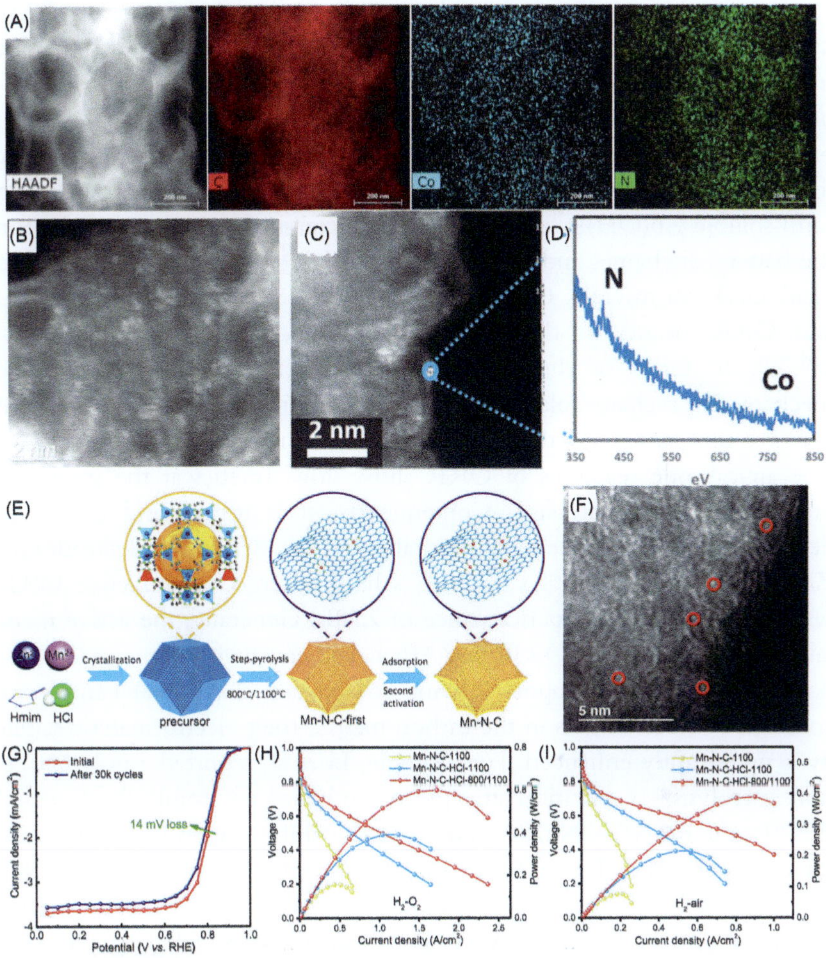

Figure 4.8 (A) HAADF-STEM image and STEM-EDS elemental maps of Co—N—PCNFs sample. (B—D) Aberration-corrected HAADF-STEM images with EELS point spectra of Co—N—PCNFs catalyst. (E) Aqueous synthesis scheme for Mn—N—C catalysts. (F) STEM images of Mn—N—C—HCl-800/1100. (G) Steady-state ORR polarization curves of Mn—N—C—HCl-800/1100 before and after potential cycling tests (0.6—1.0 V, 30,000 cycles). (H) Fuel cell performance of different catalysts under H_2—O_2. (I) Fuel cell performance of various catalysts under H_2-air (1.0 bar). *(B)—(D) Reproduced with permission Y. He, H. Guo, S. Hwang, X. Yang, Z. He, J. Braaten, et al., Single cobalt sites dispersed in hierarchically porous nanofiber networks for durable and high-power PGM-free cathodes in fuel cells, Adv. Mater. 46 (32) (2020) 2003577. Copyright 2020, Wiley-VCH. Reproduced with permission M. Chen, X. Li, F. Yang, B. Li, T. Stracensky, S. Karakalos, et al., Atomically dispersed MnN4 catalysts via environmentally benign aqueous synthesis for oxygen reduction: mechanistic understanding of activity and stability improvements, ACS Catal. 18 (10) (2020) 10523—10534. Copyright 2020, American Chemical Society.*

electrolyte solution [75]. Among them, aqueous-based rechargeable ZABs have been regarded as the most mature and reliable battery system due to their theoretically high energy density, reliable battery system, and economical price [76–78]. Under the alkaline condition, the zinc metal is oxidized at the air (anode) electrode with an electron release during the discharge process. The resulting Zn^{2+} further combines with OH^- to form soluble $Zn(OH)_4{}^{2-}$. The Zn^{2+} are increasingly supersaturated upon the battery discharge, and the soluble $Zn(OH)_4{}^{2-}$ can be decomposed to yield ZnO. Meanwhile, O_2 accepts electrons released from the Zn anode, and ORR occurs in the cathodic electrode, which produces OH^- [77,79]. In terms of the electrochemical reactions, both primary and mechanically rechargeable ZABs operate only in a discharge mode that involves the earlier-mentioned electrochemical reactions. Reactions [80]. The air-cathode reactions of ORR show slow kinetics at the gas-solid-liquid three-phase interface. Consequently, it is urgent and critical to develop high-performance ORR catalysts to be used as air electrodes for ZABs. In the past several years, SACs have emerged as attractive ORR systems to improve the performance of ZABs. Generally, the active mental centers of SACs for ORR in ZABs have been highly dispersed on N-doped carbon-based support. Owing to the fact that metal centers are confined by heteroatoms in the carbon matrix, their electrocatalytic stability is significantly enhanced. For example, Li et al. reported a polymerization–pyrolysis–evaporation strategy to synthesize an atomically dispersed $Fe-N_4$ catalyst ($Fe-N_4$ SAs/NPC) and exhibit superior performance for electrochemical HER, OER, and ORR (Fig. 4.9A). For ORR, the $Fe-N_4$ SAs/NPC catalyst exhibited an onset potential of 0.972 V and a half-wave potential of 0.885 V (vs RHE) in 0.1 M KOH (Fig. 4.9B). As a trifunctional electrocatalyst, $Fe-N_4$ SAs/NPC can be assembled into a zinc-air battery device, which has a low charge-discharge voltage gap and large power density (232 mW cm^{-2}) (Fig. 4.9C–E) [81]. Xu et al. also emphasized that the $Fe-N_4$ SAs modified overhang-eave structure could enhance the mass transfer of reactants to the accessible Fe–N4 sites. The assembled ZAB delivered a high capacity of 807.5 mAh $g_{Zn}{}^{-1}$, a high peak power density of 186.8 mW cm^{-2} as well as an energy density of 962.7 Wh $kg_{Zn}{}^{-1}$ [82]. In addition, Pennycook et al. developed cobalt (Co) single atoms anchored in porous nitrogen-doped carbon nanoflake arrays (NC–Co SA). Compared with Co nanoparticles grown on NC, the NC-Co SA electrocatalyst with a high density of $Co-N_x$ active sites demonstrates a lower OER overpotential and much higher ORR

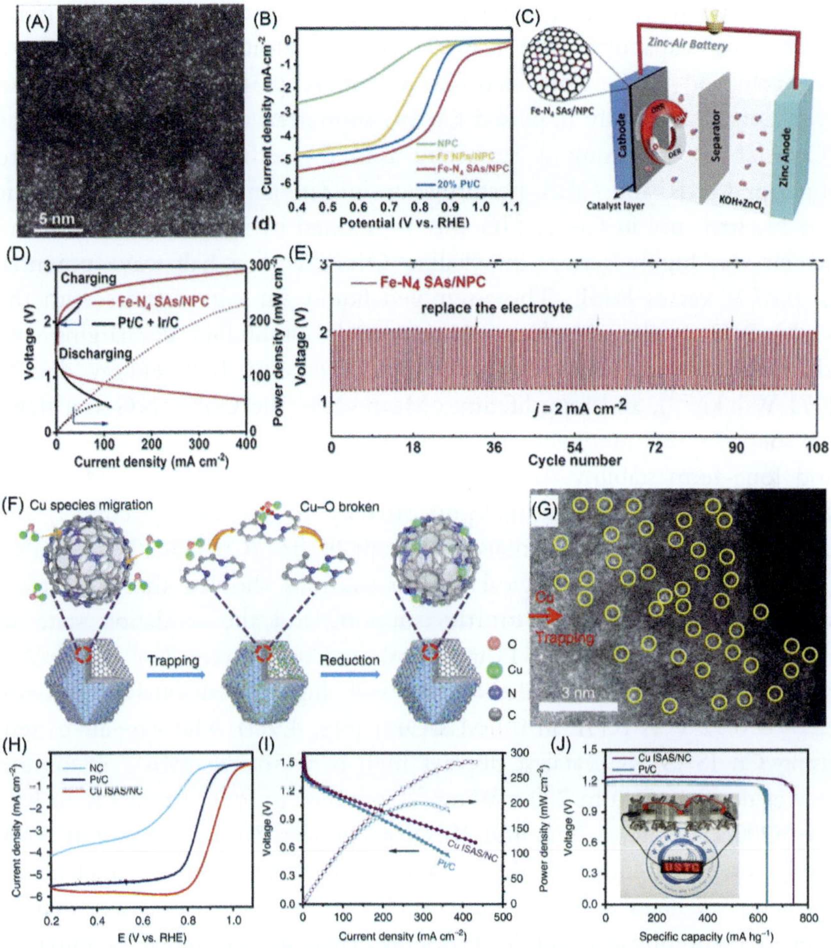

Figure 4.9 (A) HAADF-STEM of the Fe-N_4 SAs/NPC. (B) ORR polarization plots of Fe-N_4 SAs/NPC. (C) Zn–air battery. (D) Charge–discharge polarization curve and power density plot. (E) Charge–discharge cycling performance. (F) Scheme of the formation isolated copper sites (Cu ISAS/N-C) catalyst (G) AC HAADF-STEM image of Cu ISAS/NC. (H) Linear sweep voltammetry (LSV) curves of NC, Cu ISAS/NC, and Pt/C catalysts in 0.1 M KOH solution with a sweep rate of 10 mV s^{-1} and rotation rate of 1600 rpm. (I) Polarization and corresponding power density plots of Cu ISAS/NC and Pt/C-based Zn-air batteries. (J) The specific capacity of Cu ISAS/NC and Pt/C-based Zn-air batteries at 50 mA cm^{-2}. (Inset: a photograph showing a light-emitting diode panel powered by three Cu ISAS/NC-based Zn-air batteries). *(E) Reproduced with permission Y. Pan, S. Liu, K. Sun, X. Chen, B. Wang, K. Wu, et al., A bimetallic Zn/Fe polyphthalocyanine derived single-atom Fe-N4 catalytic site: a superior trifunctional catalyst for overall water splitting and Zn-Air batteries, Angew. Chem. Int. Ed. 28 (57) (2018) 8614–8618. Copyright 2018, Wiley-VCH. Reproduced with permission Z. Yang, B. Chen, W. Chen, Y. Qu, F. Zhou, C. Zhao, et al., Directly transforming copper (I) oxide bulk into isolated single-atom copper sites catalyst through gas-transport approach, Nat. Commun. 1 (10) (2019) 3734. Copyright 2019, Springer Nature.*

saturation current. Thus, the NC–Co SA presented excellent charging-discharging capability, and a peak power density of 20.9 mW cm^{-3} was achieved in the assembled Zn-air battery [83]. Cao and coworkers synthesized atomically dispersed Co-N_4 supported on N-doped graphene (CoN_4/NG) by using a surfactant-assisted method, which exhibited superior ORR and OER performance in Zn–air batteries. The atomic Co-N_4 structure in CoN_4/NG was ascertained by XAS characterizations, which was highly active for alkaline ORR with a half-wave potential of 0.87 V versus RHE. The assembled liquid Zn–air batteries with the CoN_4/NG as the air electrode presented outstanding discharging and charging properties including large power density, high energy density (671 W h kg^{-1}), and long lifetime. Meanwhile, the CoN_4/NG as a flexible solid Zn-air battery also achieved a high power density (28 mW cm^{-2}) and long-term stability (lasting 6 h) at flat and folded states [84]. Yang et al. reported a gas gas-transport strategy to successfully convert the bulk Cu_2O into isolated single-atom copper sites (Cu ISAS/NC) catalyst (Fig. 4.9F–G). Morphological characterizations showed that Cu atoms were uniformly dispersed on the support, and the oxidation state of individual Cu atoms in Cu ISAS/NC should be between Cu^0 and Cu^{2+}. The as-obtained Cu ISAS/NC showed high electrocatalytic activity ($E_{1/2}$ = 0.92 V vs RHE in 0.1 M KOH) (Fig. 4.9H). The Zn–air battery using Cu ISAS/NC catalyst displays high performance with a maximum power density of up to 280 mW cm^{-2}, superior to Pt/C-based Zn-air battery (200 mW cm^{-2}) (Fig. 4.9I–J). Notably, this Zn-air battery can work stably for up to 45 h, and the discharge voltage was basically unchanged, showing excellent stability. This strategy is readily scalable and the as-prepared sintering-resistant M ISAS/NC catalysts hold great potential in high-temperature applications [85]. Until now, a lot of research about transition metal SACs for ZABs has been reported. The ZABs can be considered to be used in conjunction with other energy devices (such as water splitting, FCs, etc.), which are expected to replace traditional fossil fuels with new energy technologies.

4.5 Conclusions and perspectives

Atomic-level dispersed metal single-atoms catalysts exhibit special catalytic performance in fuel cells and metal-air batteries. By summarizing past progress, we expect to pave the way for the development of next-generation high-performance ORR catalysts on single-atom levels.

Despite intense research efforts in SACs in the previous work, larger studies with improved experimental designs are still needed to cut costs and enhance ORR catalytic activity and stability in fuel cells and metal-air batteries. There are still hurdles to overcome before SACs become qualified ORR catalysts in practice. For precious-metal-based catalysts, the fatal drawback is still the prohibitively high cost. Although in ORR testing, some advanced SACs catalysts have already achieved high mass activities, few of them can also deliver such extraordinary performance in a practical application due to their complicated synthetics. However, for nonprecious metal-based ORR catalysts, it seems that SACs are the most promising candidates for the next generation of electrochemical energy devices. This is because they possess relatively high activity and lower cost. Even so, the state-of-the-art ORR SACs reported are far from practical application. Thus, to fully realize the potential for ORR in practical applications of SACs, there are still some opportunities and challenges that should be put forward as follows:

1. Optimizing the synthesis methods for SACs. In view of the fact that commercialization usually requires low synthetic cost, future research direction for ORR SACs would be to develop efficient synthetic strategies that allow us facilely prepare SACs on a larger scale. Despite the widespread availability of the pyrolysis process for the fabrication of atomically dispersed metal catalysts, the precise synthesis at the atomic level is a major challenge. Thus, developing more facile methods that avoid using high temperature or other energy-intensive treatments is vital for further practical applications.
2. Increasing metal loading with a well atomic dispersion. As practical ORR in fuel cells and metal-air batteries require high catalytic activity, one challenge is how to improve the metal loading in SACs and promote the utilization rate of metal sites. For the SACs, high metal loading means higher active site density, which is desirable for improving ORR activity. However, until now, many reported SACs may dissolve or aggregate to form nanoparticles in electrocatalytic reactions, or some side reactions may occur, leading to the loss of many active sites accompanied by decreased activity. Thus, designing an efficient synthesis method for SACs with high metal loading is essential.
3. Enhancing the stability of SACs. An inevitable issue with atomically dispersed metal atoms is their stability, especially under harsh electrochemical reaction conditions. In some cases, the tolerance to the acid of SACs is still limited. Meanwhile, the slow mass transfer, high

resistance, and poor durability in practical applications could seriously affect the performance of catalysts. The design of hierarchical structures and the introduction of heteroatoms (N, S, Se, P, etc.) could enhance the activity and stability of SACs, which can drive the development of SACs for large-scale application in fuel cells and ZABs.

References

[1] S. Chu, A. Majumdar, Opportunities and challenges for a sustainable energy future, Nature 488 (7411) (2012) 294–303.
[2] H.B. Wu, X.W. Lou, Metal-organic frameworks and their derived materials for electrochemical energy storage and conversion: promises and challenges, Sci. Adv. 3 (12) (2017) 928935.
[3] M. Zhou, H.-L. Wang, S. Guo, Towards high-efficiency nanoelectrocatalysts for oxygen reduction through engineering advanced carbon nanomaterials, Chem. Soc. Rev. 45 (5) (2016) 1273–1307.
[4] H. Xu, D. Cheng, D. Cao, X.C. Zeng, A universal principle for a rational design of single-atom electrocatalysts, Nat. Catal. 1 (5) (2018) 339–348.
[5] Y. Jiao, Y. Zheng, M. Jaroniec, S.Z. Qiao, Design of electrocatalysts for oxygen-and hydrogen-involving energy conversion reactions, Chem. Soc. Rev. 44 (8) (2015) 2060–2086.
[6] Y. Chen, S. Ji, S. Zhao, W. Chen, J. Dong, W.C. Cheong, et al., Enhanced oxygen reduction with single-atomic-site iron catalysts for a zinc-air battery and hydrogen-air fuel cell, Nat. Commun. 9 (1) (2018) 5422.
[7] R. Wang, H. Wang, F. Luo, S. Liao, Core–shell-structured low-platinum electrocatalysts for fuel cell applications, Electrochem. Energ. Rev. 1 (3) (2018) 324–387.
[8] O. Savadogo, K. Lee, K. Oishi, S. Mitsushima, N. Kamiya, K.I. Ota, New palladium alloys catalyst for the oxygen reduction reaction in an acid medium, Electrochem. Commun. 6 (2) (2004) 105–109.
[9] J. Suntivich, H.A. Gasteiger, N. Yabuuchi, H. Nakanishi, J.B. Goodenough, Y. Shao-Horn, Design principles for oxygen-reduction activity on perovskite oxide catalysts for fuel cells and metal–air batteries, Nat. Chem. 3 (7) (2011) 546–550.
[10] W. Xing, M. Yin, Q. Lv, Y. Hu, C. Liu, J. Zhang, Oxygen Solubility, Diffusion Coefficient, and Solution Viscosity, Elsevier, 2014, pp. 1–31.
[11] Y. Liang, Y. Li, H. Wang, J. Zhou, J. Wang, T. Regier, et al., Co_3O_4 nanocrystals on graphene as a synergistic catalyst for oxygen reduction reaction, Nat. Mater. 10 (10) (2011) 780–786.
[12] J. Cao, Y. Feng, B. Liu, H. Li, Carbon skeleton doped with Co, N, S and P as efficient electrocatalyst for oxygen evolution reaction, Sci. China Mater. 61 (5) (2018) 686–696.
[13] J. Zhang, T. Wang, D. Pohl, B. Rellinghaus, R. Dong, S. Liu, et al., Interface engineering of MoS_2/Ni_3S_2 heterostructures for highly enhanced electrochemical overall-water-splitting activity, Angew. Chem. 128 (23) (2016) 6814–6819.
[14] D.W. Wang, D. Su, Heterogeneous nanocarbon materials for oxygen reduction reaction, Energy Environ. Sci. 7 (2) (2014) 576–591.
[15] C. Zhu, S. Fu, Q. Shi, D. Du, Y. Lin, Single-atom electrocatalysts, Angew. Chem. Int. Ed. 56 (45) (2017) 13944–13960.
[16] A.P.O.' Mullane, From single crystal surfaces to single atoms: investigating active sites in electrocatalysis, Nanoscale 6 (8) (2014) 4012–4026.

[17] A. Morozan, B. Jousselme, S. Palacin, Low-platinum and platinum-free catalysts for the oxygen reduction reaction at fuel cell cathodes, Energy Environ. Sci. 4 (4) (2011) 1238–1254.

[18] J. Zhang, H. Yang, B. Liu, Coordination engineering of single-atom catalysts for the oxygen reduction reaction: a review, Adv. Energy Mater. 3 (11) (2021) 2002473.

[19] J. Su, R. Ge, Y. Dong, F. Hao, L. Chen, Recent progress in single-atom electrocatalysts: concept, synthesis, and applications in clean energy conversion, J. Mater. Chem. A 6 (29) (2018) 14025–14042.

[20] J. Liu, M. Jiao, L. Lu, H.M. Barkholtz, Y. Li, Y. Wang, et al., High performance platinum single atom electrocatalyst for oxygen reduction reaction, Nat. Commun. 8 (1) (2017) 15938.

[21] T. Tang, W.J. Jiang, S. Niu, N. Liu, H. Luo, Y.Y. Chen, et al., Electronic and morphological dual modulation of cobalt carbonate hydroxides by Mn doping toward highly efficient and stable bifunctional electrocatalysts for overall water splitting, J. Am. Chem. Soc. 139 (24) (2017) 8320–8328.

[22] L. Liu, A. Corma, Metal catalysts for heterogeneous catalysis: from single atoms to nanoclusters and nanoparticles, Chem. Rev. 118 (10) (2018) 4981–5079.

[23] C.X. Zhao, B.Q. Li, J.N. Liu, Q. Zhang, Intrinsic electrocatalytic activity regulation of M–N–C single-atom catalysts for the oxygen reduction reaction, Angew. Chem. Int. Ed. 60 (9) (2021) 4448–4463.

[24] M.B. Gawande, P. Fornasiero, R. Zboril, Carbon-based single-atom catalysts for advanced applications, ACS Catal. 10 (3) (2020) 2231–2259.

[25] J. Liu, B.R. Bunes, L. Zang, C. Wang, Supported single-atom catalysts: synthesis, characterization, properties, and applications, Environ. Chem. Lett. 16 (2) (2018) 477–505.

[26] J.C. Liu, Y.G. Wang, J. Li, Toward rational design of oxide-supported single-atom catalysts: atomic dispersion of gold on ceria, J. Am. Chem. Soc. 139 (17) (2017) 6190–6199.

[27] Y. Chen, S. Ji, C. Chen, Q. Peng, D. Wang, Y. Li, Single-atom catalysts: synthetic strategies and electrochemical applications, Joule 2 (7) (2018) 1242–1264.

[28] P. Lang, N. Yuan, Q. Jiang, Y. Zhang, J. Tang, Recent advances and prospects of metal-based catalysts for oxygen reduction reaction, Energy Technol. 8 (3) (2020) 1900984.

[29] M.K. Debe, Electrocatalyst approaches and challenges for automotive fuel cells, Nature 486 (7401) (2012) 43–51.

[30] C. Tang, Y. Jiao, B. Shi, J.N. Liu, Z. Xie, X. Chen, et al., Coordination tunes selectivity: two-electron oxygen reduction on high-loading molybdenum single-atom catalysts, Angew. Chem. 132 (23) (2020) 9256–9261.

[31] X.Q. Huang, Z.P. Zhao, L. Cao, Y. Chen, E.B. Zhu, Z.Y. Lin, et al., High-performance transition metal-doped Pt_3Ni octahedra for oxygen reduction reaction, Science 348 (6240) (2015) 1230–1234.

[32] K. Dong, J. Liang, Y. Wang, Z. Xu, Q. Liu, Y. Luo, et al., Honeycomb carbon nanofibers: a superhydrophilic O_2-entrapping electrocatalyst enables ultrahigh mass activity for the two-electron oxygen reduction reaction, Angew. Chem. 133 (19) (2021) 10677–10681.

[33] D.A. Kuznetsov, Z. Chen, P.M. Abdala, O.V. Safonova, A. Fedorov, C.R. Muller, Single-atom-substituted Mo_2CT_x: Fe-layered carbide for selective oxygen reduction to hydrogen peroxide: tracking the evolution of the Mxene phase, J. Am. Chem. Soc. 143 (15) (2021) 5771–5778.

[34] H.A. Hansen, V. Viswanathan, J.K. Nørskov, Unifying kinetic and thermodynamic analysis of 2 e^- and 4 e^- reduction of oxygen on metal surfaces, J. Phys. Chem. C. 118 (13) (2014) 6706–6718.

[35] S. Jin, A.V. Neste, E. Ghali, S. Boily, R. Schulz, New cathode materials for chlorate electrolysis, J. Electrochem. Soc. 144 (12) (1997) 4272.
[36] J.A. Keith, G. Jerkiewicz, T. Jacob, Theoretical investigations of the oxygen reduction reaction on Pt (111), ChemPhysChem 11 (13) (2010) 2779–2794.
[37] L. Du, L. Xing, G. Zhang, M. Dubois, S. Sun, Strategies for engineering high-performance PGM-free catalysts toward oxygen reduction and evolution reactions, Small Methods 4 (6) (2020) 2000016. Available from: https://doi.org/10.1002/smtd.202000016.
[38] V. Stamenkovic, B.S. Mun, K.J. Mayrhofer, P.N. Ross, N.M. Markovic, J. Rossmeisl, et al., Changing the activity of electrocatalysts for oxygen reduction by tuning the surface electronic structure, Angew. Chem. Int. Ed. 45 (18) (2006) 2897–2901.
[39] V. Tripković, E. Skúlason, S. Siahrostami, J.K. Nørskov, J. Rossmeisl, The oxygen reduction reaction mechanism on Pt (111) from density functional theory calculations, Electrochim. Acta 55 (27) (2010) 7975–7981.
[40] J.K. Nørskov, J. Rossmeisl, A. Logadottir, L. Lindqvist, J.R. Kitchin, T. Bligaard, et al., Origin of the overpotential for oxygen reduction at a fuel-cell cathode, J. Phys. Chem. B 108 (46) (2004) 17886–17892.
[41] H. Shen, E. Gracia-Espino, J. Ma, K. Zang, J. Luo, L. Wang, et al., Synergistic effects between atomically dispersed Fe-N-C and C-S-C for the oxygen reduction reaction in acidic media, Angew. Chem. 129 (44) (2017) 13988–13992.
[42] M. Tong, L. Wang, H. Fu, Designed synthesis and catalytic mechanisms of non-precious metal single-atom catalysts for oxygen reduction reaction, Small Methods 5 (10) (2021) 2100865.
[43] Z. Song, Y.N. Zhu, H. Liu, M.N. Banis, L. Zhang, J. Li, et al., Engineering the low coordinated Pt single atom to achieve the superior electrocatalytic performance toward oxygen reduction, Small 16 (43) (2020) 2003096.
[44] L. Zhang, T. Yang, W. Zang, Z. Kou, Y. Ma, M. Waqar, et al., Quasi-paired Pt atomic sites on Mo_2C promoting selective four-electron oxygen reduction, Adv. Sci. 8 (18) (2021) 2101344.
[45] K. Kamiya, R. Kamai, K. Hashimoto, S. Nakanishi, Platinum-modified covalent triazine frameworks hybridized with carbon nanoparticles as methanol-tolerant oxygen reduction electrocatalysts, Nat. Commun. 5 (1) (2014) 5040.
[46] M. Xiao, L. Gao, Y. Wang, X. Wang, J. Zhu, Z. Jin, et al., Engineering energy level of metal center: Ru single-atom site for efficient and durable oxygen reduction catalysis, J. Am. Chem. Soc. 141 (50) (2019) 19800–19806.
[47] X. Peng, S. Zhao, Y. Mi, L. Han, X. Liu, D. Qi, et al., Trifunctional single-atomic Ru sites enable efficient overall water splitting and oxygen reduction in acidic media, Small 16 (33) (2020) 2002888.
[48] C. Zhang, J. Sha, H. Fei, M. Liu, S. Yazdi, J. Zhang, et al., Single-atomic ruthenium catalytic sites on nitrogen-doped graphene for oxygen reduction reaction in acidic medium, ACS Nano 11 (7) (2017) 6930–6941.
[49] M. Xiao, J. Zhu, G. Li, N. Li, S. Li, Z.P. Cano, et al., A single-atom iridium heterogeneous catalyst in oxygen reduction reaction, Angew. Chem. Int. Ed. 58 (28) (2019) 9640–9645.
[50] Q. Liu, Y. Li, L. Zheng, J. Shang, X. Liu, R. Yu, et al., Sequential synthesis and active-site coordination principle of precious metal single-atom catalysts for oxygen reduction reaction and PEM fuel cells, Adv. Energy Mater. 20 (10) (2020) 2000689.
[51] F. He, K. Li, C. Yin, Y. Wang, H. Tang, Z.J. Wu, Single Pd atoms supported by graphitic carbon nitride, a potential oxygen reduction reaction catalyst from theoretical perspective, Carbon 114 (2017) 619–627.

[52] Y. Deng, J. Luo, B. Chi, H. Tang, J. Li, X. Qiao, et al., Advanced atomically dispersed metal-nitrogen-carbon catalysts toward cathodic oxygen reduction in PEM fuel cells, Adv. Energy Mater. 37 (11) (2021) 2101222.
[53] Y. He, S. Hwang, D.A. Cullen, M.A. Uddin, L. Langhorst, B. Li, et al., Highly active atomically dispersed CoN_4 fuel cell cathode catalysts derived from surfactant-assisted MOFs: carbon-shell confinement strategy, Energy Environ. Sci. 12 (1) (2019) 250–260.
[54] M. Xiao, J. Zhu, L. Ma, Z. Jin, J. Ge, X. Deng, et al., Microporous framework induced synthesis of single-atom dispersed Fe-NC acidic ORR catalyst and its in situ reduced Fe-N_4 active site identification revealed by X-ray absorption spectroscopy, ACS Catal. 8 (4) (2018) 2824–2832.
[55] N. Zhang, T. Zhou, M. Chen, H. Feng, R. Yuan, W. Yan, et al., High-purity pyrrole-type FeN_4 sites as a superior oxygen reduction electrocatalyst, Energy Environ. Sci. 13 (1) (2020) 111–118.
[56] G. Chen, P. Liu, Z. Liao, F. Sun, Y. He, H. Zhong, et al., Promoted oxygen reduction kinetics on nitrogen-doped hierarchically porous carbon by engineering proton-feeding centers, Adv. Mater. 32 (8) (2020) 1907399.
[57] M. Qiao, Y. Wang, Q. Wang, G. Hu, X. Mamat, S. Zhang, et al., Hierarchically ordered porous carbon with atomically dispersed FeN_4 for ultraefficient oxygen reduction reaction in proton-exchange membrane fuel cells, Angew. Chem. Int. Ed. 59 (7) (2020) 2688–2694.
[58] X. Fu, N. Li, B. Ren, G. Jiang, Y. Liu, F.M. Hassan, et al., Fuel cells: tailoring FeN4 sites with edge enrichment for boosted oxygen reduction performance in proton exchange membrane fuel cell, Adv. Energy Mater. 9 (11) (2019) 1970031.
[59] P. Yin, T. Yao, Y. Wu, L. Zheng, Y. Lin, W. Liu, et al., Single cobalt atoms with precise N-coordination as superior oxygen reduction reaction catalysts, Angew. Chem. 128 (36) (2016) 10958–10963.
[60] X. Han, X. Ling, Y. Wang, T. Ma, C. Zhong, W. Hu, et al., Generation of nanoparticle, atomic-cluster, and single-atom cobalt catalysts from zeolitic imidazole frameworks by spatial isolation and their use in zinc-air batteries, Angew. Chem. 131 (16) (2019) 5413–5418.
[61] Y. Guo, P. Yuan, J. Zhang, Y. Hu, I.S. Amiinu, X. Wang, et al., Carbon nanosheets containing discrete Co-N_x-B_y-C active sites for efficient oxygen electrocatalysis and rechargeable Zn-Air batteries, ACS Nano 12 (2) (2018) 1894–1901.
[62] Y. Ha, B. Fei, X. Yan, H. Xu, Z. Chen, L. Shi, et al., Atomically dispersed Co-pyridinic N-C for superior oxygen reduction reaction, Adv. Energy Mater. 10 (26) (2020) 2002592.
[63] J. Li, M. Chen, D.A. Cullen, S. Hwang, M. Wang, B. Li, et al., Atomically dispersed manganese catalysts for oxygen reduction in proton-exchange membrane fuel cells, Nat. Catal. 1 (12) (2018) 935–945.
[64] W. Li, C. Min, F. Tan, Z. Li, B. Zhang, R. Si, et al., Bottom-up construction of active sites in a Cu-N_4-C catalyst for highly efficient oxygen reduction reaction, ACS Nano 13 (3) (2019) 3177–3187.
[65] P. Song, M. Luo, X. Liu, W. Xing, W. Xu, Z. Jiang, et al., Zn single atom catalyst for highly efficient oxygen reduction reaction, Adv. Funct. Mater. 27 (28) (2017) 1700802.
[66] F. Zhang, Y. Zhu, C. Tang, Y. Chen, B. Qian, Z. Hu, et al., Zn single atom catalyst for highly efficient oxygen reduction reaction, Adv. Funct. Mater. 32 (16) (2022) 2110224.
[67] S. Liu, Z. Li, C. Wang, W. Tao, M. Huang, M. Zuo, et al., Turning main-group element magnesium into a highly active electrocatalyst for oxygen reduction reaction, Nat. Commun. 11 (1) (2020) 938.

[68] V. Das, S. Padmanaban, K. Venkitusamy, R. Selvamuthukumaran, F. Blaabjerg, P. Siano, Recent advances and challenges of fuel cell based power system architectures and control—a review, Renew. Sustain. Energy Rev. 73 (2017) 10—18.

[69] D. Wang, H.L. Xin, R. Hovden, H. Wang, Y. Yu, D.A. Muller, et al., Structurally ordered intermetallic platinum-cobalt core-shell nanoparticles with enhanced activity and stability as oxygen reduction electrocatalysts, Nat. Mater. 12 (1) (2013) 81—87.

[70] J. Liu, M. Jiao, B. Mei, Y. Tong, Y. Li, M. Ruan, et al., Carbon-supported divacancy-anchored platinum single-atom electrocatalysts with superhigh Pt utilization for the oxygen reduction reaction, Angew. Chem. 131 (4) (2019) 1175—1179.

[71] Y. He, H. Guo, S. Hwang, X. Yang, Z. He, J. Braaten, et al., Single cobalt sites dispersed in hierarchically porous nanofiber networks for durable and high-power PGM-free cathodes in fuel cells, Adv. Mater. 46 (32) (2020) 2003577.

[72] Y. Cheng, S. He, S. Lu, J.P. Veder, B. Johannessen, L. Thomsen, et al., Iron single atoms on graphene as nonprecious metal catalysts for high-temperature polymer electrolyte membrane fuel cells, Adv. Sci. 10 (6) (2019) 1802066.

[73] M. Chen, X. Li, F. Yang, B. Li, T. Stracensky, S. Karakalos, et al., Atomically dispersed MnN4 catalysts via environmentally benign aqueous synthesis for oxygen reduction: mechanistic understanding of activity and stability improvements, ACS Catal. 18 (10) (2020) 10523—10534.

[74] Z. Lu, B. Wang, Y. Hu, W. Liu, Y. Zhao, R. Yang, et al., An isolated zinc-cobalt atomic pair for highly active and durable oxygen reduction, Angew. Chem. Int. Ed. 9 (58) (2019) 2622—2626.

[75] Q. Zhang, J. Guan, Applications of single-atom catalysts, Nano Res. 15 (1) (2022) 38—70.

[76] X. Yan, Y. Ha, R. Wu, Binder-free air electrodes for rechargeable Zinc-Air batteries: recent progress and future perspectives, Small Methods 4 (5) (2021) 2000827.

[77] M. Xu, D.G. Ivey, Z. Xie, W. Qu, Rechargeable Zn-air batteries: progress in electrolyte development and cell configuration advancement, J. Power Sources 283 (2015) 358—371.

[78] W. Zhang, Y. Liu, L. Zhang, J. Chen, Recent advances in isolated single-atom catalysts for zinc air batteries: a focus review, Nanomaterials 1 (9) (2019) 1402.

[79] H. Yu, A. Fisher, D. Cheng, D. Cao, Cu, N-codoped hierarchical porous carbons as electrocatalysts for oxygen reduction reaction, ACS Appl. Mater. Inter. 33 (8) (2016) 21431—21439.

[80] Q. Zhang, J. Guan, Applications of atomically dispersed oxygen reduction catalysts in fuel cells and zinc-air batteries, Energy Environ. Mater. 3 (4) (2021) 307—335.

[81] Y. Pan, S. Liu, K. Sun, X. Chen, B. Wang, K. Wu, et al., A bimetallic Zn/Fe polyphthalocyanine derived single-atom Fe-N_4 catalytic site: a superior trifunctional catalyst for overall water splitting and Zn-Air batteries, Angew. Chem. Int. Ed. 28 (57) (2018) 8614—8618.

[82] C.C. Hou, L. Zou, L. Sun, K. Zhang, Z. Liu, Y. Li, et al., Single-atom iron catalysts on overhang-eave carbon cages for high-performance oxygen reduction reaction, Angew. Chem. Int. Ed. 19 (59) (2020) 7384—7389.

[83] W. Zang, A. Sumboja, Y. Ma, H. Zhang, Y. Wu, S. Wu, et al., Single Co atoms anchored in porous N-doped carbon for efficient zinc- air battery cathodes, ACS Catal. 10 (8) (2018) 8961—8969.

[84] L. Yang, L. Shi, D. Wang, Y. Lv, D. Cao, Single-atom cobalt electrocatalysts for foldable solid-state Zn-air battery, Nano Energy 50 (2018) 691—698.

[85] Z. Yang, B. Chen, W. Chen, Y. Qu, F. Zhou, C. Zhao, et al., Directly transforming copper (I) oxide bulk into isolated single-atom copper sites catalyst through gas-transport approach, Nat. Commun. 1 (10) (2019) 3734.

CHAPTER 5

Single-atom catalysts for electrocatalytic oxygen evolution reaction

Ajit Kumar Singh*, Baghendra Singh* and Arindam Indra
Department of Chemistry, IIT (BHU), Varanasi, Uttar Pradesh, India

5.1 Introduction

The effective utilization of sustainable and renewable energy sources is essential to limit the excess use of fossil fuels for energy production [1]. In this respect, electrochemical water splitting has emerged as an efficient technology to produce clean and sustainable hydrogen fuel [1]. The electrochemical water splitting can be divided into two half-cell reactions: (1) oxygen evolution reaction (OER) at anode and (2) hydrogen evolution reaction (HER) at cathode [1]. The electrochemical OER suffers from sluggish reaction kinetics due to the involvement of the multistep electron transfer with the formation of high-energy intermediates [1]. Additionally, a huge potential is required to overcome the thermodynamic barrier of OER [1].

In recent years, a series of catalysts have been developed with different structural, morphological, compositional, and electronic properties for electrochemical OER [2]. Additionally, the designing of low-cost and efficient electrocatalysts must be explored focusing on (1) low loading of the precious metal and (2) replacement of the noble metal catalysts with earth-abundant transition metal-based catalysts maintaining comparable OER performance [3].

In this respect, single-atom catalysts (SACs) have been demonstrated for the electrochemical OER [3–5]. SACs consist of a noble or nonnoble metal uniformly dispersed on the support material with an isolated single atom as its active center (Fig. 5.1) [3–5]. SACs possess remarkable catalytic activity and stability due to their unique geometric and electronic

* Equally contributed.

Single Atom Catalysts
DOI: https://doi.org/10.1016/B978-0-323-95237-8.00012-4

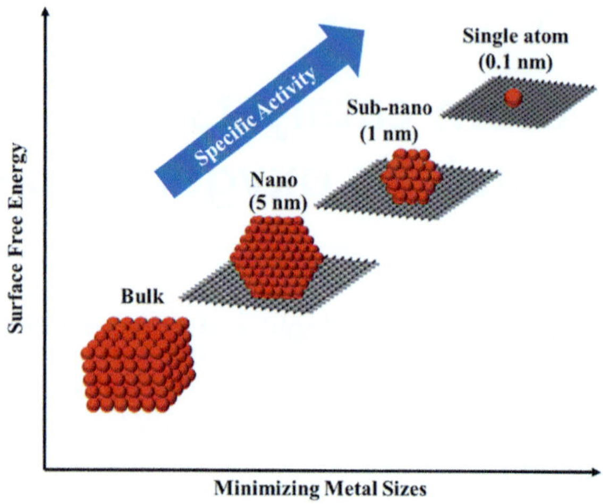

Figure 5.1 Schematic illustration showing the formation of SACs with increased specific activity.

properties. Moreover, SACs have positively charged single metal atoms. As a result, adsorption properties and binding energies of intermediates are optimized leading to the reduction in the free energy barrier [3–5]. The various supports, on which single atoms are dispersed, have diverse structures, tunable size, shape, porosity, large surface area, and high electrical conductivity. The strong metal-support interaction facilitates the uniform dispersion of the active metal atoms and enhances the stability of the catalyst [3–5].

The activity and stability of SACs can be further improved by tuning the coordination environment of the metal moieties [6–8]. The abundant heteroatoms (N, S, and O) with lone pairs of electrons are observed to act as the active coordinating sites. The unique coordination environment of the metal atom in SACs results in the extra active sites promoting the electrochemical performance [5,7,9].

The SACs have advantages like (1) isolated active species for improved catalytic activity, (2) high reusability and durability, (3) efficient atom utilization, (4) high surface free energy, and (5) strong metal atom-carrier interaction [5,7,9]. Owing to these unique properties, SACs are considered the ideal catalysts for OER. Further, the catalytic mechanisms for OER can be easily studied at the molecular level owing to their well-defined active sites and coordination properties and the mechanistic insights can provide the direction for the designing of highly active and more efficient electrocatalysts [5,7,9].

In 2011, Zhang group first reported the SAC-Pt_1/FeO_x for the CO oxidation [10]. In 2012, Sykes group developed Pd SAC on the Cu(111) surface for the selective hydrogenation of styrene and acetylene [11]. Further, Geng group reported a well-dispersed Pd single atom on TiO_2 at room temperature for the hydrogenation of benzaldehyde [12]. In further advancement, a series of SACs like Ru/CoFe-LDH, Au/NiFe-LDH, Ag/IrO_x, W-$Ni(OH)_2$, Ru-N-C, and so on have also been explored for electrochemical OER with efficient catalytic activity [5].

5.2 Synthetic strategies of single-atom catalysts

Although SACs have unique structural and electronic properties, there are several challenges associated with the designing of SACs [5]. The high surface energy of SACs results in the agglomeration of the single atoms leading to the formation of metal atom clusters and nanoparticles [5,8]. This ultimately creates a serious issue compromising the stability of SACs. Therefore, controlling the interactions between the metal atoms and the support is essential [13,14]. The strong metal-support interaction prevents agglomeration and cluster formation and also improves the charge transfer and mass transport from the support to the SACs [13]. Additionally, SACs suffer from other challenges like low efficiency, high production cost, poor thermal stability, and complicated synthetic processes [6,15]. The amount of loading has also a significant impact on the catalytic activity of SACs [13,16]. The high loading of the single atom is also a challenge for the designing of the SACs [4,13]. The synthetic processes of SACs are not economical and involve expensive precursors, high temperature, and complicated synthetic procedures [16].

Strategies like wet chemical method, high-temperature pyrolysis, atomic-layer deposition (ALD), photocatalysis, etc., have been employed to synthesize SACs [8]. The following key factors should be taken care of during the synthesis of SACs: (1) uniform distribution of the active sites on the support; (2) strong interaction between metal atoms and nearby coordination atoms; (3) improved mass and electron transport; (4) active-site accessibility; and (5) high activity, selectivity, and durability [17].

5.2.1 Wet chemical strategy

The wet chemical approach uses chemical reduction, electrochemical deposition, and photochemical reduction [8,17]. This method is straightforward and needs mild reaction conditions, which is helpful for large-scale manufacturing [8]. The wet synthesis process involves the loading of metal ions onto the support by ion exchange or impregnation method and drying via calcination followed by reduction using thermal or photo energy [5]. Selection of the suitable metal precursor, supporting material, and control over the synthetic method is essential to minimize the metal ion aggregation process. Although the wet chemical method is an efficient process for the synthesis of a single atom, the interaction of the metal with the support is weak resulting in poor stability of the SACs [18].

5.2.1.1 Coprecipitation method

The coprecipitation method is the most often used for the synthesis of SACs because it is simple and reproducible [3]. In the coprecipitation method, the combination of the precursors, support, and even promoter with an appropriate precipitant results in the formation of SACs [18]. The precipitate is further dried or calcined to obtain SACs. For example, the single-atom Ir was synthesized on NiO by the coprecipitation method and used for electrocatalytic OER in the alkaline medium [19]. The XAS studies reveal the formation of Ir single atom on NiO, and the formation of Ir-O bonding stabilizes a single atom Ir. The single atom Ru was deposited on monolayer NiFe-layered double hydroxide (LDH) by coprecipitation method [20]. The increase in the lattice parameter of NiFe-LDH from 0.8 to 0.9 nm was observed when Ru single atom is loaded on LDH.

This method can be used to deposit two or more uniformly dispersed metal elements on the support [18]. The parameters, such as pH, speed of addition, reaction time, and reaction temperature, can be controlled to tune the properties of the SAC [18]. The catalyst loading in the coprecipitation method is low. Some active sites are not exposed in the catalyst prepared by the coprecipitation method as they are buried in the bulk support.

5.2.1.2 Electrochemical deposition method

The electrochemical synthesis used electrical energy to deposit SACs on the support [8,21]. The electrochemical deposition of SACs can be controlled by tuning the parameters like voltage and current, the concentration of metal ion, pH of the medium, and deposition time [8,21]. The electrochemical deposition is facile scalable and cheap. Also, the electrochemical deposition method produces SACs that are free of binder and can be employed directly in electrocatalytic processes.

The electrochemical method leads to the deposition of the catalyst on the outermost part of the support and maximizes the utilization of the metal [21]. For example, Zhang et al. synthesized Ir SAC on the $Co_{0.8}Fe_{0.2}Se$ by anodic electrodeposition method and used it for OER [22]. Similarly, the single-atom Ru deposited on the defective NiFe-layer double hydroxide by electrochemical process showed improvement in OER activity [23]. The difference in the deposition process involving different redox processes can change the properties of the SACs [22]. However, homogeneous metal atom coating on the working electrode is difficult in the electrodeposition method.

5.2.2 High-temperature pyrolysis

The most prevalent method for the synthesis of the SACs is the high-temperature pyrolysis of the metal precursors [8,21,24,25]. This method is easy, cheap, and it does not require any sophisticated instrument. However, aggregation of single atoms occurs during pyrolysis. Hence, pretreatment of the precursor or posttreatment of the pyrolyzed product is required to get an atomically dispersed catalyst [8,21]. Adding N-rich precursors (such as melamine, dopamine, and dicyandiamide) prior to thermal treatment or calcining the precursors in an NH_3 atmosphere increases the anchoring force and enhances the metal loading [17,21]. Also, aggregation of a single atom can be prevented by modulating the kinetics of precursor decomposition [8].

The carbon matrix is widely explored as support for anchoring a single atom for the OER [25–28]. The intrinsic defects on the surface of carbon material and functional group serve as sites for the deposition of SACs in a highly dispersed state, minimize the surface agglomeration, and show good catalytic performance. Carbon material like graphitic carbon nitride [28], graphene [14,29], graphene oxide [27], and N or S-doped carbon nanotube [25], is used as a substrate for the loading of SACs. Cao et al. used phosphorous-doped carbon nitride as support to synthesize atomically dispersed Ru SACs for OER [28]. Single-atom Ni is synthesized on the defective graphene support for excellent OER activity [29].

5.2.3 Atomic-layer deposition method of deposition of single-atom catalyst

ALD is a controlled technology frequently used to deposit metal atoms in layer by layer to produce SACs [18,30]. By regulating the number of cycles, the ALD method can modify the size of catalytic species from single atoms to nanoclusters to nanoparticles [17,18]. The thickness and the metal loading can also be controlled by this method. The substrate used for the deposition of SACs generally has a functional group like -OH, -O, -N, and an organometallic compound/gaseous precursor is used as the metal precursor [17,21,31]. The Pd and Pt are mainly deposited on the N and OH functional groups containing substrate [30]. The single-atom Ru is deposited on the Co_3O_4 by the ALD method and shows a large decrease in the overpotential compared to RuO_2-Co_3O_4 for electrocatalytic OER [32]. ALD method of SAC synthesis has drawbacks like (1) difficulties in high loading of catalyst, (2) formation of clusters, (3) use of a sophisticated instrument, (4) slow rate of deposition, and (5) high-cost precursor [18].

5.3 Characterization of single-atom catalysts

The characterization of the SACs by conventional methods like powder X-ray diffraction (PXRD) and HR-TEM gives indirect evidence because of their limited sensitivity [5]. Hence, advanced techniques, including synchrotron-based X-ray absorption spectroscopy (XAS) like X-ray absorption near edge structure (XANES) and EXAFS, Mössbauer spectroscopy, scanning tunneling microscopy (STM), X-ray photoelectron spectroscopy (XPS), and scanning transmission electron microscopy (STEM), etc., have been employed to characterize and identify SACs [14,17,33,34]. These characterization techniques provide direct evidence of the formation of the SACs and their morphology. When these improved, characterization techniques are combined with the density functional theory (DFT) calculations, the structure—activity relationship may be clarified, allowing for the precise, logical design of tailored SACs [9].

The PXRD can provide indirect evidence for the formation of the SACs [5]. The absence of diffraction peaks related to the metal indicates the formation of SACs. For example, NiO-supported single-atom iridium (Ir-NiO) does not show the metallic iridium peak in the PXRD, indicating the presence of Ir in the atomic structure rather than nanoparticles (Fig. 5.2A) [19]. However, the absence of diffraction peaks related to the metal can be originated from the low metal loading (<2 wt.%), ultrasmall particle size (<4 nm), and the amorphous nature of the metal [17]. Therefore, advanced characterization techniques are required to identify SACs.

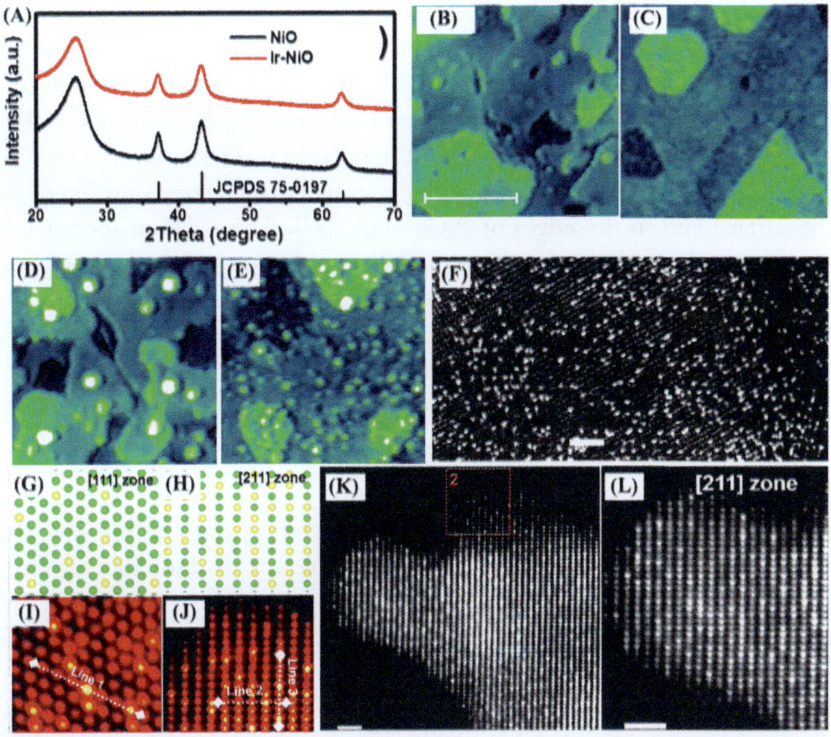

Figure 5.2 (A) Powder X-ray diffraction (PXRD) of Ir single atom on NiO (*red color*) and NiO support (*black color*). (B–C) STM image of bare ceria support and (D–E) STM image showing the deposition of Pt single-atom on the ceria support. (F) HDAAF-STEM image of atomically dispersed Ir single atom on NiO matrix; (G–H) Atomic models of the Ir-NiO along [111] and [211] zone axes; (I–L) Atomic STEM images of Ir-NiO along the [111] and [211] zone axis. *Reproduced with permission from (A) Q. Wang, X. Huang, Z.L. Zhao, M. Wang, B. Xiang, J. Li, et al., Ultrahigh-loading of Ir single atoms on NiO matrix to dramatically enhance oxygen evolution reaction, J. Am. Chem. Soc. 142 (2020) 7425–7433. https://doi.org/10.1021/jacs.9b12642, with American chemical society; (D–E) F. Dvořák, M. Farnesi Camellone, A. Tovt, N.-D. Tran, F.R. Negreiros, M. Vorokhta, et al., Creating single-atom Pt-ceria catalysts by surface step decoration, Nat. Commun. 7 (2016) 10801. https://doi.org/10.1038/ncomms10801, with Springer Nature; (I–L) Q. Wang, X. Huang, Z.L. Zhao, M. Wang, B. Xiang, J. Li, et al., Ultrahigh-loading of Ir single atoms on NiO matrix to dramatically enhance oxygen evolution reaction, J. Am. Chem. Soc. 142 (2020) 7425–7433. https://doi.org/10.1021/jacs.9b12642, with American chemical society.*

5.3.1 Scanning tunneling microscopy

STM is an ultra-high-resolution technique used to investigate surface structure at the atomic level. [36] A bias voltage is applied to the substrate surface by a conducting tip. Therefore, a tunneling current is generated. The structure of the substrate surface can be observed by monitoring the current as the tip scans across it. [17,33] The STM study reveals the formation of single-atom Pt on the ceria support (Fig. 5.2B–E) [35]. The single-atom Pt exists as PtO_4 unit in the edge of the monolayer ceria support [35]. Also, the STM detects single atom Pt, randomly distributed on the surface and in the edges of PtCu alloy [37]. The STM images of Pd/Cu alloys/Al_2O_3 show the range of Pd coverages on Cu(111) at different concentrations of Pd [38]. Pd is deposited at low concentration in the form of a single atom, while Pd forms a metal island on the copper surface at high concentration.

5.3.2 Aberration-corrected scanning transmission electron microscopy

Recent advances in Aberration-corrected scanning transmission electron microscopy (ACSTEM) resolution have enabled it to distinguish isolated single atoms (~0.5 Å) [17]. Based on Z-contrast imaging, the ACSTEM image provides the most effective approach to detecting individual metal atoms directly on the support [17,33]. Z-contrast images can be obtained by concentrating an electron beam on material and collecting electrons that scatter to an annular dark-field detector. The conversion of Pd, Pt, Au nanoparticles into single Pd, Pt, Au atoms at high temperatures on N-doped carbon support was studied by ACSTEM [39]. Other techniques like high-angle annular dark-field scanning TEM (HAADF-STEM) are also used for the characterization of the SACs [10]. HAADF-STEM directly revealed the electroscopic image of Ir/NiO_x single atoms (Fig. 5.2F) [19]. In the Ir/NiO_x, ACSTEM shows Ir single atoms on NiO were projected into [111] and [211] zone and found to be exclusively located in the same columns of Ni atoms (Fig. 5.2G–L).

The EDX mapping is also used to detect a single atom on the support [33,40,41]. The formation of a single atom is confirmed by the higher peak-to-background ratio in the EDX [40–42]. The single dispersion state of Pt on defective $Ni(OH)_x$ was described using EDX mapping analysis and HAADF-STEM [40].

5.3.3 X-ray absorption spectroscopy

Although electron microscopy allows to viewing of single atoms, the visual range of this technology is confined to a local area [5,17]. One of the most powerful techniques to confirm the atomic and electronic structure of a single atom is XAS [14,17]. The signals obtained from XAS give exact information about the structure of SACs including oxidation state, coordination number, and coordination environment [17]. Hence, XAS can be used to determine the chemical state and local atomic structure of isolated atoms. The investigation of detailed electronic structure and coordination environment is crucial for analyzing the specific structure of SACs by extended X-ray absorption fine structure (EXAFS) [27]. The isolated atomic sites from clusters or particles can be distinguished by detecting the radial distance between the active metal centers and the coordinated atoms using the radial structure-function (R space) obtained from the Fourier-transformed EXAFS (FT-EXAFS) [5,7,27]. A single major peak in R space can be attributed to the coordination of the metal with nonmetallic elements (N, C, O, B, and P) without the existence of metal-metal coordination in the SACs [7]. Further, the XANES is also utilized to determine the valence state of the atoms in SACs [5].

The single-atom XANES spectrum of Pt/CeO_x reveals distinct characteristic peaks between 11570 and 11630 eV [43]. The XANES of Pt supported on ceria shows a white line corresponding to the 11567 eV corresponding to SACs [43]. The presence of a peak at 1.8 Å and the absence of a peak above the 2.5 Å in the EXAFS studies of Cu_1-N_2/HCNS show no appreciable contribution from metallic Cu or oxidized CuO_x clusters in Cu-N_2/HCNS confirm the formation of SACs [44]. The peak at about 8979 eV in XANES shows that Cu is present in the +1.5 oxidation state in the Cu_1-N_2/HCNS (Fig. 5.3A–C) [44]. The EXAFS and XANES studies reveal the formation of Ir single atom on the cobalt oxide support [46]. The Co K-edge of $Ir_{0.06}Co_{2.94}O_4$ exhibits an enhanced valence state of Co cations compared to pure Co_3O_4 in the XANES while EXAFS of Ir L3-edge of $Ir_{0.06}Co_{2.94}O_4$ shows the presence of high Ir valence state of $\sim +4$, unlike nanoparticles or supported single atoms [46].

5.3.4 Resonant inelastic X-ray scattering

Like XANES, resonant inelastic X-ray scattering (RIXS) is another powerful method to determine the valence d state characteristics of the SACs [34]. The RIXS is used to determine the low-energy excitations in atoms and molecules [34,47]. In RIXS, atomic core electrons are excited into unfilled localized electronic valence states. The important characteristics of

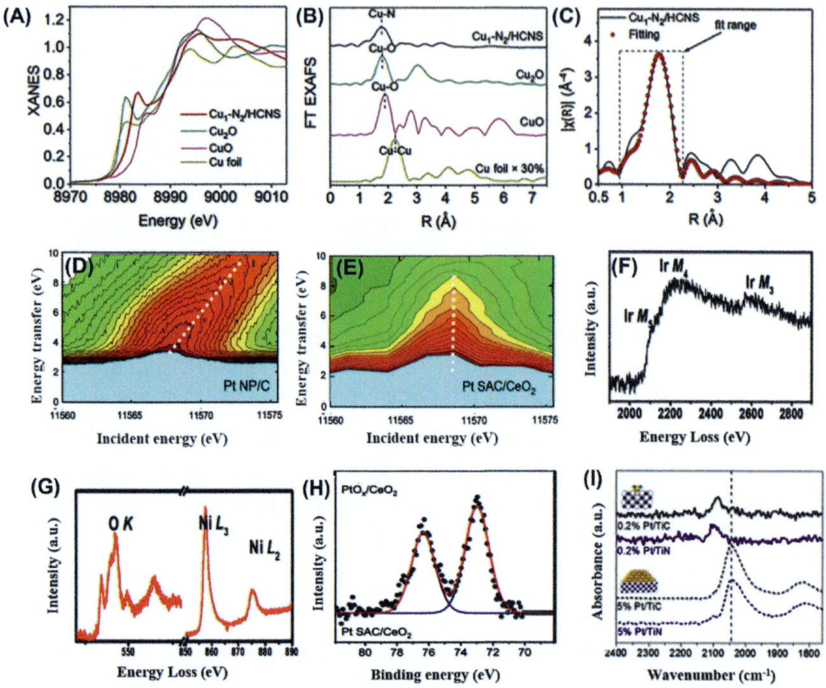

Figure 5.3 (A) XANES spectra of Cu K-edge of Cu_1-N_2/HCNS, (B) Fourier transform spectra of Cu K-edge EXAFS of Cu_1-N_2/HCNS, Cu_2O, CuO, and Cu foil; (C) EXAFS fitting curves of Cu_1-N_2/HCNS; (D–E) RIXS of the Pt nanoparticle and single atom Pt on CeO_2. (F) EELS spectra of Ir M-edge of single Ir atom and Ni L-edge and O K-edge of NiO, (H) Pt 4 f XPS spectra of Pt single atom with Pt^{2+} and Pt^0. (I) CO absorbance FTIR spectra of Pt single atoms (*solid lines*) and Pt nanoparticles (*dashed lines*). *(C) Reproduced permission from T. Zhang, X. Nie, W. Yu, X. Guo, C. Song, R. Si, et al., Single atomic Cu-N2 catalytic sites for highly active and selective hydroxylation of benzene to phenol, IScience. 22 (2019) 97–108. https://doi.org/10.1016/j.isci.2019.11.010, with Elsevier; (D–E) M. Kottwitz, Y. Li, R.M. Palomino, Z. Liu, G. Wang, Q. Wu, et al., Local structure and electronic state of atomically dispersed Pt supported on nanosized CeO2, ACS Catal. 9 (2019) 8738–8748. https://doi.org/10.1021/acscatal.9b02083, with American chemical society; (F) Q. Wang, X. Huang, Z.L. Zhao, M. Wang, B. Xiang, J. Li, et al., Ultrahigh-loading of Ir single atoms on NiO matrix to dramatically enhance oxygen evolution reaction, J. Am. Chem. Soc. 142 (2020) 7425–7433. https://doi.org/10.1021/jacs.9b12642, with American chemical society; (H) M. Kottwitz, Y. Li, R.M. Palomino, Z. Liu, G. Wang, Q. Wu, et al., Local structure and electronic state of atomically dispersed Pt supported on nanosized CeO2, ACS Catal. 9 (2019) 8738–8748. https://doi.org/10.1021/acscatal.9b02083, with American chemical society; (I) S. Yang, Y. J. Tak, J. Kim, A. Soon, H. Lee, Support effects in single-atom platinum catalysts for electrochemical oxygen reduction, ACS Catal. 7 (2017) 1301–1307. https://doi.org/10.1021/acscatal.6b02899, with American chemical society.*

the RIXS planes are attributed to the changes between occupied and unoccupied valance states of the atom/element [34,47]. As the metal valence states face a change with the external environment, variation in the RIXS spectra is observed [34,47]. For example, RIXS can detect the Pt 5d to Pt 2p d band tails for Pt metal and Pt nanoparticles while this tail is absent for single atom Pt [43]. In the Pt single atom, the d band tail arises due to the Pt 5d-O 2p transitions that appear at higher energy (Fig. 5.3D–E) [43].

5.3.5 Electron energy loss spectroscopy

The chemical composition, stoichiometry, energy levels, and electronic structure of SACs can be determined by Electron energy loss spectroscopy (EELS) [33,48,49]. EELS is recorded during TEM or STEM as it requires electrons to hit the catalyst surface [42,49]. When an electron beam with known kinetic energy interacts with an atom, it generates inelastically scattered electrons that lose energy and deviate from the pathways [33]. The EELS spectrum is obtained by distributing the number of energy-loss electrons on the material according to the energy [33,48,49]. The energy loss of a single atom state is different from other chemical states, which helps in the characterization of the catalyst [33,48,49]. Under the appropriate conditions of electronic source, accelerated voltage, and orbital electrons, a resolution of 0.1 nm can be reached in EELS [33]. The resolution of EELS is higher than XPS and Auger electron spectroscopy [33]. For example, in the Ir/NiO_x, the Ir M edges are detected using EELS, while on the substrate, the EELS showed O K and Ni L edges of NiO_x (Fig. 5.3F–G) [19].

5.3.6 X-ray photoelectron spectroscopy

As the metal in SACs coordinates with the nearby heteroatoms, the charge transfer from the metal to the coordinated heteroatoms takes place [17]. XPS can detect this oxidation of the core metal in SACs [34]. Compared to single atom Pt on *g*-CN, Pt nanoparticles loaded *g*-CN showed only minor BE shifts of C 1s, N 1s, and Pt 4f peaks, with no obvious change in peak shape in synchronous illumination XPS (SI-XPS) [50]. These results show that metallic Pt has a weaker charge-separation capability than single-atom Pt-N bonding in *g*-CN, which is matched with catalytic performance. Similarly, XPS is used to determine the oxidation state of single-atom Pt on monolayer ceria [43]. The XPS studies reveal that the binding energies of Pt atoms to the ceria step edges (6.7 eV) are greater than that of the binding energy of bulk metallic Pt, and this Pt is present in Pt^{2+} oxidation state in the single atom Pt/CeO_2. The absence of Pt^0 contributions confirms the homogeneity of Pt^{2+} sites in the SAC (Fig. 5.3H) [43]. The Ni 2p of NiFe@g-CN/CNTs shows the presence of Ni^{2+} and Ni^{3+} corresponding to NiO [51]. The absence of a peak corresponding to Ni^0 in Ni 2p spectra of NiFe@g-CN/CNTs confirmed that the Ni atom is present in a cationic form in the framework of g-CN rather than the cluster form [43]. Similarly, the Fe2p XPS spectra showed that Fe exists in the Fe^{2+} and Fe^{3+} states in the NiFe@g-CN/CNTs.

5.3.7 Fourier-transformed infrared spectroscopy

Monitoring the highly sensitive vibration mode of the adsorbed CO molecules on the SACs surface Fourier-transformed infrared spectroscopy (FTIR) can detect the absence of nanoclusters and nanoparticles [10,36,45,52]. For Pt SACs on FeO_x, TiC, and TiN support, only linearly adsorbed CO molecules at 2080 cm^{-1} were detected [45]. On the other hand, Pt nanoclusters and nanoparticles showed bridging CO molecules at 1820 cm^{-1} (Fig. 5.3I).

5.3.8 Mössbauer spectroscopy

Mössbauer spectroscopy is the study of recoil-free nuclear gamma-ray emission and absorption in materials [17]. Three distinct types of nuclear interactions are observable in Mössbauer spectra: (1) isomer shift caused by variations in neighboring electron populations, (2) quadrupole splitting caused by atomic-scale electric field gradients, and (3) magnetic Zeeman splitting caused by nonnuclear magnetic fields [17]. The Mössbauer spectroscopy is generally used to characterize Fe-based SACs to find out the coordination environment and electronic configuration of the core Fe ions in complex materials [53–55]. The Mössbauer spectra of FeN_4 SACs consist of three doublets (D1, D2, and D3) [53]. The D1 belongs to the ferrous low-spin, while D2 and D3 doublets belong to two ferrous with intermediate spin might be assigned to FeN_4.

5.4 Electrocatalytic water splitting

The OER process is kinetically sluggish than HER, involving a four-electron and proton transfer reaction [2]. Hence, the overpotential of

OER is higher than HER due to the kinetic hindrance of the individual elementary steps of OER [2]. Also, Gibb's free energy change of OER is higher than HER. The OER process depends on the pH of the medium, and it follows different mechanisms at the different pH. Alkaline OER is more facile than acidic medium [2].

The electrochemical OER is a surface phenomenon, and the active sites present on the surface play a key role to control the OER activity [2]. Therefore, the designing of the electrocatalysts with ultrathin nature and more active sites on the surface is highly essential to promote OER activity. As the SACs have exposed active metal sites on the surface, it also improves the OER activity [4,9]. The OER activity and stability of SACs can be boosted by the introduction of heteroatoms like N, O, C, S, etc. [56–60]. These coordinating atoms play a crucial role in modulating the electronic properties of the SACs. The coordinating heteroatoms are observed to act as anchoring sites for single atoms [56–60]. The electronegative N, S, and O atoms have lone-pair electrons, which can interact with the empty d-orbitals of transition metals. As a result, stable and abundant anchoring sites become available without the aggregation of single atoms [56–60]. Moreover, the strong interaction of single atoms with heteroatoms is attributed to their high electronegativity. The strong interaction regulates the electronic properties of the active sites [56–60]. Due to the modulation of electronic properties, the OER activity and stability of SACs are improved [56–60]. In addition, the coordinating heteroatoms also optimize the binding energy of the reaction intermediates.

The effect of coordinating atoms on the OER activity and stability was explained in Ni single atoms on graphene support [57]. This study showed the modulation of the electronic properties due to the coordinating C and N atoms. Further, Ni single atoms coordinated to O sites of graphene-like carbon (Ni–O–G) were reported for efficient OER activity [57]. The DFT calculations revealed that the coordinating O-atoms assisted in the optimization of the binding energy of the reaction intermediates during OER.

Still now, noble metal-based SACs have been widely employed in electrochemical oxygen evolution for efficient catalytic activity [7,60,61]. The DFT has revealed that the SACs based on noble metals, such as Ir, Pt, Ru, Rh, etc., have shown excellent OER activity and stability [7,60,61]. However, the high price and rare existence of these noble metals are the major restrictions to explore the noble metal SACs for practical applications [7,60,61]. On the other hand, the high loading of the single metal atoms is also a major concern. Therefore, noble metals are replaced with transition metals to avoid the high loading and high-cost issue of the noble metals [7,60,61]. The transition metal-based SACs exhibited excellent OER activity and stability, comparable to the noble metal-based SACs.

As the turnover frequency (TOF) is an important parameter to evaluate the intrinsic activity of the electrocatalysts for OER, TOF of the SACs is also determined to evaluate the intrinsic OER performance [7,60,61]. The TOF of electrocatalysts can be enhanced by increasing the atomic utilization as well as increasing the surface active sites, which can be achieved by dispersing the active metal sites at an atomic level. The SACs have maximum atomic utilization due to well-dispersed atoms, and the active sites are more exposed on the surface of the SACs [7,60,61]. Therefore, the SACs have been observed to possess high TOF compared to the other bulk electrocatalysts. For example, the bulk $NiCo_2O_4$ and the iridium single-atoms (Ir − SAs) on ultrathin $NiCo_2O_4$ porous nanosheets were developed for efficient OER activity by different research groups [61]. It has been observed the bulk catalyst $NiCo_2O_4$ exhibited the TOF of 0.0366 s^{-1} for OER while the SAC Ir- $NiCo_2O_4$ showed TOF of 6.70 s^{-1} for OER [61]. Similarly, Ir-NiFeS SAC also showed a high TOF of 9.85 s^{-1} indicating the high intrinsic OER activity compared to the bulk catalysts [62].

The amount of loading has also a significant impact on the catalytic activity of SACs [9,56,63–65]. It has been observed that the loading of the single atom is also a challenge for the designing of the SACs. The increment in the amount of loading of the single atoms and the dispersion of active single atomic sites also play an important role in the designing of SACs for OER [9,56,63–65]. SACs can be employed for efficient OER activity at a practical level with a sufficient amount of loading. The OER activity of SACs can be enhanced by increasing the metal mass loading. The mass loading increases the density of active centers and exposed active site area [9,56,63–65]. As a result, the water adsorption properties are improved promoting the OER activity. For example, Wang et al. developed a unique approach for achieving high loadings (up to ≈ 18 wt.%) of single Ir atoms in a nickel oxide (NiO) matrix [19]. The high loading of Ir resulted in the efficient catalytic activity for OER due to the increased number of single atoms and exposed number of surface active sites.

The single metal atoms are not stable because of the high surface energy [9,56,63–65]. Therefore, intensive studies have been performed focusing on the stabilization of single metal atoms. The enhancement of metal-support interactions is observed as an efficient route for the stabilization of single metal atoms. The metal-support interaction results in the tuning of electronic properties around the active sites in SACs [9,56,63–65]. The metal-support interaction modulates the coordination properties to tune the atomic structure and also enhances the charge transfer properties to promote the OER activity. In addition, the metal-support interaction results in structural defects in the catalyst structure improving the OER activity [9,56,63–65]. The metal-support interaction also optimizes the adsorption energies between the catalytically active sites and the reaction intermediates to promote the OER activity [9,56,63–65].

During the OER, transition metal-based electrocatalysts undergo electrochemical reconstruction to form metal-(oxy)hydroxide under the applied anodic potential in alkaline electrolyte [66]. In addition, the leaching of metals is also observed under electrochemical conditions. Therefore, it is most important that the SACs should maintain their original structure during OER under applied anodic potential [20,56,67–69]. The metal-support interaction in SACs prevents the anodic oxidation of active centers and the leaching of metal ions. Moreover, the metal-support interaction increases the stability of the SACs during OER [20,56,67–69].

As the amount of loading also plays a key role in the catalytic activity and stability of SACs, the replacement of noble metals with earth-abundant transition metals can be an alternative to design efficient, economic, and active SACs with high metal loading for OER [20,56,67–69]. Therefore, huge progress has been achieved in designing the transition metal-based SACs with high OER activity and stability. As the single metal atoms are dispersed on a support, the SACs can be categorized into three groups based on their support [20,56,67–69].

5.4.1 Carbon-supported single-atom catalysts

Carbon-supported SACs are widely utilized for OER in which metal single atoms are anchored on a carbon matrix [4,9,58]. The metal atoms can be successfully deposited on carbon materials due to their intrinsic structural defects and surface functionality. Moreover, metal atoms can be easily immobilized with high dispersion preventing undesirable aggregation [4,9,58]. When the metal atoms are deposited on the carbon matrix, they form a bond with the carbon matrix or the surface functional groups producing single-atom active centers [4,9,58]. The carbon-supported SACs have been employed for electrochemical OER with excellent catalytic performance and stability.

For example, the Ni atoms were entrapped on defective graphene to form the atomically dispersed Ni on graphene (A-Ni@DG) with 1.24 wt. % of Ni loading [70]. The electrochemical OER performance of A-Ni@DG was recorded at 270 mV attaining 10 mA cm^{-2} current density. Moreover, the improved TOF of 13.4 s was also determined at 300 mV overpotential. The improved electrochemical performance of A-Ni@DG was attributed to the unique configurations of atomic Ni in the graphene defects, which modulated the electronic properties of the Ni atoms optimizing the binding energies of OER intermediates [70]. The DFT calculations revealed that the electronic density of the states was modulated around the Ni due to the defective graphene. As a result, the energy barrier for the adsorption of intermediates was optimized promoting the OER activity.

The introduction of heteroatom like N, S, etc., in carbon support has been observed to enhance the availability of sites for metal atom deposition [56–60]. Moreover, the electronic conductivity is also modulated to improve the catalytic performance. Specifically, nitrogen atoms are more electronegative compared to carbon atoms.

For example, SACs based on 3D metals supported by nitrogen-doped holey graphene frameworks (M–NHGFs, M = Fe, Co, or Ni) have been developed for the electrochemical OER [27]. The XAS revealed the MN_4C_4 functionality in the M–NHGFs catalyst structure. The STEM images were further shown a uniform dispersion of the metal atoms throughout the graphene matrix [27]. The DFT calculations revealed that the catalysts followed the trend of Ni–NHGF > Co–NHGF > Fe–NHGF for OER activity [27]. This trend of OER activity was confirmed by the electrochemical results. The Ni–NHGF catalyst delivered 10 mA cm^{-2} current density at 331 mV overpotential for OER.

It should be mentioned here that the metal atoms in most of the SACs are bonded with covalent bonds [56–60]. However, the SACs with metal-oxygen (M-O) coordination have also been developed for electrochemical OER [56–60]. These SACs can modulate the electronic properties of the metal single atoms improving the electron transfer properties. In this regard, the Ni SACs supported on oxygenated sites of reduced graphene oxide were designed for electrochemical OER [57]. The Ni SACs required the overpotential of 328 mV to reach the current density of 10 mA cm^{-2} for OER. Further, Ni SACs with Ni–O coordination on graphene-like carbon (Ni–O–G SACs) were also synthesized and employed for efficient OER activity. The microscopic studies revealed the presence of a 3D porous framework in Ni–O–G SACs having graphene sheets with uniformly dispersed single Ni atom sites [57]. The Ni K-edge XAS studies showed the Ni–O coordination where each Ni atom was bonded to the six oxygen atoms. The catalyst attained 10 mA cm^{-2} current density at a low overpotential of 224 mV for OER. Moreover, TOF of 1.44 s^{-1} at 300 mV was also determined indicating excellent intrinsic OER activity [57]. The DFT calculation indicated that the Ni–O coordination reduced the activation energy barrier for the oxygen adsorbates to promote the OER activity of Ni–O–G SACs.

Due to unique electronic properties, SACs have also shown excellent activity and durability in acidic medium. Mostly, noble metal-based SACs have been employed for acidic OER. For example, Pt–C_2N_2 single site catalyst was explored for the durable acidic OER [71]. The Pt–C_2N_2 moieties were developed by the carbonization of polyaniline. The Pt–C_2N_2 showed a large current density of 120 mA cm^{-2} at a low overpotential of 405 mV for OER in 0.5 M H_2SO_4 [71]. Moreover, the catalyst exhibited stability of more than 12 h of in an acidic medium. Further, an atomically dispersed Ru-N_4 site anchored on nitrogen-carbon support (Ru-N-C) was synthesized for the highly efficient OER in an acidic medium [28]. The catalyst demonstrated a low overpotential of 267 mV to deliver a current density of 10 mA cm^{-2} (Fig. 5.4A). The Ru-N-C showed high OER activity after the normalization with electrochemically active surface area (Fig. 5.4). The Ru-N-C catalyst showed good mass activity as high as 3571 A g^{-1} and TOF of 3348 O_2 h^{-1} (Fig. 5.4C). The Tafel plots also indicated the lowest Tafel slope value for Ru-N-C indicating the faster OER kinetics (Fig. 5.4D). The catalyst exhibited excellent stability of 30 h for OER in acidic medium (Fig. 5.4E). Moreover, the Ru-N-C showed produced H_2 exactly two times higher than the oxygen (Fig. 5.4F).

5.4.2 Metal compound-supported single-atom catalysts

The metal compounds having surface defects can be utilized for the synthesis of metal-supported SACs [64,67,72–74]. It has been observed that SACs supported on metal oxides and hydroxides show excellent OER

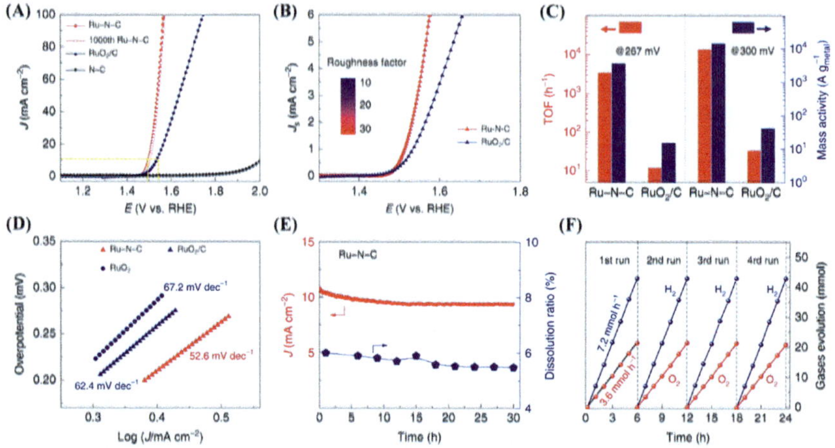

Figure 5.4 (A) LSV profiles for OER with Ru-N-C and commercial RuO_2/C in 0.5 M H_2SO_4 showing the best activity for Ru-N-C; (B) normalized LSV curves against electrochemical active surface area showing the improved OER activity for Ru-N-C; (C) TOF and mass activities for Ru-N-C and RuO_2/C electrocatalysts showing the highest TOF and mass activity for Ru-N-C; (D) Tafel plots for Ru-N-C and RuO_2/C electrocatalysts showing the lowest Tafel slope for Ru-N-C; (E) plot of current density and Ru dissolved mass ratio versus time for Ru-N-C at a constant anode voltage of 1.49 V versus RHE in 0.5 M H_2SO_4; and (F) time profiles of O_2 and H_2 evolutions in overall water splitting. *Reproduced with permission from L. Cao, Q. Luo, J. Chen, L. Wang, Y. Lin, H. Wang, et al., Dynamic oxygen adsorption on single-atomic Ruthenium catalyst with high performance for acidic oxygen evolution reaction, Nat. Commun. 10 (2019) 4849. https://doi.org/10.1038/s41467-019-12886-z with Springer Nature.*

performance due to electronic interactions between the metal atoms and the support. For example, Au single-atom-decorated $CoSe_2$ nanobelts (Au−$CoSe_2$) were synthesized for electrochemical OER [75]. The enhanced OER activity of Au−$CoSe_2$ was recorded compared to the pure $CoSe_2$ nanobelts. The DFT calculations indicated that the introduction of isolated Au atoms led to the shifting of the d-band center of Co active sites closer to the Fermi level. As a result, the H_2O adsorption energy was lowered promoting the OER activity. In another study, an Au@Ni_2P SAC was also developed for enhanced OER activity [76].

An amorphous $NiFeIr_x$/Ni core-shell nanowire@nanosheets ($NiFeIr_x$/Ni-NW@NSs) was designed for efficient OER activity [77]. The atomically isolated Ir atoms were observed to be decorated on a NiFe-based core. The $NiFeIr_x$/Ni NW@NSs catalyst required 200 mV to reach 10 mA cm^{-2} current density for OER. The DFT studies revealed that the single-atom Ir tuned the electronic properties of the Ni active site improving the intrinsic OER activity of the $NiFeIr_x$/Ni NW@NSs [77]. Song's group investigated the Ir SAC deposited on defect-rich cobalt-based hydroxide nanosheets for the electrochemical OER [78]. The CoIr catalyst having 9.7 wt.% Ir exhibited excellent OER activity outperforming the noble metal-based IrO_2 catalyst. The XAS studies revealed that the oxidation state of Co was increased, and the Co hydroxide nanosheets were converted into β-CoOOH active catalyst during OER [78]. In addition, the oxidation state of Ir was also increased during OER.

Single-atom Pt was also introduced into porous NiO nanocubes (0.5 wt.% Pt/NiO) to attain efficient OER activity [79]. The Pt introduction facilitated the electrochemical transformation of NiO to an active NiOOH catalyst. As a result, superior OER activity of Pt/NiO was recorded at lower overpotential compared to NiO. The DFT calculations showed that the improved charge transfer was observed from Pt to Ni through O resulting in the weaker local Ni-O bonding. Similarly, tungsten single-atom-doped $Ni(OH)_2$ nanosheets, NiO supported support Ir SACs were also developed for excellent OER activity and stability [80].

The SACs have been also utilized for oxygen evolution in acidic medium and have shown excellent activity and stability for hours. For example, an Ag/IrO_x SAC with Ag single atoms embedded in an IrO_2 matrix was developed for acidic OER [81]. The Ag/IrO_x SAC attained a current density of 10 mA cm^{-2} at the overpotential of 224 mV. The DFT calculation revealed that six Ir atoms possessed higher-valence Ir^{x+} ($x > 4$) optimizing the adsorption free energy of reaction intermediates.

Recently, metal SACs supported with metal alloys have been explored and employed for electrochemical OER. These catalysts also show excellent OER activity at very low overpotentials. For example, single atom Ru was decorated onto PtCu alloys for efficient electrochemical OER [82]. The study revealed a volcano-type relationship between the OER activity and lattice constant of the PtCu alloys. The $Ru-Pt_3Cu$ catalyst recorded a very low overpotential of 90 mV for a current density of 10 mA cm^{-2} in acidic media with high stability. The DFT calculations suggested that the electronic structure was modulated due to the compressive strain in the Pt shell, whereas the redox behavior of the Ru single atoms optimized the binding of reactive oxygen adsorbates for OER [82].

The SACs are not only supported on oxides, sulfides, selenides, and alloys but are developed utilizing LDHs as support [20,23,67]. LDHs are 2D materials consisting of positively charged edge-sharing MO_6 octahedral layers where M denotes the divalent and trivalent metal ions. A series of LDHs like NiFe-LDH, NiCo-LDH, CoFe-LDH, and CoMn-LDH have been studied for electrochemical OER [20,23,67]. Recently, LDHs have been utilized for the anchoring of single atoms to develop the SACs for OER. The single metal atoms are observed to be anchored on the top surface of LDHs [20,23,67]. The single atoms are bonded with the O atoms in MO_6 moieties to form M−O−M bonds [20,23,67]. The SACs supported on LDHs can optimize the electronic structure at active sites on LDHs and improves in-plane active sites. Moreover, the charge transfer properties are also enhanced to improve the OER activity. Therefore, NiFe-LDH-supported Au SACs [67], CoCr-LDH-supported Ru SACs [83], NiFe-LDH-supported Ru SACs [23], $Co(OH)_2$ nanosheet-supported Ir SACs [78] were developed and employed for efficient OER activity. It can be concluded that SACs can be developed with metal alloy support improving the overpotential as well as stability for OER.

5.5 Mechanism of oxygen evolution reaction with single-atom catalysts

When the SACs are employed for electrochemical OER, their electrochemical reconstruction is expected under applied anodic potential [5,9,20,32,33,56,69]. The XANES studies revealed that most of the single atoms are converted into the high oxidation state (lower than 4+) at higher potential (~1.6 V) and the reversibility of the oxidation state is also observed. The single atoms return back into their original oxidation state after the potential is removed. For example, a single-atom Ru catalyst anchored on a CoFe-LDH (Ru/CoFe-LDH) was developed, and the operando XANES study was carried out [73]. The oxidation state of Ru was increased at the applied potential of 1.6 V during the OER. However, the oxidation state of Ru was lower than that of RuO_2 (Ru^{4+}). Under open circuit potential (OCP) after the OER, the Ru was observed to return to its prior oxidation state showing the reversibility of a single Ru atom [73]. In contrast, the high oxidation state was maintained by Co and Fe under OCP conditions during the OER [73]. A similar conclusion was drawn from the operando XAS measurements of the Ru−N−C catalyst during electrochemical OER [28]. The results revealed the reversibility of the oxidation state of Ru to maintain the stability of the SACs. When the potential was increased to 1.6 V versus RHE, a new peak was originated at 764 cm^{-1} assigned for the single oxygen adsorption (O*) intermediates. This result showed the O−Ru−N_4 configuration of the Ru−N−C catalyst under the OER conditions [28].

The formation of the active catalyst structure was determined by Zhang et al. during the OER of Au/NiFe LDH SAC [67]. The EXAFS and in situ Raman studies were carried out to detect the active catalyst structure. The in situ Raman spectra revealed that during the anodic half sweep from 1.12 to 1.32 V, the two peaks were observed at 462 and 535 cm^{-1}, assigned for the Ni-O vibrations of defective or disordered NiFe LDH. When the potential was increased to 1.44 V, these two peaks were disappeared and new peaks were observed at 475 and 559 cm^{-1}, assigned to the characteristic bands in NiOOH [67]. This result indicated that the LDHs were electrochemically transformed into oxy-hydroxides under the applied anodic potential during OER. This result was also consistence with the DFT calculations [67]. The origin of the high OER activity was attributed to the formation of active metal-(oxy)hydroxide catalyst. Moreover, the single atom Au assisted in the charge redistribution promoting the OER activity [67].

The DFT study showed that the rate-determining step for Ru/CoFe-LDHs is the formation of *OOH from *O [73]. The DFT calculation shows that in the case of Ru/CoFe-LDHs structure, Ru single atom lowers the Gibbs free energy of rate-determining step for the formation of *OOH group from *O group than Fe active site of CoFe-LDH. Hence, Ru/CoFe-LDHs are showing better OER activity [73]. Similarly, the formation of O-Ru-N during OER facilitates the lowering of overpotential for OER process. In situ FTIR shows that at 1.2 V versus RHE, there was no noticeable peak between 600−900 cm^{-1} observed in FTIR [73]. When the potential is increased to 1.6 V versus RHE, a new peak appears at 764 cm^{-1}, indicating the presence of single oxygen (O*) intermediates adsorbed on the Ru-NC that is absent at lower potential 1.6 V versus RHE due to no OER activity.

5.6 Conclusions

In this chapter, the basic principle of designing SACs with special reference to electrochemical OER has been discussed. The importance of SACs and their merits and demerits have been explained with suitable examples. The different strategies for designing the SACs, such as wet chemical strategy, coprecipitation strategy, electrodeposition method, high-temperature pyrolysis, and ALD method, have been described in detail. The various characterization techniques employed for the investigation of structural, coordination, electronic, and morphological properties have been explained. Interestingly, the application of the various SACs for electrochemical OER has been discussed with suitable examples focusing on the stability, mechanism, mass loading, as well as structure–activity relationship.

Further, the mechanism of OER with SACs has been well established with the help of XAS studies and DFT calculations. The recent progress and challenges have been summarized in this chapter for a better understanding of the electrochemical OER with SACs.

References

[1] B. Singh, A. Indra, Prussian blue- and Prussian blue analogue-derived materials: progress and prospects for electrochemical energy conversion, Mater. Today Energy. 16 (2020) 100404. Available from: https://doi.org/10.1016/j.mtener.2020.100404.

[2] B. Singh, A. Singh, A. Yadav, A. Indra, Modulating electronic structure of metal-organic framework derived catalysts for electrochemical water oxidation, Coord. Chem. Rev. 447 (2021) 214144. Available from: https://doi.org/10.1016/J.CCR.2021.214144.

[3] Y. Wang, D. Wang, Y. Li, Rational design of single-atom site electrocatalysts: from theoretical understandings to practical applications, Adv. Mater. 33 (2021) 2008151. Available from: https://doi.org/10.1002/adma.202008151.

[4] Y. Yang, Y. Yang, Z. Pei, K.-H. Wu, C. Tan, H. Wang, et al., Recent progress of carbon-supported single-atom catalysts for energy conversion and storage, Matter 3 (2020) 1442–1476. Available from: https://doi.org/10.1016/j.matt.2020.07.032.

[5] W.H. Lee, Y.-J. Ko, J.-Y. Kim, B.K. Min, Y.J. Hwang, H.-S. Oh, Single-atom catalysts for the oxygen evolution reaction: recent developments and future perspectives, Chem. Commun. 56 (2020) 12687–12697. Available from: https://doi.org/10.1039/D0CC04752J.

[6] K. Liu, X. Zhao, G. Ren, T. Yang, Y. Ren, A.F. Lee, et al., Strong metal-support interaction promoted scalable production of thermally stable single-atom catalysts, Nat. Commun. 11 (2020) 1263. Available from: https://doi.org/10.1038/s41467-020-14984-9.

[7] L. Zhou, S.-Y. Lu, S. Guo, Recent progress on precious metal single atom materials for water splitting catalysis, SusMat 1 (2021) 194–210. Available from: https://doi.org/10.1002/sus2.15.

[8] Y. Zhou, J. Li, X. Gao, W. Chu, G. Gao, L.-W. Wang, Recent advances in single-atom electrocatalysts supported on two-dimensional materials for the oxygen evolution reaction, J. Mater. Chem. A. 9 (2021) 9979–9999. Available from: https://doi.org/10.1039/D1TA00154J.

[9] C. Zhu, Q. Shi, S. Feng, D. Du, Y. Lin, Single-atom catalysts for electrochemical water splitting, ACS Energy Lett. 3 (2018) 1713–1721. Available from: https://doi.org/10.1021/acsenergylett.8b00640.

[10] B. Qiao, A. Wang, X. Yang, L.F. Allard, Z. Jiang, Y. Cui, et al., Single-atom catalysis of CO oxidation using Pt1/FeOx, Nat. Chem. 3 (2011) 634–641. Available from: https://doi.org/10.1038/nchem.1095.

[11] G. Kyriakou, M.B. Boucher, A.D. Jewell, E.A. Lewis, T.J. Lawton, A.E. Baber, et al., Isolated metal atom geometries as a strategy for selective heterogeneous hydrogenations, Science 335 (2012) 1209−1213.

[12] L. Kuai, Z. Chen, S. Liu, E. Kan, N. Yu, Y. Ren, et al., Titania supported synergistic palladium single atoms and nanoparticles for room temperature ketone and aldehydes hydrogenation, Nat. Commun. 11 (2020) 1−9. Available from: https://doi.org/10.1038/s41467-019-13941-5.

[13] B. Zhang, T. Fan, N. Xie, G. Nie, H. Zhang, Versatile applications of metal single-atom @ 2D material nanoplatforms, Adv. Sci. 6 (2019) 1901787. Available from: https://doi.org/10.1002/advs.201901787.

[14] H. Fei, J. Dong, D. Chen, T. Hu, X. Duan, I. Shakir, et al., Single atom electrocatalysts supported on graphene or graphene-like carbons, Chem. Soc. Rev. 48 (2019) 5207−5241. Available from: https://doi.org/10.1039/C9CS00422J.

[15] J. Zheng, K. Lebedev, S. Wu, C. Huang, T. Ayvalı, T.-S. Wu, et al., High loading of transition metal single atoms on chalcogenide catalysts, J. Am. Chem. Soc. 143 (2021) 7979−7990. Available from: https://doi.org/10.1021/jacs.1c01097.

[16] Z. Chen, S. Zhang, J. Yang, C. Chen, Y. Song, C. Xu, et al., Rational design and controllable synthesis of polymer aerogel-based single-atom catalysts with high loading, Mater. Adv. 2 (2021) 6885−6900. Available from: https://doi.org/10.1039/D1MA00720C.

[17] J. Han, J. Bian, C. Sun, Recent advances in single-atom electrocatalysts for oxygen reduction reaction, Research 2020 (2020) 9512763. Available from: https://doi.org/10.34133/2020/9512763.

[18] J. Guo, H. Liu, D. Li, J. Wang, X. Djitcheu, D. He, et al., A minireview on the synthesis of single atom catalysts, RSC Adv. 12 (2022) 9373−9394. Available from: https://doi.org/10.1039/D2RA00657J.

[19] Q. Wang, X. Huang, Z.L. Zhao, M. Wang, B. Xiang, J. Li, et al., Ultrahigh-loading of Ir single atoms on NiO matrix to dramatically enhance oxygen evolution reaction, J. Am. Chem. Soc. 142 (2020) 7425−7433. Available from: https://doi.org/10.1021/jacs.9b12642.

[20] Z. Wang, S.-M. Xu, Y. Xu, L. Tan, X. Wang, Y. Zhao, et al., Single Ru atoms with precise coordination on a monolayer layered double hydroxide for efficient electrooxidation catalysis, Chem. Sci. 10 (2019) 378−384. Available from: https://doi.org/10.1039/C8SC04480E.

[21] P. Aggarwal, D. Sarkar, K. Awasthi, P.W. Menezes, Functional role of single-atom catalysts in electrocatalytic hydrogen evolution: current developments and future challenges, Coord. Chem. Rev. 452 (2022) 214289. Available from: https://doi.org/10.1016/j.ccr.2021.214289.

[22] Z. Zhang, C. Feng, C. Liu, M. Zuo, L. Qin, X. Yan, et al., Electrochemical deposition as a universal route for fabricating single-atom catalysts, Nat. Commun. 11 (2020) 1215. Available from: https://doi.org/10.1038/s41467-020-14917-6.

[23] P. Zhai, M. Xia, Y. Wu, G. Zhang, J. Gao, B. Zhang, et al., Engineering single-atomic ruthenium catalytic sites on defective nickel-iron layered double hydroxide for overall water splitting, Nat. Commun. 12 (2021) 4587. Available from: https://doi.org/10.1038/s41467-021-24828-9.

[24] J. Wu, H. Zhou, Q. Li, M. Chen, J. Wan, N. Zhang, et al., Densely populated isolated single Co-N site for efficient oxygen electrocatalysis, Adv. Energy Mater. 9 (2019) 1900149. Available from: https://doi.org/10.1002/aenm.201900149.

[25] P. Chen, T. Zhou, L. Xing, K. Xu, Y. Tong, H. Xie, et al., Atomically dispersed iron−nitrogen species as electrocatalysts for bifunctional oxygen evolution and reduction reactions, Angew. Chem. Int. Ed. 56 (2017) 610−614. Available from: https://doi.org/10.1002/anie.201610119.

[26] H. Xu, D. Cheng, D. Cao, X.C. Zeng, A universal principle for a rational design of single-atom electrocatalysts, Nat. Catal. 1 (2018) 339–348. Available from: https://doi.org/10.1038/s41929-018-0063-z.

[27] H. Fei, J. Dong, Y. Feng, C.S. Allen, C. Wan, B. Volosskiy, et al., General synthesis and definitive structural identification of MN_4C_4 single-atom catalysts with tunable electrocatalytic activities, Nat. Catal. 1 (2018) 63–72. Available from: https://doi.org/10.1038/s41929-017-0008-y.

[28] L. Cao, Q. Luo, J. Chen, L. Wang, Y. Lin, H. Wang, et al., Dynamic oxygen adsorption on single-atomic Ruthenium catalyst with high performance for acidic oxygen evolution reaction, Nat. Commun. 10 (2019) 4849. Available from: https://doi.org/10.1038/s41467-019-12886-z.

[29] K. Jiang, H. Wang, Electrocatalysis over graphene-defect-coordinated transition-metal single-atom catalysts, Chem 4 (2018) 194–195. Available from: https://doi.org/10.1016/j.chempr.2018.01.013.

[30] L. Xuning, L. Linghui, R. Xinyi, G. Jiajian, H. Yanqiang, L. Bin, Microenvironment modulation of single-atom catalysts and their roles in electrochemical energy conversion, Sci. Adv. 6 (2022). Available from: https://doi.org/10.1126/sciadv.abb6833. eabb6833.

[31] B. Seo, D.S. Baek, Y.J. Sa, S.H. Joo, Shape effects of nickel phosphide nanocrystals on hydrogen evolution reaction, CrystEngComm 18 (2016) 6083–6089. Available from: https://doi.org/10.1039/C6CE00985A.

[32] C. Zhao, Y. Tang, C. Yu, X. Tan, M.N. Banis, S. Li, et al., Insights into the electronic origin of enhancing the catalytic activity of Co3O4 for oxygen evolution by single atom ruthenium, Nano Today 34 (2020) 100955. Available from: https://doi.org/10.1016/j.nantod.2020.100955.

[33] P. Qi, J. Wang, X. Djitcheu, D. He, H. Liu, Q. Zhang, Techniques for the characterization of single atom catalysts, RSC Adv. 12 (2022) 1216–1227. Available from: https://doi.org/10.1039/D1RA07799F.

[34] M. Kottwitz, Y. Li, H. Wang, A.I. Frenkel, R.G. Nuzzo, Single atom catalysts: a review of characterization methods, Chemistry–Methods 1 (2021) 278–294. Available from: https://doi.org/10.1002/cmtd.202100020.

[35] F. Dvořák, M. Farnesi Camellone, A. Tovt, N.-D. Tran, F.R. Negreiros, M. Vorokhta, et al., Creating single-atom Pt-ceria catalysts by surface step decoration, Nat. Commun. 7 (2016) 10801. Available from: https://doi.org/10.1038/ncomms10801.

[36] A.J. Therrien, A.J.R. Hensley, M.D. Marcinkowski, R. Zhang, F.R. Lucci, B. Coughlin, et al., An atomic-scale view of single-site Pt catalysis for low-temperature CO oxidation, Nat. Catal. 1 (2018) 192–198. Available from: https://doi.org/10.1038/s41929-018-0028-2.

[37] M.D. Marcinkowski, M.T. Darby, J. Liu, J.M. Wimble, F.R. Lucci, S. Lee, et al., Pt/Cu single-atom alloys as coke-resistant catalysts for efficient C–H activation, Nat. Chem. 10 (2018) 325–332. Available from: https://doi.org/10.1038/nchem.2915.

[38] M.B. Boucher, B. Zugic, G. Cladaras, J. Kammert, M.D. Marcinkowski, T.J. Lawton, et al., Single atom alloy surface analogs in Pd0.18Cu15 nanoparticles for selective hydrogenation reactions, Phys. Chem. Chem. Phys. 15 (2013) 12187–12196. Available from: https://doi.org/10.1039/C3CP51538A.

[39] S. Wei, A. Li, J.-C. Liu, Z. Li, W. Chen, Y. Gong, et al., Direct observation of noble metal nanoparticles transforming to thermally stable single atoms, Nat. Nanotechnol. 13 (2018) 856–861. Available from: https://doi.org/10.1038/s41565-018-0197-9.

[40] J. Zhang, X. Wu, W.-C. Cheong, W. Chen, R. Lin, J. Li, et al., Cation vacancy stabilization of single-atomic-site Pt1/Ni(OH)x catalyst for diboration of alkynes and alkenes, Nat. Commun. 9 (2018) 1002. Available from: https://doi.org/10.1038/s41467-018-03380-z.
[41] X. Zhang, W.-X. Liu, Y.-W. Zhou, Z.-D. Meng, L. Luo, S.-Q. Liu, Single-atom nickel anchored on surface of molybdenum disulfide for efficient hydrogen evolution, J. Electroanal. Chem. 894 (2021) 115359. Available from: https://doi.org/10.1016/j.jelechem.2021.115359.
[42] R.F. Egerton, M. Watanabe, Characterization of single-atom catalysts by EELS and EDX spectroscopy, Ultramicroscopy 193 (2018) 111–117. Available from: https://doi.org/10.1016/j.ultramic.2018.06.013.
[43] M. Kottwitz, Y. Li, R.M. Palomino, Z. Liu, G. Wang, Q. Wu, et al., Local structure and electronic state of atomically dispersed Pt supported on nanosized CeO2, ACS Catal. 9 (2019) 8738–8748. Available from: https://doi.org/10.1021/acscatal.9b02083.
[44] T. Zhang, X. Nie, W. Yu, X. Guo, C. Song, R. Si, et al., Single atomic Cu-N2 catalytic sites for highly active and selective hydroxylation of benzene to phenol, IScience 22 (2019) 97–108. Available from: https://doi.org/10.1016/j.isci.2019.11.010.
[45] S. Yang, Y.J. Tak, J. Kim, A. Soon, H. Lee, Support effects in single-atom platinum catalysts for electrochemical oxygen reduction, ACS Catal. 7 (2017) 1301–1307. Available from: https://doi.org/10.1021/acscatal.6b02899.
[46] J. Shan, C. Ye, S. Chen, T. Sun, Y. Jiao, L. Liu, et al., Short-range ordered iridium single atoms integrated into cobalt oxide spinel structure for highly efficient electrocatalytic water oxidation, J. Am. Chem. Soc. 143 (2021) 5201–5211. Available from: https://doi.org/10.1021/jacs.1c01525.
[47] D.J. Higley, Z. Chen, M. Beye, M. Hantschmann, A.H. Reid, V. Mehta, et al., Stimulated resonant inelastic x-ray scattering in a solid, Commun. Phys. 5 (2022) 83. Available from: https://doi.org/10.1038/s42005-022-00857-8.
[48] F.S. Hage, G. Radtke, D.M. Kepaptsoglou, M. Lazzeri, Q.M. Ramasse, Single atom vibrational spectroscopy in the scanning transmission electron microscope, Science (80–) 367 (2020) 1124–1127. Available from: https://doi.org/10.1126/science.aba1136.
[49] M. Xu, A. Li, M. Gao, W. Zhou, Single-atom electron microscopy for energy-related nanomaterials, J. Mater. Chem. A. 8 (2020) 16142–16165. Available from: https://doi.org/10.1039/D0TA04918B.
[50] L. Zhang, R. Long, Y. Zhang, D. Duan, Y. Xiong, Y. Zhang, et al., Direct observation of dynamic bond evolution in single-atom Pt/C_3N_4 catalysts, Angew. Chem. Int. Ed. 59 (2020) 6224–6229. Available from: https://doi.org/10.1002/anie.201915774.
[51] D. Liu, S. Ding, C. Wu, W. Gan, C. Wang, D. Cao, et al., Synergistic effect of an atomically dual-metal doped catalyst for highly efficient oxygen evolution, J. Mater. Chem. A. 6 (2018) 6840–6846. Available from: https://doi.org/10.1039/C8TA00550H.
[52] D. Kunlun, G. Ahmet, A.M. Johnson, N.M. Schweitzer, G.D. Stucky, L.D. Marks, et al., Identification of active sites in CO oxidation and water-gas shift over supported Pt catalysts, Science (80–) 350 (2015) 189–192. Available from: https://doi.org/10.1126/science.aac6368.
[53] U.I. Kramm, I. Herrmann-Geppert, J. Behrends, K. Lips, S. Fiechter, P. Bogdanoff, On an Easy way to prepare metal–nitrogen doped carbon with exclusive presence of MEN4-type sites active for the ORR, J. Am. Chem. Soc. 138 (2016) 635–640. Available from: https://doi.org/10.1021/jacs.5b11015.

[54] J. Li, S. Ghoshal, W. Liang, M.-T. Sougrati, F. Jaouen, B. Halevi, et al., Structural and mechanistic basis for the high activity of Fe−N−C catalysts toward oxygen reduction, Energy Environ. Sci. 9 (2016) 2418−2432. Available from: https://doi.org/10.1039/C6EE01160H.

[55] J. Wang, Z. Huang, W. Liu, C. Chang, H. Tang, Z. Li, et al., Design of N-coordinated dual-metal sites: a stable and active Pt-free catalyst for acidic oxygen reduction reaction, J. Am. Chem. Soc. 139 (2017) 17281−17284. Available from: https://doi.org/10.1021/jacs.7b10385.

[56] Q. Zhang, J. Guan, Single-atom catalysts for electrocatalytic applications, Adv. Funct. Mater. 30 (2020) 1−53. Available from: https://doi.org/10.1002/adfm.202000768.

[57] Y. Li, Z.S. Wu, P. Lu, X. Wang, W. Liu, Z. Liu, et al., High-valence nickel single-atom catalysts coordinated to oxygen sites for extraordinarily activating oxygen evolution reaction, Adv. Sci. 7 (2020). Available from: https://doi.org/10.1002/advs.201903089.

[58] X. Li, P. Cui, W. Zhong, J. Li, X. Wang, Z. Wang, et al., Graphitic carbon nitride supported single-atom catalysts for efficient oxygen evolution reaction, Chem. Commun. 52 (2016) 13233−13236. Available from: https://doi.org/10.1039/c6cc07049c.

[59] G. Gao, S. Bottle, A. Du, Understanding the activity and selectivity of single atom catalysts for hydrogen and oxygen evolution via ab initial study, Catal. Sci. Technol. 8 (2018) 996−1001. Available from: https://doi.org/10.1039/c7cy02463k.

[60] G.M. Tomboc, T. Kim, S. Jung, H.J. Yoon, K. Lee, Modulating the local coordination environment of single-atom catalysts for enhanced catalytic performance in hydrogen/oxygen evolution reaction, Small 2105680 (2022) 1−36. Available from: https://doi.org/10.1002/smll.202105680.

[61] P. Xi, B. Huang, J. Yin, J. Jin, M. Lu, H. Zhang, et al., Iridium single atoms coupling with oxygen vacancies boosts oxygen evolution reaction in acid media, J. Am. Chem. Soc. 142 (2020) 18378−18386. Available from: https://doi.org/10.1021/jacs.0c05050.

[62] Z. Lei, W. Cai, Y. Rao, K. Wang, Y. Jiang, Y. Liu, et al., Coordination modulation of iridium single-atom catalyst maximizing water oxidation activity, Nat. Commun. 13 (2022) 1−10. Available from: https://doi.org/10.1038/s41467-021-27664-z.

[63] L. Peng, L. Shang, T. Zhang, G.I.N. Waterhouse, Recent advances in the development of single-atom catalysts for oxygen electrocatalysis and zinc−air batteries, Adv. Energy Mater. 10 (2020) 1−22. Available from: https://doi.org/10.1002/aenm.202003018.

[64] Q. Wang, Z. Zhang, C. Cai, M. Wang, Z.L. Zhao, M. Li, et al., Single iridium atom doped Ni2P catalyst for optimal oxygen evolution, J. Am. Chem. Soc. 143 (2021) 13605−13615. Available from: https://doi.org/10.1021/jacs.1c04682.

[65] B. Singh, V. Sharma, R.P. Gaikwad, P. Fornasiero, R. Zbořil, M.B. Gawande, Single-atom catalysts: a sustainable pathway for the advanced catalytic applications, Small 17 (2021) 1−27. Available from: https://doi.org/10.1002/smll.202006473.

[66] A. Indra, B. Singh, A. Yadav, Realizing electrochemical transformation of metal-organic framework precatalyst into metal hydroxide-oxy(hydroxide) active catalyst during alkaline water oxidation, J. Mater. Chem. A. (2022). Available from: https://doi.org/10.1039/D1TA09424F.

[67] J. Zhang, J. Liu, L. Xi, Y. Yu, N. Chen, S. Sun, et al., Single-atom Au/NiFe layered double hydroxide electrocatalyst: probing the origin of activity for oxygen evolution reaction, J. Am. Chem. Soc. 140 (2018) 3876−3879. Available from: https://doi.org/10.1021/jacs.8b00752.

[68] C. Chen, G. Chen, X. Kong, Enhanced oxygen evolution reaction for single atomic co catalyst via support modification: a density functional theory design predication, Inorg. Chem. 57 (2018) 13020–13026. Available from: https://doi.org/10.1021/acs.inorgchem.8b02294.

[69] Z. Zhang, C. Feng, X. Li, C. Liu, D. Wang, R. Si, et al., In-situ generated high-valent iron single-atom catalyst for efficient oxygen evolution, Nano Lett. 21 (2021) 4795–4801. Available from: https://doi.org/10.1021/acs.nanolett.1c01335.

[70] L. Zhang, Y. Jia, G. Gao, X. Yan, N. Chen, J. Chen, et al., Graphene defects trap atomic Ni species for hydrogen and oxygen evolution reactions, Chem 4 (2018) 285–297. Available from: https://doi.org/10.1016/j.chempr.2017.12.005.

[71] H. Su, M.A. Soldatov, V. Roldugin, Q. Liu, Platinum single-atom catalyst with self-adjustable valence state for large-current-density acidic water oxidation, EScience 2 (2021) 102–109. Available from: https://doi.org/10.1016/j.esci.2021.12.007.

[72] Y. Xu, W. Zhang, Y. Li, P. Lu, Z.S. Wu, A general bimetal-ion adsorption strategy to prepare nickel single atom catalysts anchored on graphene for efficient oxygen evolution reaction, J. Energy Chem. 43 (2020) 52–57. Available from: https://doi.org/10.1016/j.jechem.2019.08.006.

[73] P. Li, M. Wang, X. Duan, L. Zheng, X. Cheng, Y. Zhang, et al., Boosting oxygen evolution of single-atomic ruthenium through electronic coupling with cobalt-iron layered double hydroxides, Nat. Commun. 10 (2019) 1–11. Available from: https://doi.org/10.1038/s41467-019-09666-0.

[74] L. Yang, B. Zhang, B. Fang, L. Feng, A comparative study of NiCo2O4 catalyst supported on Ni foam and from solution residuals fabricated by a hydrothermal approach for electrochemical oxygen evolution reaction, Chem. Commun. 54 (2018) 13151–13154. Available from: https://doi.org/10.1039/c8cc08251k.

[75] X. Zhao, P. Gao, Y. Yan, X. Li, Y. Xing, H. Li, et al., Gold atom-decorated CoSe2 nanobelts with engineered active sites for enhanced oxygen evolution, J. Mater. Chem. A 5 (2017) 20202–20207. Available from: https://doi.org/10.1039/c7ta06172b.

[76] C. Cai, S. Han, Q. Wang, M. Gu, Direct observation of yolk-shell transforming to gold single atoms and clusters with superior oxygen evolution reaction efficiency, ACS Nano 13 (2019) 8865–8871. Available from: https://doi.org/10.1021/acsnano.9b02135.

[77] X. Luo, X. Wei, H. Zhong, H. Wang, Y. Wu, Q. Wang, et al., Single-atom Ir-anchored 3D amorphous NiFe nanowire@nanosheets for boosted oxygen evolution reaction, ACS Appl. Mater. Interfaces. 12 (2020) 3539–3546. Available from: https://doi.org/10.1021/acsami.9b17476.

[78] Y. Zhang, C. Wu, H. Jiang, Y. Lin, H. Liu, Q. He, et al., Atomic iridium incorporated in cobalt hydroxide for efficient oxygen evolution catalysis in neutral electrolyte, Adv. Mater. 30 (2018) 1–8. Available from: https://doi.org/10.1002/adma.201707522.

[79] C. Lin, Y. Zhao, H. Zhang, S. Xie, Y.F. Li, X. Li, et al., Accelerated active phase transformation of NiO powered by Pt single atoms for enhanced oxygen evolution reaction, Chem. Sci. 9 (2018) 6803–6812. Available from: https://doi.org/10.1039/c8sc02015a.

[80] J. Yan, L. Kong, Y. Ji, J. White, Y. Li, J. Zhang, et al., Single atom tungsten doped ultrathin α-$Ni(OH)_2$ for enhanced electrocatalytic water oxidation, Nat. Commun. 10 (2019) 1–10. Available from: https://doi.org/10.1038/s41467-019-09845-z.

[81] F.F. Zhang, C.Q. Cheng, J.Q. Wang, L. Shang, Y. Feng, Y. Zhang, et al., Iridium oxide modified with silver single atom for boosting oxygen evolution reaction in acidic media, ACS Energy Lett. 6 (2021) 1588–1595. Available from: https://doi.org/10.1021/acsenergylett.1c00283.

[82] Y. Yao, S. Hu, W. Chen, Z.Q. Huang, W. Wei, T. Yao, et al., Engineering the electronic structure of single atom Ru sites via compressive strain boosts acidic water oxidation electrocatalysis, Nat. Catal. 2 (2019) 304–313. Available from: https://doi.org/10.1038/s41929-019-0246-2.
[83] C. Dong, X. Zhang, J. Xu, R. Si, J. Sheng, J. Luo, et al., Ruthenium-doped cobalt–chromium layered double hydroxides for enhancing oxygen evolution through regulating charge transfer, Small 16 (2020) 1–7. Available from: https://doi.org/10.1002/smll.201905328.

CHAPTER 6

Single atom catalysts for electrocatalytic hydrogen evolution reaction

Tianwei Wang[1], Haoxuan Liu[1] and Xijun Liu[2]
[1]State Key Laboratory of Featured Metal Materials and Life-cycle Safety for Composite Structures, MOE Key Laboratory of New Processing Technology for Nonferrous Metals and Materials, School of Resources, Environment and Materials, Guangxi University, Nanning, P.R. China
[2]MOE Key Laboratory of New Processing Technology for Non-Ferrous Metals and Materials, and Guangxi Key Laboratory of Processing for Non-Ferrous Metals and Featured Materials, School of Resource, Environments and Materials, Guangxi University, Nanning, P.R. China

6.1 Fundamentals of the HER

6.1.1 The mechanism of HER

Hydrogen evolution reaction (HER) is a multistep electrochemical reaction process taking place on the surface of an electrode. The reaction can be described by the following equations:

$$2H^+ + 2e^- \rightarrow H_2 \tag{6.1}$$

(Under acidic condition)

$$2H_2O + 2e^- \rightarrow H_2 + 2OH^- \tag{6.2}$$

(Under neutral or alkaline conditions)

The HER mechanism depends on the type of electrolyte environment and generally consists of two steps. The first step is the generation of hydrogen adsorption (H_{ads}), known as the Volmer reaction; the second step is a hydrogen desorption process, including Heyrovsky reaction or Tafel reaction (Fig. 6.1).

In acidic solution, the Volmer step is a process of an electron capturing a proton at the electrode surface (denoted as M) to form an absorbed hydrogen atom (M_H_{ads}): [2]

$$M + H^+(aq) + e^- \rightarrow M\_H_{ads} \tag{6.3}$$

Once H_{ads} is formed, the second step can take place via two different ways. One is Heyrovsky reaction, where the adsorbed hydrogen reacts

Single Atom Catalysts
DOI: https://doi.org/10.1016/B978-0-323-95237-8.00002-1

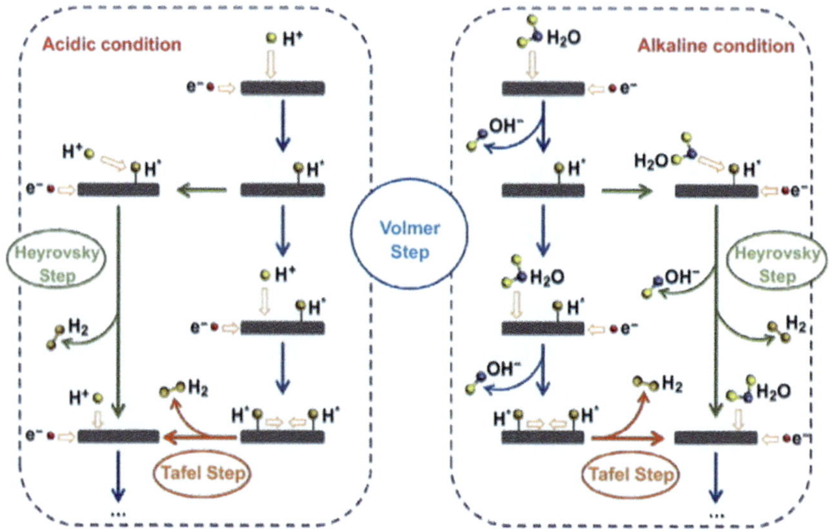

Figure 6.1 HER reaction diagram [1]. *Reproduced with permission from J. Wei, M. Zhou, A. Long, Y. Xue, H. Liao, C. Wei, et al., Nano-Micro. Lett. 10 (2018) 75.*

with an electron in the presence of a proton to release H_2 through electrochemical desorption:

$$M\_H_{ads} + H^+(aq) + e^- \rightarrow M + H_2 \tag{6.4}$$

The other one is Tafel reaction, which two adjacent adsorbed hydrogen intermediates chemically react with each other to produce H_2 through chemical desorption:

$$2M\_H_{ads} \rightarrow 2M + H_2 \tag{6.5}$$

In alkaline solution and neutral solution, the HER follows the Volmer-Heyrovsky or Volmer-Tafel pathway as well. However, the strong $H-O-H$ covalent bond in water molecules requires more energy to break, which means it is hard to obtain protons to form $M-H^*$ (Volmer step) [3].

In the Volmer step, a water molecule combines with an electron on the catalyst active surface to generate an adsorbed hydrogen atom: [2]

$$M + H_2O + e^- \rightarrow M\_H_{ads} + OH^- \tag{6.6}$$

In the following Heyrovsky reaction, the adsorbed hydrogen atom reacts with a water molecule and electron simultaneously to generate a hydrogen molecule.

$$M\_H_{ads} + H_2O + e^- \rightarrow M + H_2 + OH^- \tag{6.7}$$

The principle of the Tafel reaction under this condition is the same as the acidic condition counterpart, so it will not be stated here.

6.1.2 Electrochemical characterization parameters for HER

The HER electrochemical performance can be evaluated by three main aspects: activity, kinetic, and stability. Several important evaluation parameters are introduced in the following section [3].

6.1.2.1 Overpotential

Overpotential is the difference value between the electrode potential when an electrode reaction deviates from equilibrium and the equilibrium potential, which is one of the most commonly used parameters to evaluate catalyst activity in electrochemical catalysis [4]. The relationship between the overpotential (η) and required overall potential (E_i) for an electrochemical reaction can be described as follow formula: [5]

$$E_i = E_{\text{reversible}} + iR + \eta \qquad (6.8)$$

where $E_{\text{reversible}}$ is the thermo-dynamic potential, iR is the ohmic potential drop derived from the flow of current in ionic electrolytes.

In this formula, η is the variable part while the other two quantities are fixed in certain electrochemical scenarios and specific reactions. The polarization curves, recorded by the linear sweep voltammetry (LSV) method, are usually used to obtain the η (vs reversible hydrogen electrode). The smaller required η indicates the lower total energy consumption. Generally, the onset η (η_{onset}, η at $j = 0.5$ or 1 mA cm^{-2}) and the η at $j = 10$ mA cm^{-2} (η_{10}) are two common parameters for assessing the catalytic activity [5]. Under the same j, the higher activity of electrocatalysts exhibit, the lower η is performed.

6.1.2.2 Tafel slope and exchange current density

As for HER, the Tafel slope (b) and the exchange current density (j_0) are two important parameters to judge the reaction mechanism, such as the elementary steps, the rate-determining steps (RDS), or describe the catalytic activity. The relationship between these two parameters can be described as the following Eqs. (6.9) and (6.10) [6].

The Tafel empirical equation:

$$\eta = a + b\log(j) \qquad (6.9)$$

where η is the overpotential, j is the current density and b is the Tafel slope.

For the simple electrochemical redox reaction as HER, it can be depicted by the Butler–Volmer equation: [7]

$$j = j_0\left\{\exp\left(-\alpha\frac{F}{RT}\eta\right) - \exp\left[(1-\alpha)\frac{F}{RT}\eta\right]\right\} \tag{6.10}$$

where j_0 is exchange current density, α is symmetry coefficient, F is the Faraday constant; R is the universal gas constant, and T is absolute temperature.

The following statement describes the derivation process of Tafel formula under different reaction mechanisms.

1. Volmer reaction:

Under strong polarization conditions, it can be directly obtained:

$$\eta = -\frac{RT}{\alpha F}\ln(j_0) + \frac{RT}{\alpha F}\ln(j) = -\frac{2.3RT}{\alpha F}\lg(j_0) + \frac{2.3RT}{\alpha F}\lg(j) \tag{6.11}$$

Take α equal to 0.5: [7]

$$\eta = -\frac{2.3\times 2RT}{F}\lg(j_0) + \frac{2.3\times 2RT}{F}\lg(j) \tag{6.12}$$

Comparing (6.9) and (6.12) formulas to get the parameters a and b.

$$a = -\frac{2.3\times 2RT}{F}\lg(j_0)$$

$$b = \frac{2.3\times 2RT}{F}$$

When T equal to 25°C, $b \approx 118$ mv dec^{-1}.

2. Heyrovsky reaction: [8]

If the activity of hydrogen atoms is expressed by the surface coverage (θ_{MH}) of hydrogen atoms adsorbed on the electrode, the HER equilibrium potential expression of (6.3) can be written by the Nernst equation:

$$\varphi_{eq} = \varphi_H^0 + \frac{RT}{F}\ln\left(\frac{a_{H^+}}{\theta_{MH}^0}\right) \tag{6.13}$$

where φ_{eq} is the equilibrium electrode potential of the HER; φ_H^0 is the standard electrode potential; R is the universal gas constant; T is the reaction temperature; F is the Faraday constant; a_{H^+} is the hydrogen ion activity on the electrode surface, and θ_{MH}^0 is surface coverage of hydrogen atoms adsorbed on the electrode without current.

When there is a static current passing through, due to the fast discharge step, it can be considered that Eq. (6.3) is still in equilibrium and can be described by the Nernst equation:

$$\varphi = \varphi_H^0 + \frac{RT}{F}\ln\left(\frac{a_{H^+}}{\theta_{MH}}\right) \tag{6.14}$$

Therefore, hydrogen evolution overpotential is

$$\eta = \varphi_{eq} - \varphi = \frac{RT}{F}\ln\left(\frac{\theta_{MH}}{\theta_{MH}^0}\right) \tag{6.15}$$

$$\theta_{MH} = \theta_{MH}^0 \exp\left[\frac{F}{RT}\eta\right] \tag{6.16}$$

Under strong polarization conditions, according to (6.4), the kinetic equation of the entire electrode reaction can be written:

$$j = 2Fk' c_{H^+}\theta_{MH}\exp\left[\frac{\alpha F}{RT}\eta\right] \tag{6.17}$$

where k' is reaction rate constant in this reaction; c_{H^+} is hydrogen ion concentration.

According to (6.16) and (6.17), hydrogen evolution overpotential can be expressed by the following formula:

$$\eta = a + \frac{2.3RT}{(1+\alpha)F}\lg(j) \tag{6.18}$$

When T equal to 25°C, $b \approx 39$ mv dec^{-1}.

3. Tafel reaction:

In this reaction, it is still following the same first step—Heyrovsky reaction. The previous equilibrium of (6.3) and adsorption assumptions (θ_{MH}) still hold here.

Under strong polarization conditions, the reversible reaction of (6.5) can be ignored. Based on this reaction, the hydrogen evolution current of the entire electrode reaction can be expressed as:

$$j = 2Fk\theta_{MH}^2 \tag{6.19}$$

where the k is reaction rate constant.

According to (6.16) and (6.19), hydrogen evolution overpotential can be expressed by the following formula:

$$\eta = a + \frac{2.3RT}{2F}\lg(j) \tag{6.20}$$

When T equal to 25°C, $b \approx 29.5$ mv dec^{-1}.

The earlier-mentioned equations can link Tafel slope with HER mechanism, which could provide a guide for finding the potential HER mechanism of electrocatalysts via obtaining Tafel slope of the corresponding electrocatalysts. As for obtaining the Tafel slope, there are three methods: (1) calculation from the polarization curves; (2) direct evaluation via an electrochemical workstation; and (3) analysis from the electrochemical impedance data [9]. The first one is the widely used method because the Tafel slopes could be received from the linear fitting plots of log$j$ versus η. This method is commonly to select low η range to get the accurate Tafel slope because the log$j$ diverges further from the linearity if the bubbles forming [5]. Comparing the experimental value and the theoretical value of the Tafel slope, it can confirm the RDS in the current reaction [6].

As for the exchange current density (j_0), it can be calculated from the polarization curves, in Eq. (6.9) when the $\eta = 0$. Since the value of j_0 only depends on the electrolyte, temperature, and composition of the electrocatalyst, it has become a key parameter describing the inherent attribute of electrode reactions. The ability of electrocatalysts to transfer electrons and describe the reaction rates at equilibrium potentials can be better revealed by the j_0 [10,11]. The better the catalytic activity of the catalyst, the larger j_0 exhibit.

6.1.2.3 Electrochemical impedance spectroscopy

In electrochemical impedance spectroscopy (EIS) analysis, the charge transfer resistance (R_{ct}) and adsorption resistance (R_{ad}) are two important metrics that provide information on reaction kinetics at the electrode/electrolyte interface. R_{ct} is related to the interfacial charge transfer process of the electrodes and is usually calculated from the semicircles in the high-frequency zones [10–12]. When electron transfer at the interface is fast, the R_{ct} value will be low. Besides, the larger η could generate smaller R_{ct} value [2]. In addition, R_{ad} represents the adsorption of oxidizing species on the electrode, which is usually obtained from a semicircle in the low-frequency zones. The value of R_{ad} is related to the onset potential, a more positive onset potential leads to a smaller R_{ad} [13].

6.1.2.4 Electrochemical active surface area

As for the assessment of intrinsic electrocatalytic activity on different electrocatalysts, Electrochemical active surface area (ECSA) is an important parameter. Generally, three common methods could be applied to obtain the ECSA value, including CO stripping technology, hydrogen underpotential deposition, and Faraday double-layer capacitance method [14–17]. The first two methods are more suitable for ECSA evaluation of noble metals, such as platinum (Pt) and rhodium (Rh) [16,18,19]. For the third method, since there is a proportional relationship between the effective electrochemical surface area of the electrode and the double-layer capacitance (C_{dl}), it is suitable for high-conductivity samples [9]. Cyclic voltammetry (CV) is applied to calculate C_{dl}, which involves the calculation of scan rate (v) and current density j:

$$C_{dl} = \frac{j}{v} \tag{6.21}$$

ECSA is calculated according to formula (6.22):

$$\mathrm{ECSA} = \frac{C_{dl}}{C_S} \tag{6.22}$$

Where C_S is the specific capacitance of the electrode with a flat surface [17,20]. The value of A generally ranges from 20 to 60 mF cm^{-2}, with an intermediate value of 40 being used when evaluating nonprecious metal-based catalysts. However, for carbon-based electrocatalysts, this approach is not applicable as the non-Faradaic current derived from high surface area carbon could lead to the overestimation of the corresponding ECSA value [16,19]. Highly active materials have abundant active sites, corresponding to larger ECSA values [17].

6.1.2.5 Turnover frequency

The definition of turnover frequency (TOF) is the number of reactant molecules participating in each active site in unit time under the specified potential. It is a key indicator for evaluating the intrinsic activity of electrocatalysts, TOF can be calculated by the following formula: [5]

$$\mathrm{TOF} = \frac{jA}{2nF} \tag{6.23}$$

where j is the current density, A is the area of the working electrode, n is the number of moles of the electrocatalyst, and F is the Faraday constant.

The coefficient 1/2 in the equation indicates that the formation of a hydrogen molecule requires two electrons to participate in the transfer [21]. Higher TOF value means that the electrocatalyst has better inherent activity.

6.1.2.6 Faradaic efficiency

Faradaic efficiency (FE) is an important physical quantity to evaluate the selectivity of catalysts to products, which describes the relationship between the number of moles of reaction products and the total amount of electrons consumed by the external circuit [13]. In simple terms, FE can be obtained by calculating the ratio of the actual obtained amount of the product to the theoretically calculated amount. In HER, although its product is only H_2, the FE of HER is hard to reach 100% due to the non-Faradaic capacitive current in the total measured current [3].

6.1.2.7 Stability

Stability test is a key measurement for assessing the practical applications of electrocatalytsts. One method of the stability test is to perform CV cycles or LSV scan within a certain voltage range. Through the thousands of CV cycles, the stability of the catalyst was judged by comparing the η values of the same j before and after cycling. The Small differences indicate excellent stability [5]. Another method of testing the stability of electrocatalysts is $i-t$ test at a constant potential (chronoamperometry) or the $V-t$ test at a constant current density (chronopotentiometry). This method can confirm the stability of electrocatalyst by studying the fluctuation of the potential (or j) at a constant j (or potential) over time. The higher the fluctuation, the worse the stability. In HER, a j of 10 mA/cm^2 is often judged as the constant current density in the stability test [3].

6.1.3 Thermodynamic model: Gibbs free energy of hydrogen adsorption

Theoretically, the Gibbs free energy (ΔG) can indicate the reaction pathways. For HER, the H-binding energy could be used to predict the activity and kinetics of an electrocatalyst. The Gibbs free energy change for H* adsorption on the catalyst surface (ΔG_{H^*}) reflects the H-binding strength, which can be used to study the H* adsorption and H_2 desorption behaviors [3]. It can be obtained by theoretical calculation and is always used to judge the RDS of the HER process.

According to the Sabatier principle, an optimal adsorption/desorption strength would give rise to the best electrochemical properties [21]. The positive value of ΔG_{H^*} ($\Delta G_{H^*} > 0$) represents that the adsorption of hydrogen on the active site is weak, which means H-binding strength is too feeble to support the formation of stable intermediates on the electrocatalyst surface. At this point, the Volmer step is the RDS and it affects the overall reaction kinetics of HER. Contrarily, the negative value of ΔG_{H^*} ($\Delta G_{H^*} < 0$) indicates the strong adsorption of hydrogen, the H-binding strength is intense that makes it difficult to release H_2. Therefore, the desorption step (Heyrovsky or Tafel step) can limit the HER rate. According to these two situations, only when the value of ΔG_{H^*} close to zero ($\Delta G_{H_*} \sim 0$), which suggests the H-binding strength is suitable for adsorption/desorption of the hydrogen, the activity of HER could achieve the best performance [21,22]. The ΔG_{H^*} values of different metals can be described by the volcano-type plot based on the microkinetic model study. The X-coordinate is ΔG_{H^*}, and it can be calculated by quantum chemistry calculated; Y-coordinate is the exchange current density (in log form, log j_0) [23]. The theoretical activity of the catalyst in HER can be judged by ΔG_{H^*} and j_0 relationship in the volcano plot. (Fig. 6.2)

Theoretical studies have revealed that the value of ΔG_{H^*} on the active metal center is strongly influenced by the position of the *d*-bond center of

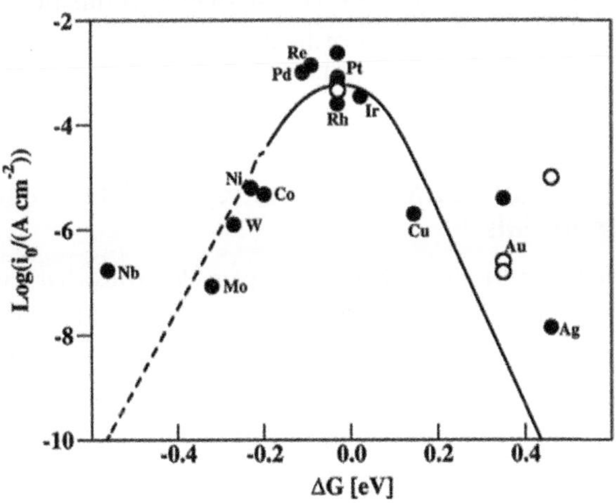

Figure 6.2 The volcano plot shows the relationship between exchange current density and the ΔG_{H^*} at U = 0 V [24]. *Reproduced with permission from E. Skúlason, V. Tripkovic, M. E. Björketun, S. Gudmundsdóttir, G. Karlberg, J. Rossmeisl, et al., J. Phys. Chem. C. 2010, 114, 18182–18197.*

the metal atom [25]. In most cases, the *d*-band center can be served as a good descriptor for the adsorption energy variations since it can correlate well with the position of the upper *d*-band edge, which basically controls the position and filling of the antibonding states. To be specific, the closer *d*-band center near to Femi level, the stronger the adsorption energy it has, which means ΔG_{H^*} is much more negative. On the contrary, the *d*-band center stays away from Femi level, the adsorption energy is weak and ΔG_{H^*} becomes more positive. When the *d*-band center and Fermi level of the catalyst are in the most suitable position, that is, when ΔG_{H^*} is closer to zero, the best catalytic effect can be obtained [26]. However, owing to the nonperiodic structure of SACs, the *d*-band center theory cannot be applied to the SACs, and herein, the shift of *d*-orbital of the metal atoms is recommended as a parameter to semiquantitatively estimate the activity of corresponding SACs [27–29].

6.2 SACs for HER

Controlling the coordination relationship between central metal atoms and support atoms could be served as an effective way to mediate the HER electrocatalytic activity of SACs. The coordination number mainly depends on the valence state of metal atoms and their coordination environment [30]. The direct coordination between central metal atoms and support atoms could be classified as first-shell coordination [30]. In this section, supporting atoms and coordination numbers in first-shell coordination will be introduced, including N-coordination, O-coordination, and S-coordination.

6.2.1 Single metal atoms coordination with N in carbon substrate materials

The electronegative N atoms have lone-pair electrons that can bond with the metal empty *d*-orbitals, and thus provide stable and abundant anchoring sites for single atoms on carbon materials [25]. Under a strong coordination of the metal atoms with the surrounding nitrogen and carbon atoms, N-doped carbon-based SACs can exhibit the enhanced electronic conduction, desirable charge distribution, and a modulated electronic structure of the active sites [31]. Fig. 6.3 shows the proposed molecular model and corresponding Gibbs free energy calculation for HER, which are constructed by considering all possible active sites for single atoms on N-doped graphene (NG) surface, including graphitic-N, graphitic-C,

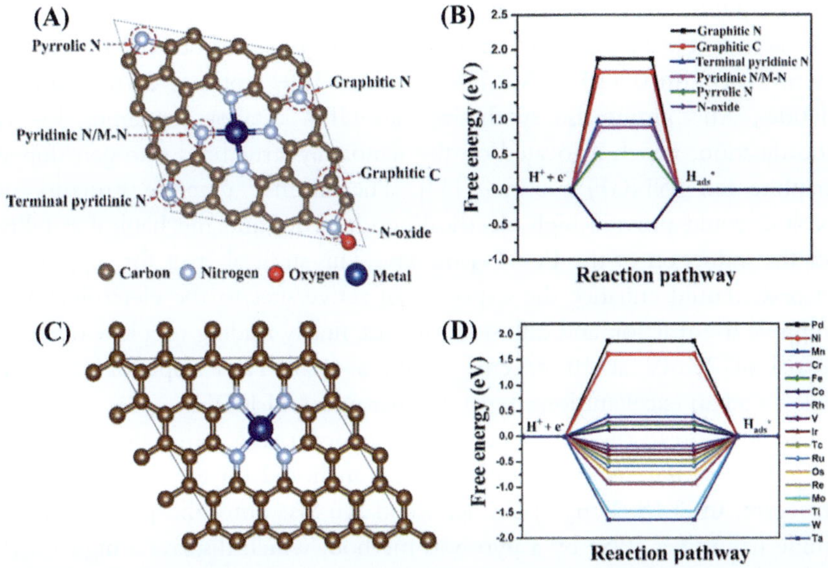

Figure 6.3 N-coordinating carbon-based materials molecular model (A) without and (C) with metal. The Gibbs free energy of hydrogen adsorption (ΔG_{H*}) (B) without and (D) with metal atoms. *Reproduced with permission from M.D. Hossain, Z. Liu, M. Zhuang, X. Yan, G.L. Xu, C.A. Gadre, et al., Adv. Energy Mater., 9 (2019). 1803689.*

pyridinic-N, terminal pyridinic-N, pyrrolic-N, and N-oxide species [25]. Hossain et al. pointed out that the substrate had almost no activity for HER (Fig. 6.3B), while after introducing a single metal atom as the active center, M-N_4 structure could be formed, and its catalytic HER activity is greatly improved (Fig. 6.3C,D) [25]. In addition to M-N_4 coordination, other numbers of N coordination could also be fabricated and display the promoted HER activity, which will be covered in the following introduction.

6.2.1.1 M − N_4 structure

In most cases of SACs, the coordination number of metal atoms with N atoms is 4, which could be attributed to the valence state and electronic structures of central atoms [30].

In a recent study, Li et al. obtained equally distributed isolated Pt SAs on N-doped porous carbon (Pt_1/NPC) via the modified photochemical solid-phase reduction method, where could achieve a high loading of about 3.8 wt.% for Pt SAs [32]. Due to the stable Pt-N_4 structure and the presence of more unoccupied Pt 5*d* orbitals for facilitating the fast

H adsorption/desorption kinetics, the electrocatalyst finally shows an excellent activity for HER, which could afford a low overpotential (25 mV at 10 mA cm^{-2} in 0.5 M H_2SO_4) with a low Tafel slope of 28 mV dec^{-1}. Besides, Xu's group also synthesized an HER catalyst containing Pt-N_4 coordination, which is located on the nanoarray-structured nitrogen-doped graphite foil (NNGF) substrate [33]. The retained graphite structure of NNGF could provide high electrical conductivity and mechanical stability for the induction of the $Pt-N_4$ moiety. This surficial layer for supporting Pt SACs could enhance the exposure of active sites to the electrolyte and improve the reaction and diffusion kinetics, finally leading to a low overpotential of 23 mV at 10 mA cm^{-2} with a small Tafel slope of 29.1 mV dec^{-1} and an excellent long-term durability in 1 M KOH.

In addition to the excellent $Pt-N_4$ coordination, other noble metals, for example, Ru and Rh, could also be anchored on the similar M—N_x structure. In 2018, Yang et al., deposited Ru SAs onto phosphorus nitride imide nanotubes (PN) by a pyrolysis method, which displays a high HER activity in 0.5 M H_2SO_4 (a minimal η_{10} of 24 mV) and superior stability [34]. Its excellent catalytic activity comes from the Ru—N_4 tetrahedral structure in the PN matrix, where charge transfer occurred from Ru to N, leading to positively charged Ru SAs. Besides, the DFT calculation shows the Ru SAs@PN have the lowest ΔG_{H^*}. In other research, Chen and coworkers studied a variety of metal SACs with a structure of TM—N_4 supported on a C_8N_8 monolayer material [35]. Through theoretical calculations of the reaction-free energy of HER and overpotential of OER, the Rh-N_4 catalyst was screened out as the best with low hydrogen adsorption free energy ($\Delta G_{H_*} = 0.08$ eV) and overpotential ($\eta = 0.49$ V) at pH = 7. Moreover, the first-principle molecular dynamics simulation proves it can be stable at a high temperature (1000 K).

Although the superior HER electrocatalytic performance of noble SACs has been revealed in many reports, their further wide application has been limited by their scarceness and high price. Therefore, the single transition-metal atom electrocatalysts have attracted more attention recently and have been tried to serve as efficient alternatives for the noble ones in the field of HER [36]. Until now, many transition metals, such as Co, Ni, Fe, have been successfully proved to be outstanding SACs for HER, and most of them are also coordinated with four nitrogen atoms.

In 2018, Hossain and coworkers used DFT calculations and electrochemical measurements to describe the relationship of TM—N_4 coordination that exists in NG supports for HER activity [25]. The calculated the

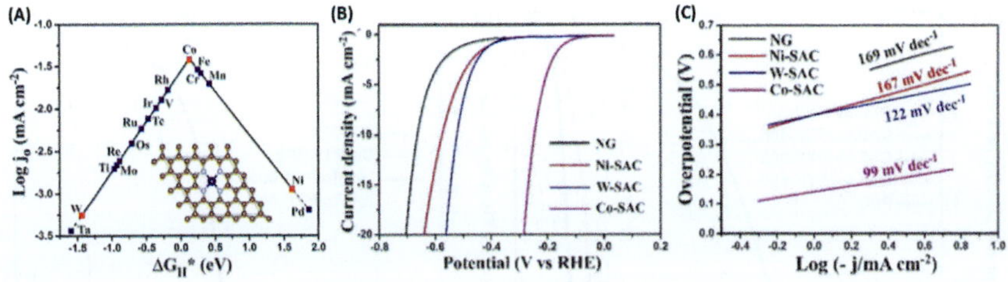

Figure 6.4 (A) Volcano diagram for SACs activity towards HER, (B) LSV curves of SACs for HER at a scan rate of 10 mV s^{-1}. (C) Corresponding Tafel plots. *Reproduced with permission from M.D. Hossain, Z. Liu, M. Zhuang, X. Yan, G.L. Xu, C.A. Gadre, et al., Adv. Energy Mater. 9 (2019). 1803689.*

ΔG_{H^*} of various TM-N_4 structure shows in Fig. 6.4. The result shows that Co has the best activity of HER with overpotential of 230 mV at 10 mA cm^{-2} in 0.5 M H_2SO_4, because the antibonding state orbital is neither completely empty nor fully filled in the case of Co-SAC. In detail, the lower and higher valence d_z^2 orbitals, respectively, could result in weaker and stronger interactions with hydrogen, while the Co SACs with active valence d_z^2 orbitals near the Fermi level partially filled the antibonding state orbital (σ^*), causing a smaller ΔG_{H^*} value.

In 2019, Yi and coworkers synthesized a cobalt single-atoms located on a porous porphyrin-triazine framework, named as CoSAs/PTF-600, which shows a superior HER activity than other nonnoble metals with a very low overpotential of 94 mV at 10 mA cm^{-2} and small onset potential of 21 mV in 0.5 M H_2SO_4 (Fig. 6.5) [37]. The excellent catalytic performance comes from the highly coordinated unsaturated active site Co−N_4, as well as its high porosity and high electrical conductivity.

Besides, Pan et al. put the isolated Fe atoms on NPC support using bimetallic Zn/Fe polyphthalocyanine (ZnFe-BPPC) as a precursor to get an efficient HER electrocatalyst, showing high catalytic activity with the overpotential of 202 mV at 10 mA cm^{-2} in 1 M KOH (Fig. 6.6) [38]. The excellent performance is contributed by the presence of Fe-N_4 catalytic active site that significantly reduce the energy barriers and ΔG_{H^*} as well as porous and conductive NPC.

Interestingly, the structure of dual metal atoms coordinating with N_4, respectively, can mutually promote catalytic performance. Rong and coworkers reported an atomically dispersed Ru/Co dual-sites catalyst on N-doped carbon (Ru/Co−N−C) with excellent HER and OER activity in both acidic and alkaline electrolytes, realizing pH-universal overall

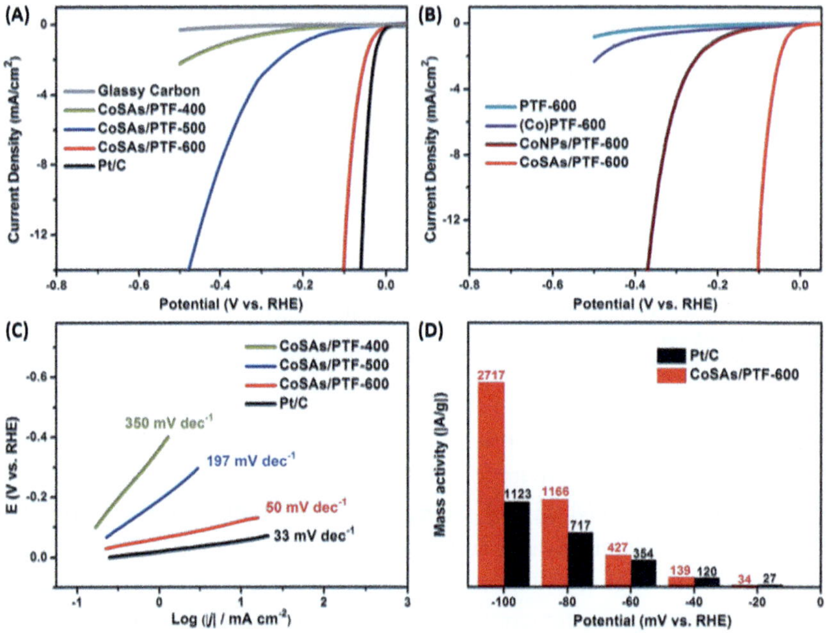

Figure 6.5 Comparison of LSV test data (A, B), Tafel slope (C) and mass activity (D) of different control samples and CoSAs/PTF-600. *Reproduced with permission from J.-D. Yi, R. Xu, G.-L. Chai, T. Zhang, K. Zang, B. Nan, et al., J. Mater. Chem. A 7 (2019) 1252–1259.*

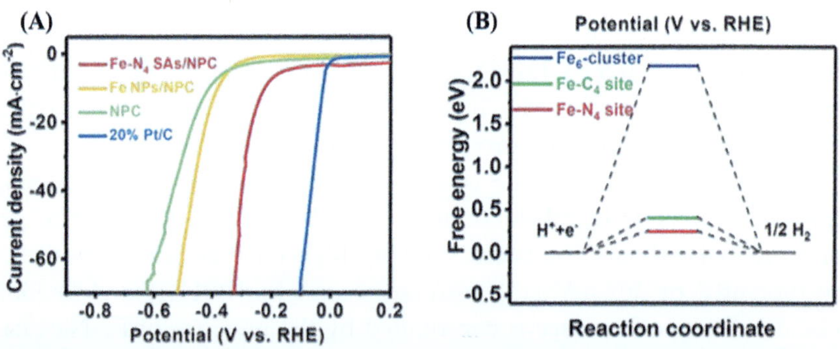

Figure 6.6 (A) The LSV test and (B) free energy diagrams for HER. *Reproduced with permission from Y. Pan, S. Liu, K. Sun, X. Chen, B. Wang, K. Wu, et al., Angew. Chem. Int. Ed. Engl., 57 (2018) 8614–8618.*

water splitting [39]. Experimental results and theoretical calculations reveal that the main activity center of Ru/Co−N−C-800°C is Ru−N_4 sites, while Co−N_4 structure plays a key role in adjusting the electronic

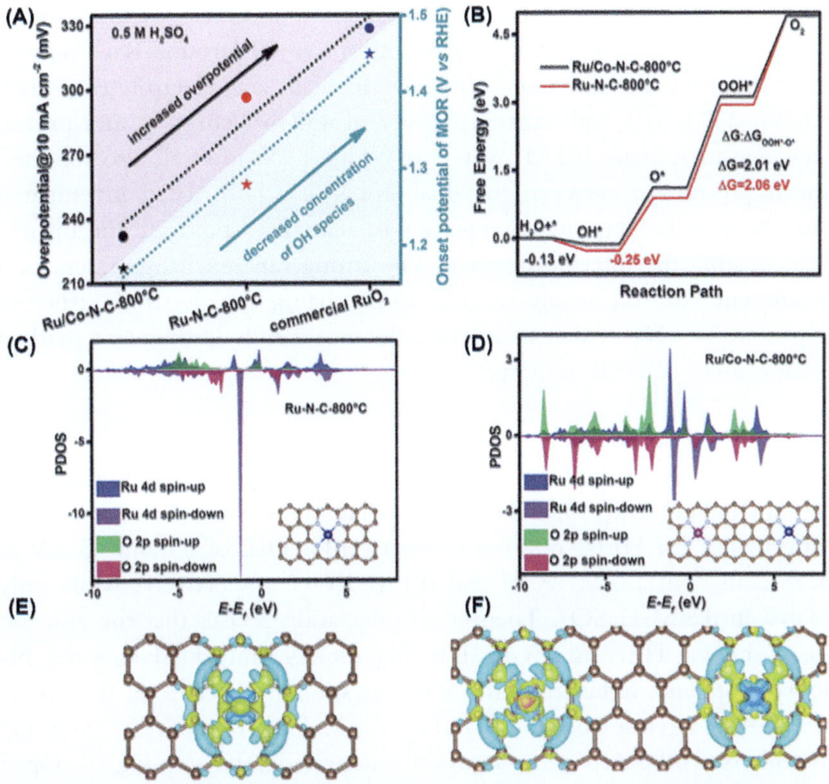

Figure 6.7 (A) Relationship between the concentration of OH intermediates and OER performance in different Ru-containing catalysts. (B) OER free energy diagram. Partial electronic density of states (PDOS) of (C) Ru−N−C-800°C and (D) Ru/Co−N−C-800°C. Differential charge density at the atomical Ru centers between Ru and neighboring C/N atoms in (E) Ru−N−C-800°C and (F) Ru/Co−N−C-800°C. *Reproduced with permission from C. Rong, X. Shen, Y. Wang, L. Thomsen, T. Zhao, Y. Li, et al., Adv. Mater. 34 (2022) e2110103.*

structure and bonding strength between oxygen/hydrogen intermediate species and $Ru-N_4$. The promotion effects of $Co-N_4$ structure on Ru single atoms can be reflected in two aspects. On one hand, $Co-N_4$ sites can change the electronic structure of Ru by elongating Ru−O covalency and increase the electron density of Ru, which optimizes bonding strength between oxygen/hydrogen intermediate species with Ru sites, thereby enhancing HER and OER performance (Fig. 6.7). Electrochemical test results show that the Ru/Co−N−C catalyst can get low overpotential of 13 mV for HER and 232 mV for OER at 10 mA cm^{-2} in 0.5 M H_2SO_4, while 23 and 247 mV counterparts in 1 M

KOH, outperforming benchmark catalysts Pt/C and RuO_2. On the other hand, the optimized electron distribution of $Co{-}N_4$ around $Ru{-}N_4$ sites could improve its corrosion resistance in acid-base electrolytes, which exhibits 330 h at a high current density of 450 mA cm_{geo}^{-2} in a proton exchange membrane (PEM) water. In addition, Shan et al. also reported similar promotion between the dual-atoms sites [40]. They investigated $Mn{-}N_4$ and $Fe{-}N_4$ site interactions to rearrange spintronic occupancy by regulating the intersite distance. The strong superexchange interaction of adjacent $Mn{-}N_4$ groups can effectively change the electronic structure of pristine $Fe{-}N_4$ to alter its intersite distance, finally leading to a marked enhancement of HER activity.

6.2.1.2 $M{-}N_3$ structure

Guan et al. used a facile pyrolysis method to anchor an atomically dispersed rhodium with three N atoms (Rh@NG), which showed superior performance for HER, offering a low overpotential of 33 and 45 mV at 10 mA cm^{-2} in 1.0 M KOH and 0.1 M KOH, respectively, while only 29 mV in 0.5 M H_2SO_4. Theoretical calculation reveals that the efficient active site for HER might be $Rh{-}N_3$ moiety embedded into the N-doped graphene, rather than the $Rh{-}N_4$ structure because of the ΔG_{H^*} of $Rh{-}N_3$ is close to Pt (111) (Fig. 6.8A) [41]. Moreover, Zang and coworkers synthesized a Ni atomically dispersed catalyst using N-doped carbon material as a substrate (named as Ni-SA/NC) with $Ni{-}N_3$ and $Ni{-}N_4$ coordination, which exhibits the low overpotential of 102 mV in alkaline electrolyte and 139 mV in seawater to achieve the current density

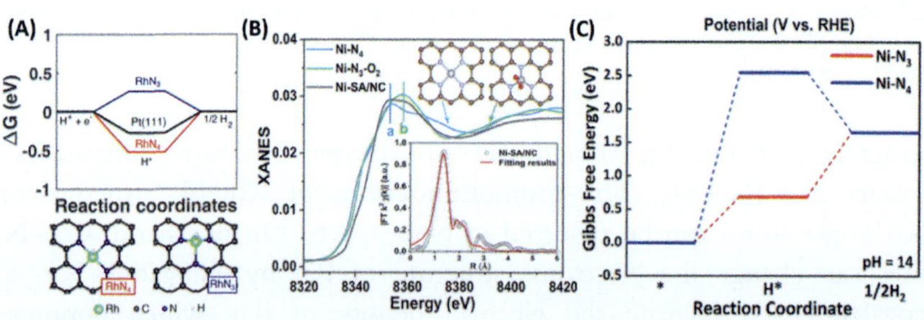

Figure 6.8 (A) Free energy diagram of HER on RhN_4, RhN_3, and Pt (111). (B) Calculated curves of the proposed Ni-N4 and $Ni\text{-}N_3\text{-}O_2$ coordination and experimental XANES curve of Ni-SA/NC. (C) Comparison of Gibbs Free Energy of $Ni\text{-}N_3$ and $Ni\text{-}N_4$. *(A) Reproduced with permission from J. Guan, X. Wen, Q. Zhang, Z. Duan, Carbon 164 (2020) 121–128. (C) Reproduced with permission from W. Zang, T. Sun, T. Yang, S. Xi, M. Waqar, Z. Kou, et al., Adv. Mater. 33 (2021) e2003846.*

of 10 mA cm^{-2} [42]. The excellent HER performance most comes from the Ni−N_3 structure. The evolution of HER in alkaline or seawater depends on the adsorption/dissociation of H_2O. The theoretical analysis demonstrates that the adsorption/dissociation kinetics of Ni−N_4 for water is slow, while the Ni−N_3 sites show a better performance, which can be judged that Ni-N_3 played a greater role in HER activity (Fig. 6.8B,C).

6.2.1.3 M-N_2 structure

Early in 2017, Peng et al. embedded ruthenium ions into the molecular skeletons of graphitic carbon nitride (C_3N_4) nanosheets and gained good HER activity when the ratio of Ru atoms and pyridine nitrogen atoms is close to 1:2, finally exhibiting a low overpotential of 140 mV at 10 mA cm^{-2} with a low Tafel slope of 57 mV dec^{-1} in 0.5 M H_2SO_4 [43]. DFT calculations proved that the good performance of the catalyst depends on the formation of Ru−N_2 structure, which promotes the H adsorption. Similar studies have also been reported by Lu and coworkers [44]. The group used ruthenium and nitrogen codoped carbon nanowires to synthesize effective SACs for HER, which only needs a low η_{10} of 12 and 47 mV in 1 M KOH and 0.1 M KOH, respectively (Fig. 6.9A,B). In addition, it displays outstanding stability with no attenuation after a long 10,000 CV cycles (Fig. 6.9C). The first principle of computations demonstrated that Ru−C_2N_2 is the most favorable active center where both the Ru and adjacent carbon atoms could act as the active sites showing nearly zero value of ΔG_{H^*} (<0.2 eV) (Fig. 6.9D).

6.2.2 Single metal atoms anchored in non-carbon substrate materials

6.2.2.1 S-coordinating

Transition metal chalcogenide is broadly investigated as support materials for single-atom catalysis, for example, 2D molybdenum disulfide (MoS_2), because of its similarities with graphene [45]. Zhu and coworkers fixed single Pt atoms on the MoS_2 nanosheets (NSs) to prepare the hybrid SA-Pt/MoS_2 catalysts, which display excellent HER activity [46]. In acidic media (0.5 M H_2SO_4), SA-Pt/MoS_2 shows a minimal overpotential of 44 mV at 10 mA cm^{-2} with the Tafel slope of 34.83 mV dec^{-1}, while in alkaline counterpart (1 M KOH) it still exhibits the overpotential of 123 mV with 76.71 mV dec^{-1} Tafel slope. Theoretical simulations reveal that Pt could improve the absorption behavior of H atoms and the charge-transfer kinetics of MoS_2, significantly enhancing HER activity.

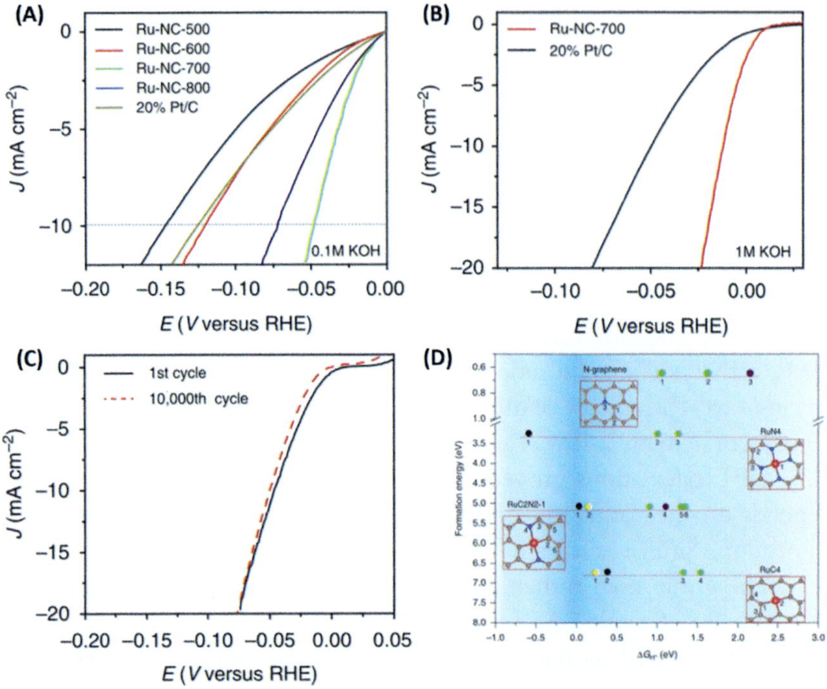

Figure 6.9 LSV comparison of Ru-NC-700 and several other comparative samples in (A) 0.1 M KOH and (B) 1 M KOH. (C) The stability of Ru-NC-700. (D) The Gibbs free energy of different RuC_xN_y structures. *Reproduced with permission from B. Lu, L. Guo, F. Wu, Y. Peng, J.E. Lu, T.J. Smart, et al., Nat. Commun. 10 (2019) 631.*

In another study, Guan et al. prepared the Pt SACs based on MoS_2/NiS_2 for HER, which displays a low η_{10} of 34 mV in 0.5 M H_2SO_4 and excellent stability of only 5% loss in the overpotential at the current density of 10, 20, and 50 mA cm^{-2} after 72 h by chronoamperometry test [47]. The excellent performance of the catalyst comes from the coordination formed between Pt and S. Under this coordination, the adsorption-free energy of S atom and Pt atom alone for H is significantly reduced, which effectively improves the catalytic activity for HER (Fig. 6.10). Moreover, support of Pt@MoS_2/NiS_2 is metallic rather than semiconducting behavior, which enhanced charge transfer and conductivity, thereby promoting the HER.

6.2.2.2 O-coordinating

Metal oxide supports could also be served as substrates for SACs owing to the multiple defect sites (steps, corners, vacancies) and ample -OH groups

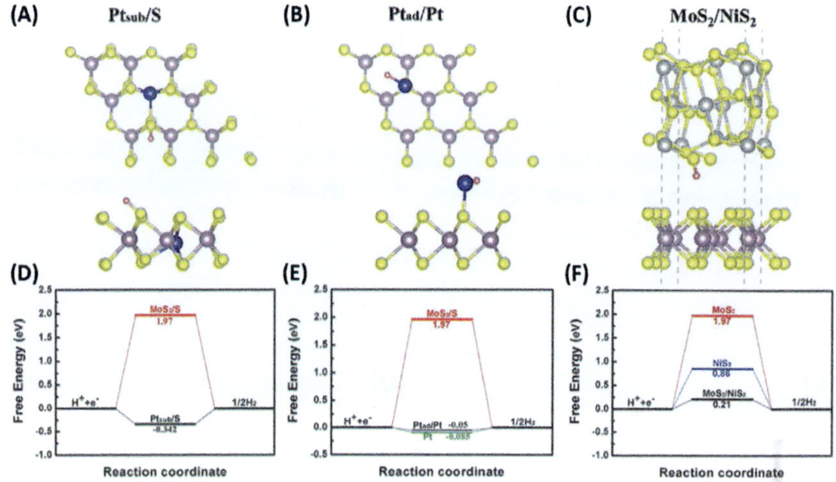

Figure 6.10 The Model and Gibbs free energy of H adsorption of (A) (D) Pt_{sub}/S, (B) (E) Pt_{as}/Pt, (C) (D) MoS_2/NiS_2. *Reproduced with permission from Y. Guan, Y. Feng, J. Wan, X. Yang, L. Fang, X. Gu, et al., Small, 14 (2018) e1800697.*

on its surface, all of which could efficiently anchor single metal atoms [48–50]. Besides, due to the robust and stable properties at high temperatures, the metal oxide substrates could exhibit good mechanical and thermal stability in catalytic reactions [31,51].

Generally, the first shell coordination of M–O could be formed for those metal oxide-based SACs. Recently, Zhou et al. develop a single-atom platinum immobilized on the NiO/Ni heterostructure (Pt_{SA}-NiO/Ni) as an alkaline HER catalyst with a quite low overpotential of 26 mV at 10 mA cm^{-2} and high mass activity of 20.6 A $mg^{-1}{}_{Pt}$ in 1 M KOH [52]. Through the DFT calculation and HAADF-STEM image, the structure of Pt coordination is proved to be Pt–O and Pt–Ni_6. After introducing single Pt atoms on NiO/Ni heterostructure, the electronic structure of Pt SAs could be redistributed at the interfaces of the heterostructure because of the different electronegativity. In detail, the electrons could be shifted from Pt to coordinating O atoms, while electrons of adjacent Ni could be shifted to the coordinating Pt atoms, generating an enhanced local electric field with a half-moon shape area around the Pt site (Fig. 6.11) and thus accelerating the adsorbed H conversion and H_2 evolution [53,54].

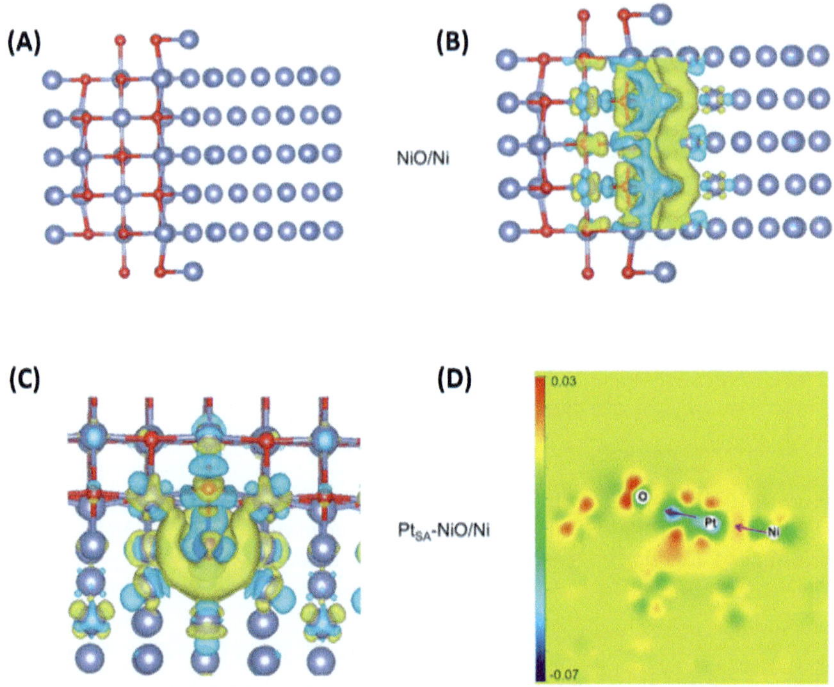

Figure 6.11 (A) Computational models and (B) the localized electric field distribution of NiO/Ni heterostructure. (C) The localized electric field distribution with Pt atom at NiO/Ni phase junction. (D) 2D isosurface map of the cross section consisting of Pt, Ni, and O atoms in Pt_{SA}-NiO/Ni. *Reproduced with permission from K.L. Zhou, Z. Wang, C.B. Han, X. Ke, C. Wang, Y. Jin, et al., Nat. Commun. 12 (2021) 3783.*

6.2.2.3 P-coordinating

Transition metal phosphide catalysts have potentially advantageous for HER due to their distinct physicochemical properties, such as the desirable electronic properties, polytropic compositions, and structures [55]. In 2020, Wu et al. reported single-atom W-doped CoP nanoarrays (W-CoP) as a robust pH-universal electrocatalyst toward HER [56]. The W-CoP catalyst presents the minimal overpotentials of 40, 36, and 48 mV in 1 M KOH, 1 M phosphate buffer, and 0.5 M H_2SO_4 solutions, respectively. Besides, it also displays decent stability of only 17% loss of current density after 20 h at 75 mV, comparable to commercial Pt/C with about 25% decline in current density under the same conditions. The researchers established the module of W-CoP (111) surface with W atoms placed at the sites of Co and revealed that the ΔG_{H^*} of P on W-CoP decreased obviously after doping with W atoms, indicating the positive influence of W-doping to the H adsorption of P.

6.2.3 Single nonmetallic elements atoms as active centers

In addition to single metal atoms as active sites, the nonmetallic single atoms could also be served as the SAC active sites and exhibit decent activity for HER.

In 2019, Zhao et al. reported the single-atom nickel iodide (SANi-I) catalysts for electrocatalytic HER with atomically dispersed nonmetal iodine atoms as the active sites, affording a 60 mV overpotential at 100 mA cm^{-2} in 1 M KOH [57]. They found that the single iodine atoms can promote the water dissociation and H adsorption through I-H_{ads} bond formation and thus enhance the HER kinetics, reaching a Tafel slop of 34.6 mV dec^{-1}. In addition, Fu et al. firstly anchored the single P atoms on the unique single crystal Mo_2C hexagonal nanosheet arrays with a (001) plane supported by carbon sheet (SAP-Mo_2C-CS) and realized efficient HER performance [58]. In detail, it exhibits superior HER activity close to Pt/C in 0.5 M H_2SO_4 solution with a low overpotential of 36 mV at 10 mA cm^{-2} and TOF of 0.23 H_2 s^{-1}, which also surpasses other reported nonprecious metal catalysts [59,60]. The DFT calculations demonstrated that the synergistic effects between P and Mo could accelerate H* desorption and cause zero-closing ΔG_{H*} to show excellent HER kinetics.

6.3 Outlook on SACs for HER

Facing the urgent demand of high-activity and long-duration HER electrocatalysts for establishing green and economical energy conversion and storage systems, SACs have been displaying strong potential for future large-scale application. In detail, the ultrahigh electrocatalytic HER activity and ultralow content of metal atoms offer huge space for reducing synthesis cost; and their simple structure provides more chances for modifying their physical and chemical properties to adapt to various HER electrochemical environment. However, there are still some challenges to be focused and solved in the future.

1. Advanced synthesis method. Until now, most of SACs for HER or even other electrochemical reactions have been synthesized via the combination of wet-chemistry method and high-temperature calcination. Although the high-temperature process could facilitate the

formation of $M-N_x$ structure, bring better electronic conductivity, and introduce more defects for significantly enhancing the HER catalytic performance, the aggregation of single atoms, higher production cost, and energy consumption should also be noted, all of which sever the large-scale SACs synthesis [2,61–64]. Therefore, the easier, more greener, and cost- and energy-effective SACs methods should be furtherly paid attention. So far, some researchers have been trying to discard the high-temperature process and stepp into more economic SACs preparation. For example, Peng et al. successfully developed a pyrolysis-free path and applied the van der Waals interaction to link $Fe-N_x-C$ moieties and the COF network, finally realizing the fabrication of Fe-based SACs, while the loading of a single atom is only 0.94 wt.% [65]. Therefore, how to increase the metal atom loading is another challenge for developing advanced SACs method in the future.

2. Deeper mechanism research. Due to the clear and simple structure of SACs, the various modification could be easily applied to the single-atom metal center and corresponding coordination environment for purusing better HER performance, such as the nonmetal/metal atoms dopant, vacancies fabrication, and metal particles/clusters adjustment [66–71]. Although researchers have studied the earlier-mentioned modifications via different theoretical calculation methods and recognized some regular patterns in the modifications, more detailed and deeper mechanism research are still limited for SACs in the field of HER. For example, the intersite distance effect of the single metal atoms for HER SACs has not been deeply research yet. A high loading of single metal sites has been usually served as a potential goal for achieving higher catalytic activity, while under the high site density, the effect caused by these adjacent single metal atoms in SACs is largely overlooked in the current research. Recently, Yu, Xiao, and coworkers found the effect of strong interaction between the adjacent $Fe-N_4$ moieties on the electronic property of the single-atom sites, which could directly affect the catalytic activity of oxygen electroreduction reaction [72]. And similarly, Li groups also reported the site distance effect of Cu-based SACs on the organic oxidation reaction [73]. However, this effect could not be applied to other catalytic systems due to the coadsorption of their different reactant product on the active sites [73]. Therefore, the effect should also be investigated in the HER field, favoring to construct deeper catalytic mechanisms on

SACs. Moreover, the rapid development of machine learning (ML) could realize the quick and easier investigation for the SACs HER mechanism. While, how to construct sufficient and consistent datasets and ensure data credibility are challenges for current ML.

3. Deeper stability research. Although much effort has been devoted to developing better HER catalytic activity for current SACs research, fewer work has been paid attention to increase the HER stability for SACs. In the practical application, stability is an important parameter to assess the reliability of selected electrocatalysts during long-term and large-scale production activity, which plays a vital role in the commercialization of SACs in hydrogen industry. As for the SACs, good stability means less aggregation and loss of the single metal sites during the electrocatalysis process, which requires stable physical and chemical coordination for those metal atoms. Therefore, in situ characterization of the whole electrochemical process for SACs is highly desirable, which could facilitate the further understanding of the structure—durability relationship, laying a foundation for designing highly stable SACs for HER [74].
4. Wider selection of water resources. Owing to the global shortage of fresh water, the electrochemical seawater splitting becomes a research trend for future low-cost hydrogen production [42,75]. In the aspect of reaction principle, the seawater electrocatalysis could be basically based on the freshwater counterpart, while its corrosive and complex environment has negative impacts on the activity and stability of HER electrocatalysts [76,77]. For example, during the HER process in the seawater, its pH values could be fluctuation, and the carbonate buffers are not hinder the local pH changes in the cathode, leading to a huge volatility window from 5 to 9 in pH value (even at under the low current density) [78,79]. Such a dramatic pH-value change on the cathode surface could easily cause the degradation of the cathode and the slight Na, Mg, and Ca could be deposited on the cathode, resulting in an inferior catalytic performance [78,80]. Therefore, furtherly optimizing SACs for realizing highly-effective HER in seawater is a meaningful way to commercialize the HER.
5. Advanced water-splitting electrolyzer. Although some achievements have been reached in reducing production costs and enhancing HER performance for SACs, the effective and cheap electrolyzer has not been paid more attention. Currently, PEM and anion exchange membrane (AEM) have been considered the main electrolyzer techniques

for overall water splitting [81,82]. However, the cathode choice of PEM has been limited in the noble metals, such as Pt, Ir, and related oxides, under the acidic electrolyte;[83] besides, the Nafion-containing PEM and the titanium stack components furtherly raise the capital cost for the electrolysis process, hindering its wide application [84]. Owing to combining the advantages of alkaline electrolysis and conventional PEM, AEM has attracted more attention recently [84]. Thus, designing suitable AEM electrolyzer for SACs is a crucial step for hydrogen economy commercialization. For example, the flow of liquid electrolytes in the AEM electrolyte should be optimized for the SACs to higher hydrogen production efficiency [84]. During working status, numerous H_2 bubbles could be formed on the cathode, surface, which could directly block the active sites, decreasing the HER performance. And those bubbles could be removed by the electrolyte flow, and thus finding a suitable electrolyte flow rate for the corresponding SACs becomes a key point in constructing a commercial AEM electrolyzer.

Looking ahead, although facing many challenges and problems during the development of SACs in the field of HER, we still believe that SACs hold huge potential to accelerate the realization and development of the hydrogen economy.

References

[1] J. Wei, M. Zhou, A. Long, Y. Xue, H. Liao, C. Wei, et al., Nano-Micro. Lett. 10 (2018) 75.
[2] H. Liu, X. Peng, X. Liu, ChemElectroChem 5 (2018) 2963–2974.
[3] L. Li, P. Wang, Q. Shao, X. Huang, Chem. Soc. Rev. 49 (2020) 3072–3106.
[4] J. Wang, F. Xu, H. Jin, Y. Chen, Y. Wang, Adv. Mater. 29 (2017) 1605838.
[5] H. Liu, Y. Liu, M. Li, X. Liu, J. Luo, Mater. Today Adv. 7 (2020) 2590. 0498.
[6] T. Shinagawa, A.T. Garcia-Esparza, K. Takanabe, Sci. Rep. 5 (2015) 13801.
[7] A.J. Brad, L.R. Faulkner, Electrochemical Methods: Fundamentals and Applications, Wiley, 2000.
[8] S. Sun, The Frontier of Nanomaterials–Electrocatalytic Nanomaterials (Chinese Edition), Chemical Industry Press, Beijing, 2018.
[9] H. Vrubel, T. Moehl, M. Gratzel, X. Hu, Chem. Commun(Camb) 49 (2013) 8985–8987.
[10] H. Liu, X. Peng, X. Liu, G. Qi, J. Luo, ChemSusChem 12 (2019) 1334–1341.
[11] H. Liu, X. Liu, Z. Mao, Z. Zhao, X. Peng, J. Luo, et al., J. Power Sources 400 (2018) 190–197.
[12] Y. Hou, J. Wang, J. Liu, C. Hou, Z. Xiu, Y. Fan, et al., Adv. Energy Mater. 9 (2019) 901751.
[13] A.D. Carbó, J. Solid. State Electrochem. 13 (2008) 505–506.
[14] M. Escudero-Escribano, A. Verdaguer-Casadevall, P. Malacrida, U. Gronbjerg, B.P. Knudsen, A.K. Jepsen, et al., J. Am. Chem. Soc. 134 (2012) 16476–16479.

[15] T.J. Schmidt, H.A. Gasteiger, G.D. Stab, P.M. Urban, D.M. Kolb, R.J. Behm, J. Electrochem. Soc. 145 (1998) 2354–2358.
[16] D. Chen, Q. Tao, L.W. Liao, S.X. Liu, Y.X. Chen, S. Ye, Electrocatalysis 2 (2011) 207–219.
[17] D. Voiry, M. Chhowalla, Y. Gogotsi, N.A. Kotov, Y. Li, R.M. Penner, et al., ACS Nano 12 (2018) 9635–9638.
[18] C. Wei, S. Sun, D. Mandler, X. Wang, S.Z. Qiao, Z.J. Xu, Chem. Soc. Rev. 48 (2019) 2518–2534.
[19] W. Sheng, M. Myint, J.G. Chen, Y. Yan, Energy. Environ. Sci. 6 (2013) 1509.
[20] N. Bhuvanendran, S. Ravichandran, Q. Xu, T. Maiyalagan, H. Su, Int. J. Hydrog. Energ. 47 (2022) 7113e7138.
[21] J.K. Norskov, T. Bligaard, J. Rossmeisl, C.H. Christensen, Nat. Chem. 1 (2009) 37–46.
[22] P. Aggarwal, D. Sarkar, K. Awasthi, P.W. Menezes, Coordin. Chem. Rev. 452 (2022) 214289.
[23] Y. Zheng, Y. Jiao, M. Jaroniec, S.Z. Qiao, Angew. Chem. Int. Ed. Engl. 54 (2015) 52–65.
[24] E. Skúlason, V. Tripkovic, M.E. Björketun, S. Gudmundsdóttir, G. Karlberg, J. Rossmeisl, et al., J. Phys. Chem. C. 114 (2010) 18182–18197.
[25] M.D. Hossain, Z. Liu, M. Zhuang, X. Yan, G.L. Xu, C.A. Gadre, et al., Adv. Energy Mater. 9 (2019) 1803689.
[26] J.K. Norskov, F. Abild-Pedersen, F. Studt, T. Bligaard, Proc. Natl. Acad. Sci. U S A 108 (2011) 937–943.
[27] Z. Fu, B. Yang, R. Wu, Phys. Rev. Lett. 125 (2020) 156001.
[28] M.T. Gorzkowski, A. Lewera, J. Phys. Chem. C. 119 (2015) 18389–18395.
[29] M.T. Greiner, T.E. Jones, S. Beeg, L. Zwiener, M. Scherzer, F. Girgsdies, et al., Nat. Chem. 10 (2018) 1008–1015.
[30] X. Li, H. Rong, J. Zhang, D. Wang, Y. Li, Nano Res. 13 (2020) 1842–1855.
[31] G.M. Tomboc, T. Kim, S. Jung, H.J. Yoon, K. Lee, Small 18 (2022) e2105680.
[32] T. Li, J. Liu, Y. Song, F. Wang, ACS Catal. 8 (2018) 8450–8458.
[33] J. Xu, R. Li, R. Zeng, X. Yan, Q. Zhao, J. Ba, et al., ACS Appl. Mater. Interfaces 12 (2020) 38106–38112.
[34] J. Yang, B. Chen, X. Liu, W. Liu, Z. Li, J. Dong, et al., Angew. Chem. Int. Ed. 57 (2018) 9495–9500.
[35] H. Chen, C. Zhu, C. Wen, M. Wang, M. Zhang, Y. Geng, et al., Appl. Surf. Sci. 549 (2021) 149320.
[36] N. Mahmood, Y. Yao, J.W. Zhang, L. Pan, X. Zhang, J.J. Zou, Adv. Sci. (Weinh.) 5 (2018) 1700464.
[37] J.-D. Yi, R. Xu, G.-L. Chai, T. Zhang, K. Zang, B. Nan, et al., J. Mater. Chem. A 7 (2019) 1252–1259.
[38] Y. Pan, S. Liu, K. Sun, X. Chen, B. Wang, K. Wu, et al., Angew. Chem. Int. Ed. Engl. 57 (2018) 8614–8618.
[39] C. Rong, X. Shen, Y. Wang, L. Thomsen, T. Zhao, Y. Li, et al., Adv. Mater. 34 (2022) e2110103.
[40] Y. Shan, Z. Wu, T. Li, Mater. Chem. Phys. 279 (2022) 125748.
[41] J. Guan, X. Wen, Q. Zhang, Z. Duan, Carbon 164 (2020) 121–128.
[42] W. Zang, T. Sun, T. Yang, S. Xi, M. Waqar, Z. Kou, et al., Adv. Mater. 33 (2021) e2003846.
[43] Y. Peng, B. Lu, L. Chen, N. Wang, J.E. Lu, Y. Ping, et al., J. Mater. Chem. A 5 (2017) 18261–18269.
[44] B. Lu, L. Guo, F. Wu, Y. Peng, J.E. Lu, T.J. Smart, et al., Nat. Commun. 10 (2019) 631.

[45] H. Guo, L. Li, X. Wang, G. Yao, H. Yu, Z. Tian, et al., ACS Appl. Mater. Interfaces 11 (2019) 36506–36514.
[46] J. Zhu, Y. Tu, L. Cai, H. Ma, Y. Chai, L. Zhang, et al., Small 18 (2022) e2104824.
[47] Y. Guan, Y. Feng, J. Wan, X. Yang, L. Fang, X. Gu, et al., Small 14 (2018) e1800697.
[48] S. Cai, L. Wang, S. Heng, H. Li, Y. Bai, D. Dang, et al., J. Phys. Chem. C. 124 (2020) 24566–24579.
[49] P. Matczak, C. R. Chim. 21 (2018) 669–675.
[50] K. Qi, M. Chhowalla, D. Voiry, Mater. Today 40 (2020) 173–192.
[51] R. Lang, X. Du, Y. Huang, X. Jiang, Q. Zhang, Y. Guo, et al., Chem. Rev. 120 (2020) 11986–12043.
[52] K.L. Zhou, Z. Wang, C.B. Han, X. Ke, C. Wang, Y. Jin, et al., Nat. Commun. 12 (2021) 3783.
[53] D. Liu, X. Li, S. Chen, H. Yan, C. Wang, C. Wu, et al., Nat. Energy 4 (2019) 512–518.
[54] K.L. Zhou, C. Wang, Z. Wang, C.B. Han, Q. Zhang, X. Ke, et al., Energy Environ. Sci. 13 (2020) 3082–3092.
[55] W. Jin, J. Chen, B. Liu, J. Hu, Z. Wu, W. Cai, et al., Small 15 (2019) e1904210.
[56] J. Wu, N. Han, S. Ning, T. Chen, C. Zhu, C. Pan, et al., ACS Sustain. Chem. Eng. 8 (2020) 14825–14832.
[57] Y. Zhao, T. Ling, S. Chen, B. Jin, A. Vasileff, Y. Jiao, et al., Angew. Chem. Int. Ed. Engl. 58 (2019) 12252–12257.
[58] W. Fu, Y. Wang, W. Tian, H. Zhang, J. Li, S. Wang, et al., Angew. Chem. Int. Ed. Engl. 59 (2020) 23791–23799.
[59] H. Duan, D. Li, Y. Tang, Y. He, S. Ji, R. Wang, et al., J. Am. Chem. Soc. 139 (2017) 5494–5502.
[60] J. Yin, Q. Fan, Y. Li, F. Cheng, P. Zhou, P. Xi, et al., J. Am. Chem. Soc. 138 (2016) 14546–14549.
[61] X. Peng, Y. Mi, X. Liu, J. Sun, Y. Qiu, S. Zhang, et al., J. Mater. Chem. A 10 (2022) 6134.
[62] W. Wei, X. Shi, P. Gao, S. Wang, W. Hu, X. Zhao, et al., Nano Energy 52 (2018) 29–37.
[63] S. Liu, M. Jin, J. Sun, Y. Qin, S. Gao, Y. Chen, et al., Chem. Eng. J. 437 (2022) 135294.
[64] J. Xu, S. Lai, D. Qi, M. Hu, X. Peng, Y. Liu, et al., Nano Res. 14 (2021) 1374.
[65] P. Peng, L. Shi, F. Huo, C. Mi, X. Wu, S. Zhang, et al., Sci. Adv. 5 (2019) eaaw2322.
[66] H. Zhang, Y. Qiu, S. Zhang, Q. Liu, J. Luo, X. Liu, Ionics 28 (2022) 3927–3934.
[67] H. Zhang, Y. Luo, P. Chu, Q. Liu, X. Liu, S. Zhang, et al., J. Alloy. Compd. 922 (2022) 166113.
[68] T. Sun, W. Zang, H. Yan, J. Li, Z. Zhang, Y. Bu, et al., ACS Catal. 11 (2021) 4498–4509.
[69] J. Park, S. Lee, H.-E. Kim, A. Cho, S. Kim, Y. Ye, et al., Angew. Chem. Int. Ed. 58 (2019) 16038.
[70] G. Meng, M. Jin, T. Wei, Q. Liu, S. Zhang, X. Peng, et al., Nano Res. 15 (2022) 8890–8896. Available from: https://doi.org/10.1007/s12274-022-4747-y.
[71] Q. He, D. Tian, H. Jiang, D. Cao, S. Wei, D. Liu, et al., Adv. Mater. 32 (2020) 1906972.
[72] Z. Jin, P. Li, Y. Meng, Z. Fang, D. Xiao, G. Yu, Nat. Catal. 4 (2021) 615–622.
[73] B. Wang, C. Cheng, M. Jin, J. He, H. Zhang, W. Ren, et al., Angew. Chem. Int. Ed. (2022) e202207268.

[74] M. Tamtaji, H. Gao, M.D. Hossain, P. Galligan, H. Wong, Z. Liu, et al., J. Mater. Chem. A 10 (2022) 15309–15331.
[75] Y. Ma, C. Wu, X. Feng, H. Tan, L. Yan, Y. Liu, et al., Energy Environ. Sci. 10 (2017) 788–798.
[76] S. Ye, W. Xiong, P. Liao, L. Zheng, X. Ren, C. He, et al., J. Mater. Chem. A 8 (2020) 11246–11254.
[77] S. Wang, M. Wang, Z. Liu, S. Liu, Y. Chen, M. Li, et al., ACS Appl. Mater. Interfaces 14 (2022) 15250–15258.
[78] X. Lu, J. Pan, E. Lovell, T. Tan, Y. Ng, R. Amal, Energy Environ. Sci. 11 (2018) 1898–1910.
[79] M. Auinger, I. Katsounaros, J. Meier, S. Klemm, P. Biedermann, A. Topalov, et al., Phys. Chem. Chem. Phys. 13 (2011) 16384–16394.
[80] J. Bennett, Int. J. Hydrog. Energy 5 (1980) 401–408.
[81] D. Li, E. Park, W. Zhu, Q. Shi, Y. Zhou, H. Tian, et al., Nat. Energy 5 (2020) 378–385.
[82] M. Rashid, M. Mesfer, H. Naseem, M. Danish, Int. J. Eng. Adv. Technol. 4 (2015) 2249–8958.
[83] M. Carmo, D. Fritz, J. Mergel, D. Stolten, Int. J. Hydrog. Energy 38 (2013) 4901–4934.
[84] I. Vincent, E.C. Lee, H.M. Kim, Sci. Rep. 11 (2021) 293.

CHAPTER 7

Single-atom catalysts for electrocatalytic carbon dioxide reduction

Subhash Chandra Shit[1] and Indranil Mondal[2,3]
[1]Department of Chemistry, Gwangju Institute of Science and Technology (GIST), Gwangju, Republic of Korea
[2]Department of Chemistry, Technische Universität Berlin, Berlin, Germany
[3]School of Chemistry, IISER Thiruvananthapuram, Kerala, India

7.1 Introduction

The unusual climate change and global warming are majorly caused by the exponentially increasing atmospheric carbon dioxide (CO_2) level. Over a couple of years, almost 13 gtc/year of CO_2 emission occurs because of anthropogenic emissions by man-made sector, including electricity generation, agriculture and land usage, synthetic industries, transportation, etc [1]. The roadmap for tackling this issue was prepared in Paris Climate Accord meeting, where the participating countries agreed that net CO_2 emission needs to be reduced immediately and meet that cross zero around the year 2050, while targeting to constrain global warming to a safe level of 1.5°C above preindustrial levels [2].

While putting significant effort toward decarbonization, researchers are desperately searching for alternative renewable energy resources such as solar, wind, tidal, photothermal, etc., which can minimize fossil fuel utilization and hence CO_2 emission. However, these resources are weather dependent and require complicated setup for storing and transportation. In addition, industrial raw materials production such as cement, steel, plastic, and ammonia cannot get rid of CO_2 emission completely. Simply, it is almost impossible to run these industries even if the energy sector fully relies on renewable energy sources.

CO_2 is the highest oxidized (with a formal oxidation state of + IV) and oxygenated form of carbon. Therefore, the production of high-energy carrier from CO_2 requires protons and electrons. These classes of reactions simply are called CO_2 reduction reaction (CO_2RR) where the

Single Atom Catalysts
DOI: https://doi.org/10.1016/B978-0-323-95237-8.00003-3

oxidation state of carbon is reduced. The combustion of hydrocarbons in the presence of oxygen either in biological system (e.g., glucose) or non-biologial system (e.g fossil fuel) releases H_2O. Therefore, the most feasible way would involve H_2O as proton and electron source to reduce CO_2. In the biological system, this process can be done using photosynthesis (Eq. 7.1) [3]; however, in the nonbiological system, it is called reverse water-gas shift or syngas formation reaction (Eq. 7.2)

$$xCO_2 + yH_2O \rightarrow \text{Glucose} + zO_2 \tag{7.1}$$

$$CO_2 + H_2O \rightarrow CO + H_2 \tag{7.2}$$

The syngas mixture can be converted into several forms of hydrocarbons depending on the heterogeneous catalysts (such as pure metals, alloys, metal/semiconductor composite, etc.) employed in the reaction mixture [4–6]. However, these processes require high temperatures (more than 250°C) and pressure (150 bar), which makes the method energy inefficient for large-scale production. In this regard, electrochemical splitting of H_2O to produce H_2 using renewable energy sources (such as solar, photothermal, wind, etc.) and using it for CO_2 reduction is one of the sustainable approaches to solve it. However, the direct electrochemical reduction of CO_2 has several advantages over the hydrogenation of CO_2 using H_2 gas produced from electrochemical water splitting [7]. Mainly, (1) it allows to occur electrochemical water splitting and CO_2 hydrogenation within a single electrochemical process, (2) it enables easy formation of hydrocarbons and related products, which cannot be generated easily using thermo-catalytic pathway, (3) most of the electrochemical reactions undergo at room temperature and ambient pressure. These features assure that CO_2 reduction is more sustainable and much easier to scale up using electrochemical processes as compared to thermal processes [8]. It is worth noting here that in a few electrochemical CO_2RR the CO forms as the major product, which is also a valuable intermediate that allows further reduction to form multicarbon hydrocarbons or alcohols [9].

The prime challenge for CO_2 reduction is its reactivity (e.g., adsorption to product formation) with different catalysts and product selectivity. This requires deep learning of the thermodynamics and kinetics of the electrochemical CO_2RR. CO_2RR involves multielectron or proton (2–18 e^-/H^+) pathway depending on the potential applied, working electrode (WE), type of electrolyzer (static or flow setup), reaction medium, including temperature, pH, supporting electrolyte ions, etc.

Moreover, CO_2RR simultaneously occurs with hydrogen evolution reaction (HER) within the same potential window and is considered as one of the competing reactions. Therefore, designing catalysts and electrochemical reaction conditions selectively for CO_2RR and suppressing HER is of high interest. The standard redox potential for the conversion of CO_2 to different products is tabulated in Table 7.1 [10]. Furthermore, it is important to understand the surface reaction mechanism (e.g., adsorption-desorption, rate-determining step (RDS)), which defines the chemical phase space of the products, especially for complex C_2 or C_3 molecules (Fig. 7.1) [11].

The electro-reduction of CO_2 on inorganic materials has been evolved with the discovery of a different class of materials, such as pure metals, alloys, semiconductor, and their varied combination in the heterostructure (or nanocomposite) (Fig. 7.2). In addition, the morphology, elemental

Table 7.1 Standard reduction potential values for the formation of C-containing products from electrochemical CO_2RR [10].

Reaction	E^{θ}/V (vs SHE)	Reaction	E^{θ}/V (vs SHE)
$CO_2 + e^- \rightarrow CO_2^{-\cdot}$	−1.90	$CO_2 + 4H^+ + 4e^- \rightarrow C + 2H_2O$	−0.200
$CO_2 + 2H^+ + 2e^- \rightarrow CO + H_2O$	−0.530	$CO_2 + 2H_2O + 4e^- \rightarrow C + 4OH^-$	−1.040
$CO_2 + 2H_2O + 2e^- \rightarrow CO + 2OH^-$	−1.347	$2CO_2 + 12H^+ + 12e^- \rightarrow C_2H_4 + 4H_2O$	−0.340
$CO_2 + 2H^+ + 2e^- \rightarrow HCOOH$	−0.610	$2CO_2 + 8H_2O + 12e^- \rightarrow C_2H_4 + 12OH^-$	−1.177
$CO_2 + H_2O + 2e^- \rightarrow HCOO^- + OH^-$	−1.491	$2CO_2 + 14H^+ + 14e^- \rightarrow C_2H_6 + 4H_2O$	−0.270
$CO_2 + 4H^+ + 4e^- \rightarrow HCHO + H_2O$	−0.480	$2CO_2 + 2H^+ + 2e^- \rightarrow H_2C_2O_4$	−0.913
$CO_2 + 3H_2O + 4e^- \rightarrow HCHO + 4OH^-$	−1.311	$2CO_2 + 2e^- \rightarrow C_2O_4^{2-}$	−1.003
$CO_2 + 6H^+ + 6e^- \rightarrow CH_3OH + H_2O$	−0.380	$2CO_2 + 12H^+ + 12e^- \rightarrow C_2H_5OH + 3H_2O$	−0.330
$CO_2 + 5H_2O + 6e^- \rightarrow CH_3OH + 6OH^-$	−1.225	$2CO_2 + 9H_2O + 12e^- \rightarrow C_2H_5OH + 12OH^-$	−1.157
$CO_2 + 8H^+ + 8e^- \rightarrow CH_4 + 2H_2O$	−0.240	$3CO_2 + 18H^+ + 18e^- \rightarrow C_3H_7OH + 5H_2O$	−0.320
$CO_2 + 6H_2O + 8e^- \rightarrow CH_4 + 8OH^-$	−1.072	$2H^+ + 2e^- \rightarrow H_2$	−0.420

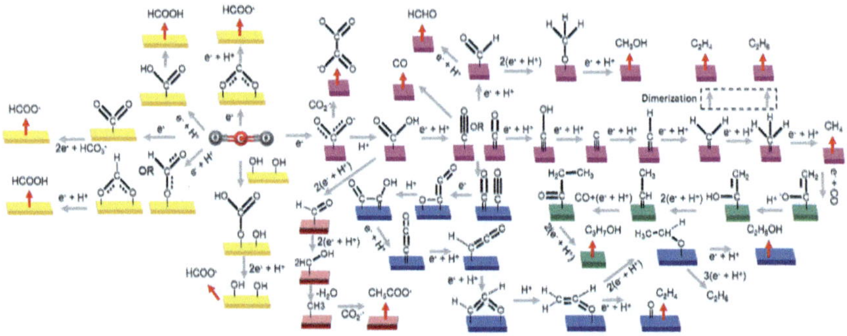

Figure 7.1 Mechanistic pathway for the CO_2 reduction to different products.

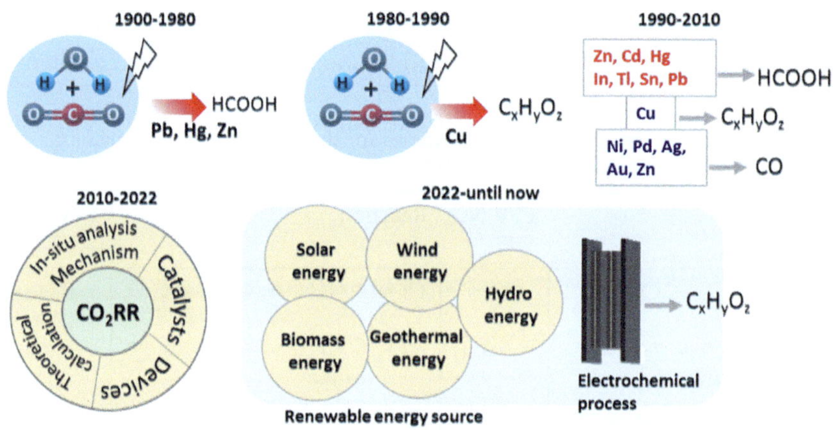

Figure 7.2 The time course for the evolution of CO_2 reduction research.

composition, and their respective electronic states have tremendous influence on the catalytic turnover frequency (TOF) and product selectivity. For example, during late 1900, the CO_2RR was mostly explored with metal catalysts such as Zn, Pb, Hg, Cu amalgam, which can selectively produce formate [12–14]. Following which, during 1980s, the CO_2RR using Ru, Mo, and few semiconductors including their heterojunctions, such as n/p-GaAs, n/p-Si, etc., received popularity for producing methanol [15,16]. Later, the study on the Cu-based materials has been emerged as researchers found extraordinary capability of tuning multicarbon product formation [17]. With further advancement of several spectroscopic tools, a more fundamental the studied the structure-activity properties, product formation pathway, reconstruction of catalysts using cutting edge spectroscopic techniques.

7.2 Device assembly and operation parameter

7.2.1 Electrolyzer components

In the research landscape of a conventional CO_2RR electrolysis, two electrodes (working and counter electrodes) setup are used, which are separated using a suitable membrane. The ion transportation occurs through a solid or liquid electrolyte. The supplied CO_2 in the cathode is being reduced under dissolved form (e.g to CO) and OH^- migrates [in case of anion exchange membrane (AEM)] into the anode to produce O_2. In the electrochemical cell, the catalytic performance and product selectivity largely determined by the cell configuration, electrode architecture, type of electrolyte, etc. This section highlights the technical aspect of the electrolyzer including other factors, such as temperature, pressure, and pH, on the CO_2RR performance.

The liquid electrolyte electrolyzer can be subdivided into two classes, namely, H-type and flow cell reactor, in which an ion exchange membrane (IEM) is used between cathode and anode compartment. IEM are categorized into three types, namely, (1) proton exchange membrane (PEM) (2) AEM, and (3) bipolar membrane (BPM). PEM (e.g Nafion) usually used where water oxidation is coupled with CO_2RR. The H^+ resulting from anode reaction transported using electrolyte, diffused through PEM into the cathode compartment [18]. Interestingly, the mechanical stability and ion-exchange properties of the PEM can govern the efficiency and stability of the electrolysis. The detrimental aspect of the PEM reactor is that H^+ migration from the anolyte could reduce the pH of the catholyte that partially favours HER over CO_2RR, thus limiting the Faradaic efficiency (FE) of the CO_2RR. As a result, PEM can mostly be found in the three-compartment setup specifically designed for the production of liquid products [19]. On the other hand, AEM (e.g Sustain ion) controls the flow of anion from the cathode to anode compartment [20–22]. Since the major charge carriers for the AEM electrolyzer are carbonate and bicarbonate ions, the mass transportation in AEM electrolyzer is much higher than that of PEM. However, both in the PEM and AEM electrolyzer, the overall cell efficiency decreases due to product (or ion) crossover. In particular, in AEM cell, the dissolved CO_2 reacts with OH- to form carbonate and bicarbonate species in the cathode compartment, which replace OH- ion in the membrane, resulting in a lowering of mass transportation. The rate of formation of carbonate species depends on the concentration of OH^-, mass transportation, CO_2 concentration in the cathode [23]. An alternative to reduce the crossover phenomena is the implementation of BPM, which is a hybrid of PEM and AEM. For example, PEM polymers with sulfonic or carboxylic groups and AEM polymers with amine and imidazole moiety are often used to build up a BPM. In a BPM reactor, carbonate ions cannot participate in the ion-exchange process on the membrane. Instead, OH^- and H^+ are transported through AEM and PEM respectively, which leads to inhabitation of product crossover. H^+ after originating from water dissociation, in the BPM connection layer and is transported through a PEM in the direction of the cathode and through an AEM to the anode, respectively [24].

The H-type reactor features a double-cell configuration (Fig. 7.3A), in which the cathode and anode are separated by PEM. The product formation

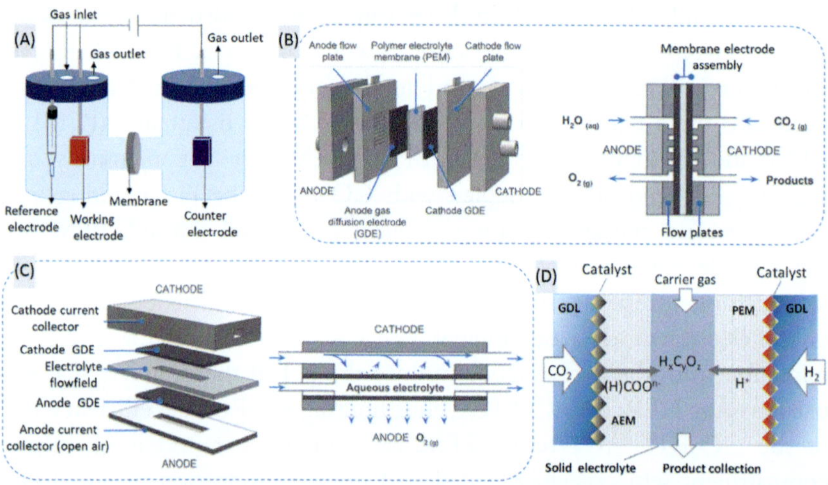

Figure 7.3 Design of several CO_2RR electrolyzers: (A) conventional double-cell H-type electrolyzer with suitable membrane (PEM or AEM); (B and C) three-dimensional view (left) and cross-sectional (right) diagrams of two common electrochemical CO_2RR flow reactors. Panel A shows a membrane-based reactor in which a membrane electrode assembly (MEA) is fixed with the anode and cathode gas diffusion electrodes (GDEs) on each side of a polymer electrolyte membrane (PEM). The MEA is assembled in between the anode and cathode current collectors and flow field plates. MEA separates the respective cathodic and anodic reactions; Panel B displays a microfluidic reactor in which an electrolyte flow channel is constructed between the anode and cathode GDE materials. The supplied $CO_{2(g)}$ in cathode side diffuses through the cathode gas diffusion layer (GDL) to the electrocatalyst, whereas OER occurs at the anode and the gas is released to air. (D) Electrolyzer having solid state electrolyte (SSE) (two types: solid oxide electrolyte (SOE) and solid polymer electrolyte (SPE)). *(B and C) Reproduced with permission from Ref. [20]. Copyright 2018, American Chemical Society.*

in the cathode is usually detected using chromatography (with or without coupled mass) technique. Even though H-type cell is widely used in academia, the long distance between the electrode WE limits mass transportation and thereby inhibit the high FE. This largely discourages the large-scale CO_2RR in the H-type cell [25]. In this regard, the mass transportation difficulties can be greatly minimized using a flow cell reactor, which is subcategorized into two classes, namely, (1) membrane-based flow cell and (2) without membrane microfluidic flow cell [20]. In both cases, the CO_2 is supplied to the cathode using either externally pumped CO_2-saturated electrolyte insertion or direct purging of CO_2 gas using a gas diffusion electrode (GDE). A typical laminar structured membrane-based flow cell and sandwich-structured microfluidic cell can be displayed as Fig. 7.3B,C [20]. Important to note that the CO_2RR performance in a flow cell can be optimized by controlling the (1) electrolyte flow rate, (2) system pressure distribution, (3) chamber/channel size, and (4) electrode structure. A more advancement in the flow technique has been realized with the discovery of zero-gap CO_2 electrolyzer. This is mainly assembled with porous anode and cathode GDE, which are in direct contact with the opposite sides of an IEM [26]. Due to narrow gap between the cathode and anode, the mass transportation is higher as compared to other flow electrolyzer. However, the major problem of these reactors are (1) carbonate salt precipitation on the membrane and (2) water flooding issue, which reduce the longevity of the electrolyzer. Later, with more advancement in the cell designing technique, researchers used a custom-build flow cell reactor with modified IEM architecture based on the oxidation reaction (e.g., water, ethanol) in the anode [27,28]. The earlier designed electrolyzers can provide useful guidance for further optimizing devices for CO_2RR studies. In the conventional liquid electrolyte electrolyzer if the liquid product form, they exist as formate, which is most often mixed with impurity ions from solutes (K^+, HCO_3^-, etc.). This mixture requires appropriate separation steps to obtain pure liquid product [29]. These steps are energy intensive and expensive, which restricts its application for a large scale. This problem can be addressed using solid-state electrolyte (SSE), which offers high ionic conductance between electrodes and easy separation of the pure liquid products. There are two types of SSE, namely, solid oxide electrolyte (SOE) and solid polymer electrolyte (SPE). In the SPE electrolyzer, catalysts are deposited on both sides of the polymer membrane, whereas catalysts are separately assembled with the SOE layer in the corresponding reactor (Fig. 7.3D) [30]. The polymer membrane usually holds good mechanical stability, avoiding electrolyte erosion and leakage; however, the CO_2 conversion efficiency is still to reach to that of SOE-based electrolyzer. The major concern for any SOE-based electrolyzer is that it cannot perform with long-term testing conditions under high temperature, which is still under intensive investigation.

7.2.2 Working electrode assembly

The WE in the cathode has a tremendous effect on the catalytic performance of the CO_2RR. In particular, WE largely control the (1) onset potential, (2) product selectivity and conversion efficiency, and (3) electrochemical stability through controlling the adsorption-desorption of the CO_2 and its intermediates, mass transportation, and local pH of the medium [31]. Based on the size, shape, dimension, and morphology, the WEs are subcategorized into several forms, such as (1) 2D planner electrodes (mostly metal electrodes), (2) mildly roughened planner electrodes, which are mostly derived from surface modification of the smooth planner surface, (3) nanostructured materials grown (e.g., using electrodeposition, plasma treatment, chemical etching, and bottom-up approach, etc.) on conductive support, and (4) externally deposition of nanostructured materials (such as oxide, alloys, etc.) using physical method such as (spin coating, electrophoretic deposition, doctors blade method, etc.) (Fig. 7.4). In case of externally deposited materials, the chemical binder plays a crucial role to provide mechanical stability and homogeneous conductivity throughout the catalyst layer. In several cases, a gas diffusion layer is introduced to facilitate CO_2 diffusion and interact strongly with the catalyst. In the case of a porous 3D nanostructured surface, the electrochemically active surface area is more as compared to the smooth planner surface. As a result, a material with a rougher surface allows mass transportation and surface utilization for the adsorption-desorption into the bulk of the material, resulting in higher conversion efficiency. Moreover, a porous rough surface-based catalyst retard the HER more than the planner electrode,

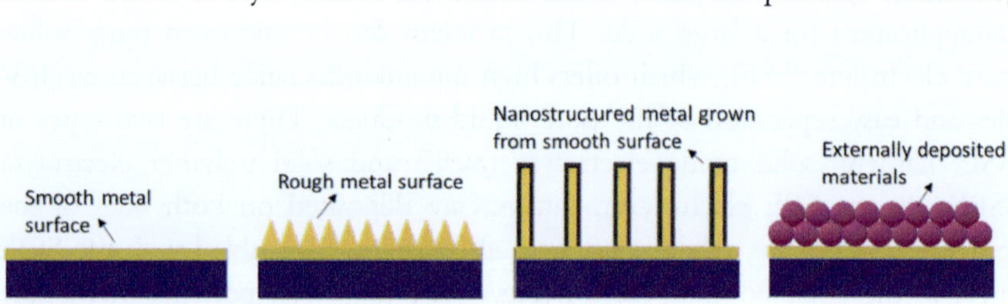

Figure 7.4 Various types of working electrode architecture for the CO_2RR.

which makes nanostructured electrodes more efficient than the planner and smoothed surface electrode.

7.2.3 Electrolyte parameters

7.2.3.1 Nature of electrolytes

The Electrolyte is responsible for the conduction of ions (and/or mass) onto respective electrodes, which greatly influence the catalytic efficiency depending on the type of cation and anion it transport, and the pH of the medium. Three types of the liquid electrolyte can be used, namely, (1) nonaqueous (e.g., polar protic solvent like dimethyl sulfoxide, *N,N*-dimethyl formamide, methanol, acetonitrile, etc.), (2) aqueous electrolyte (e.g., carbonate, KOH, bicarbonate, etc.), and (3) ionic liquids (e.g., 1-ethyl-3-methylimidazolium tetra-fluoroborate, EMIM-BF_4) [32]. Organic electrolytes usually allow better solubility of CO_2 and liquid products. In particular, methanol is a prudent choice because of low cost, low toxicity, and high conversion efficiency of CO_2 than water. On the other hand, ionic liquids are also often used as CO_2RR electrolyte, which resembles the properties of molten salts. The high ionic conductivity and wide operating electrochemical potential make the ionic liquid beneficial as compared to conventional organic solvents [33]. In several studies, a little amount of water was deliberately mixed with ionic liquid, because water can facilitate the H^+ transfer using a hydrogen bond network thereby improving mass transportation [34]. Among the aqueous electrolytes mostly bicarbonates are used in CO_2RR. The bicarbonate plays a role as (1) a proton donating agent, (2) controlling the bulk pH of the solution, (3) it helps to persist high local CO_2 concentration that involves the equilibrium between dissolved CO_2 and bicarbonate, (4) stabilizing the intermediate such as *CHO, *COOH, and bidentate COO^- [35]. Moreover, it was observed that the addition of halide ions also determines the catalytic rate and product selectivity [36]. For example, with Cu electrode, the CO_2RR rate increased with the addition of Cl^-, Br^-, I^-. However, CO is selectively generated with Cl^- and Br^-, whereas with I^- CH_4 results [37]. Similarly, the influence of the anionic species cation size present in the electrolyte also causes varied CO_2RR performance of the CO_2RR [35]. For example, with Ag and Sn cathode, with the increase in the size of alkali metal cations the CO_2RR exchange current density increase; however, it does not affect the HER. On the other hand, this is not true for Cu electrodes, where a mixture of gaseous products (H_2, CO, and CH_4) forms.

7.2.3.2 Effect of pH

pH of the electrolyte solution significantly controls the equilibrium potential of CO_2RR and determines the CO_2 conversion efficiency and product selectivity. The favorable product formation can be demonstrated using the Pourbiax diagram defined with solution pH and equilibrium potential (Fig. 7.5) [38]. The product formation is also very sensitive toward localized pH near the outer Helmholtz plane, as high local pH favors the formation of CO and thereby multicarbon product over HER and CH_4 [39,40]. The effect of local pH and buffer capacity is more meaningful in a nanostructured rough cathode surface as compared to the smooth metal surface. The local pH near a rough surface is relatively higher, thus facilitating the mass transportation and formation of multicarbon hydrocarbons.

7.2.3.3 Effect of temperature and pressure

The reaction medium temperature controls (1) the rate of CO_2RR and (2) product selectivity [41]. With a decrease in temperature, the HER rate decreases and CO_2RR increases. Additionally, few studies show that a low temperature favors CO formation over other gaseous products like CH_4 and C_2H_4. On the other hand, expectedly, the partial pressure of CO_2 also regulates the catalytic efficiency and selectivity by modeling the CO_2 solubility [42]. For instance, a study shows on group VIII metal electrodes that with CO_2 pressure increment up to a suitable level can dramatically suppress the HER and increase the product selectivity [43].

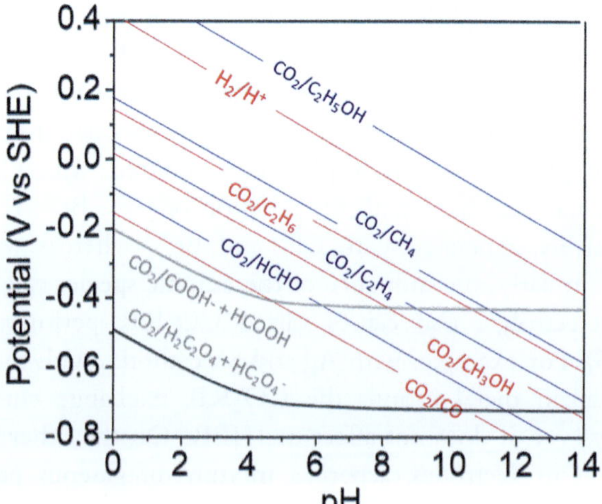

Figure 7.5 Pourbaix diagram of CO_2 and its possible reduced products.

7.3 Development of single-atom catalysts for CO_2 reduction into value-added products

Various classes of transition metal-based-catalyst, such as sulfides, phosphides, oxides, intermetallic compounds, alloys, etc., have been developed for CO_2 reduction [17,44]. SACs are one of them in which metal atoms are coordinated through electron donating sites to form a composite structure with high interatomic distances between metal atoms. Since the development of Pt_1/FeO_x SAC in 2011 [45], extensive research has been carried out on SACs for CO_2 electro-reduction [46]. In contrast to conventional catalytic systems, a single atomic site in SACs is located near the surface of the catalyst, thus achieving maximum atom utilization, long-range interactions, and the near-range coordination environment, which impacts the catalytic TOF, FE as well as selectivity for the CO_2RR. SACs are also excellent models for understanding the structure–activity relationship [46]. The mechanism and RDS of the CO_2RR on SACs can vary depending on the catalytic component of SACs, reaction conditions, and applied potential [47]. Herein, we discuss the CO_2RR efficiency, product selectivity, catalysts' own properties including the nature of the metal atom, coordination environments, etc., for the SACs.

7.3.1 HER suppression in the SACs

HER is a competitive reaction to CO_2RR that occurs within the same potential window. HER usually proceeds through a two-step mechanism either with (1) Volmer–Heyrovsky or (2) Volmer–Tafel pathway (Eqs. 7.3–7.5) [48,49].

$$H^+ + e^- \rightarrow H^*(\text{Volmer step}) \tag{7.3}$$

$$H^* + H^+ + e^- \rightarrow H_2(\text{Heyrovsky step}) \tag{7.4}$$

$$H^* + H^* \rightarrow H_2(\text{Tafel step}) \tag{7.5}$$

Both the Volmer and Heyrovsky step occurs on a mono-atomic site, whereas the Tafel step requires two adjacent metal sites to hold the H^* species, which can couple to form H_2. Due to large interatomic distances, Tafel step is energetically unfavorable for SACs, and thus could follow the Volmer–Heyrovsky step. According to previous studies on the HER mechanism, Heyrovsky step often has an activation energy twice as high as compared to Tafel step, which causes the HER to be restricted in the SACs. On the other hand, Heyrovsky mechanism is more plausible for the CO_2RR, which becomes more preferred than the HER for SACs (Eqs. 7.6–7.8).

$$CO_2 + H^+ + e^- \rightarrow COOH^* \quad (7.6)$$

$$COOH^* + H^* \rightarrow CO + H_2O \text{ (Tafel step)} \quad (7.7)$$

$$COOH^* + H^* + e^- \rightarrow CO + H_2O \text{ (Heyrovsky step)} \quad (7.8)$$

Interestingly, in several cases, the low solubility of CO_2 (34 mM) in aqueous electrolytes hinders the efficiency and selectivity of the CO_2RR [50]. At high overpotentials, the reaction rate for the CO_2RR becomes independent of the dissolved CO_2 molecules, thus leading to increased competition with the HER. In this regard, in order to adjust the adsorption energy of H^*, Liu and coworkers added Cu to the Ni-SAC and obtained CO with a 97% Faradic Efficiency (FE_{CO}). The theoretical calculation revealed that after the introduction of Cu, the d-band center was upshifted from −1.520 to −1.438 eV, which resulted in the increment of H^* adsorption energy, and consequently the HER was suppressed [51].

7.3.2 SACs for CO_2 to CO

Depending on the SACs, generally, CO_2RR to CO follows two different types of mechanisms based on the *COOH species formation. Few catalysts such as Fe SACs supported on nitrogen-doped graphene allows the direct formation of *COOH and its transformation into *CO via two-step coupled proton-electron transfer (CPET) (Eqs. 7.9 and 7.12). This requires a moderate operating potential [52]. On the other hand, Fe SAC on N-doped carbon support prefers CO production via consecutive CO_2^- and *COOH formation (Eqs. 7.10 and 7.11) [53].

$$CO_2 + {}^* + H^+ + e^- \rightarrow {}^*COOH \quad (7.9)$$

$$CO_2 + {}^* + e^- \rightarrow {}^*CO_2^- \quad (7.10)$$

$$^*CO_2^- + H^{+\,-} \rightarrow {}^*COOH \quad (7.11)$$

$$^*COOH + H^* + e^- \rightarrow {}^*CO + H_2O \quad (7.12)$$

$$^*CO \rightarrow CO + {}^* \quad (7.13)$$

Recently, theoretical calculations were conducted to understand the CO_2RR on M-N-C (porphyrin)-containing SACs containing 3d transition metals [54]. Interestingly, a volcano relationship between ΔG_{*CO}

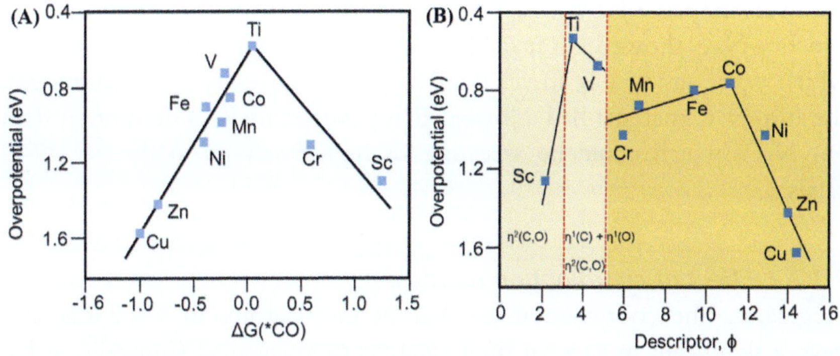

Figure 7.6 The volcano plots of electrocatalytic CO_2 reduction activities vs. descriptors and the reaction mechanism on transition metal containing single atom catalysts (TM-SACs). Overpotential for CO formation as a function of (A) bonding energy and (B) the intrinsic descriptor.

(binding energy) and η^{CO} (overpotential) was found (Fig. 7.6A), which helps to define a new descriptor (*F*) as follows (Fig. 7.6B).

$$F = VM \times E_M / r_M$$

VM, valence electron number of metals; E_M, the electronegativity of central metals; r_M, radius of metal ions.

Two volcano regions were found where the left region ($F < 3.0$) favors HER and the right ($F > 5.0$) prefers the CO_2RR. This description can be used to design M-N-C catalysts for selective CO_2RR in future. Moreover, it has been proposed that the specific charge on the metal centers might affect the activity of SACs and guide them to take alternative reduction pathways, which can be manipulated through tuning of choice of metal and coordinating atoms in the SACs as described below.

7.3.2.1 Modulation of the metal atomic center

The catalytic performance largely depends on d-band energies. The higher the d-band center of a metal atom, the stronger the binding between metal and the reaction intermediates. In this viewpoint, Jiang et al. prepared various metal (Fe, Co, Cu, and Ni) containing SACs having fixed support (N-doped carbon) (named as M_1-N-C where M indicate metal centers) through pyrolyzing metal-organic-framework (MOF) [55]. They showed that Ni_1-N-C exhibited the best activity and selectivity (TOF value 11315 h^{-1}) for the production of CO followed by Fe_1-N-C, Co_1-N-C, and Cu_1-N-C. Theoretical calculations showed that all SACs followed the same RDS which was the formation of *COOH. Ni_1-N-C and Fe_1-N-C showed comparable energy barriers for *COOH formation which was considerably lower than Co_1-N-C and Cu_1-N-C. Moreover, the Ni_1-N-C material had a lower energy barrier for CO desorption than Fe_1-N-C, which could be a reason for higher activity in the Ni_1-N-C catalyst.

7.3.2.2 Effect of coordination number

Regulating the coordination number of metal atoms in SACs can also have a significant impact on their catalytic performance. Generally, it has been observed that a lower coordination number of a metal atom facilitate stronger adsorption of the reaction intermediates. In this regard, Li and co-workers constructed Co SACs having different coordination numbers (e.g., Co-N_2, Co-N_3, and Co-N_4) through the pyrolysis of MOF precursors at various temperatures. Among all, Co-N_2 showed the highest catalytic activity for CO production [56]. Theoretical calculations explained that endergonic $^*CO_2^-$ formation was the RDS for all the Co SACs and the highest activity of Co-N_2 could be due to a higher d-band center (−0.81 eV) that could help to lower the energy barrier for $^*CO_2^-$ formation. A similar conclusion was found for Ni SACs (NiN_x; $x = 1 - 4$) reported by Li et al. [57] and Mn SACs by Zhang et al. [58].

7.3.2.3 Effect of the coordinated nitrogen type

The type of coordinating nitrogen atoms in SACs and their interactions with metal centers can greatly influence the catalytic activity. Pyridinic N and pyrrolic N are two types of nitrogen atoms that are mostly found in the carbon lattice of SACs. Pyridinic N denotes the nitrogen atoms that are part of a six-membered aromatic ring, while pyrrolic N refers to nitrogen atoms that are part of a five-membered ring. Intrinsically, a metal atom coordinated with pyrrolic-N has a lower d-band center than a metal atom coordinated with pyridinic-N. Hence, pyridinic-N coordinated metal sites are more likely to bind strongly to CO_2RR intermediates. For instance, Wang and coworkers observed that SAC with pyridinic-N coordinated Co^{II} center displayed higher activity as compared and Co-porphine where pyrrolic N coordinated with Co^{II} is present. DFT simulations showed COOH* formation is the RDS that was controlled by the d-band center. Using Bader charge analysis, they have also demonstrated that the pyrrolic N coordinated Co^{II} sites were able to receive less electronic charge from the ligand as compared to pyridinic-N coordinated Co^{II}, which directly affects the catalytic activity [59].

7.3.2.4 Effect of other heteroatoms

The doping of heteroatoms (e.g., nitrogen, sulfur, phosphorus, etc.) into the SACs composite can lead to the formation of new electronic states such as localized orbitals or impurity states in the catalyst materials. This can alter the metal *d*-band center, facilitate electron transfer, and thus improve the catalytic performance. Moreover, the presence of heteroatoms can introduce additional active sites for catalytic reactions and modulate the adsorption properties of the material, which are crucial for the CO_2RR. For example, Liu et al. observed a lowering of overpotential and high catalytic activity for sulfur doped Ni SACs ($Ni\text{-}N_3S$) as compared to undoped analog ($Ni\text{-}N_4$) [60]. Similarly, Zhang et al. showed that chlorine (Cl) doping in Mn SAC greatly improves the activity for selective CO formation [61]. According to a theoretical study, the Cl doping resulted in weak *CO adsorption, which could be the reason for improved selectivity.

7.3.2.5 Effect of dual metal atoms

The construction of dual metal atomic sites is a promising approach to improve catalytic activity and selectivity. Dual metal atomic sites refer to active sites composed of two metal atoms that are closely spaced, typically within 0.5 nm. This close proximity allows electronic interactions between two metal atoms, leading to enhanced catalytic activity and selectivity. For example, Gong et al. constructed Zn-Co-based SAC that showed enhanced performance as compared to their monometallic analogs for CO production [62]. The X-ray absorption spectroscopy confirmed the electronic interactions between Co and Zn atoms. Theoretical calculation showed that the introduction of Co atom in Zn SAC leads to the RDS step (formation of *COOH) becoming exergonic from endergonic that facilitates the adsorption of *COOH, thereby promoting CO production.

7.3.2.6 Effect of surrounding functional groups

Through the introduction of various functional groups (generally electron-withdrawing or electron-donating) into the SACs, it is possible to tune the partial charge on the metal atoms and consequently, CO_2RR performance could be manipulated. For instance, in a comparative study Huang et al. observed the electron-withdrawing nitro functionalized Co SAC (nitro-CoPc/G) exhibited higher CO production activity and selectivity as compared to electron-donating amino functionalized Co SAC (amino-CoPc/G) [63]. According to the theoretical study, nitro-CoPc/G has a lower energy barrier for $^{*}CO_2^{-}$ formation which was the RDS step. Also, it was concluded that because of the electron-withdrawing group the Co center was partially positively charged, which resulted in upshifting of the d_z^2 orbital. Hence, a large energy difference between d_z^2 and d_{xy} orbital occurs, which alters the energy barrier for $^{*}CO_2^{-}$ formation.

7.3.3 SACs for CO_2 to formic acid/formate

The conversion of CO_2 to formic acid has gained attention in recent years due to its potential applications as a feedstock chemical, hydrogen source, and hydrogen carrier. Formic acid can be used directly in various industrial processes such as leather tanning, textile dyeing, and the production of animal feed additives, and it can also be used as a fuel for fuel cells. It can also be converted to methanol, which is an important feedstock to produce chemicals such as formaldehyde, acetic acid, and olefins. Additionally, formic acid has the highest profit per mole of electrons among the products of CO_2 reduction. The electrocatalytic CO_2 reduction to formic acid follows two mechanistic pathways (Eqs. 7.14–7.17).

$$CO_2 + \left(H^+ + e^-\right) + {}^* \rightarrow {}^*COOH \tag{7.14}$$

$$CO_2 + \left(H^+ + e^-\right) + {}^* \rightarrow {}^*HCOO \tag{7.15}$$

$$^*COOH + (^*H + e^-) \rightarrow HCOOH + {}^* \tag{7.16}$$

$$^*HCOO + (^*H + e^-) \rightarrow HCOOH + {}^* \tag{7.17}$$

Au SAC on B_{36} is a potential catalyst for formic acid production because of strong selectivity towards *HCOO as opposed to *COOH. Here, the electron can easily transfer from Au surface to *HCOO as HCOO* is a far better electron withdrawing group than *COOH [64].

7.3.4 SACs for CO_2 to CH_3OH

Methanol is a promising building block to produce various chemicals and fuels, and its conversion into higher-value products through established techniques such as MTO and MTA can promote its economic value.

In general, the following two mechanisms were established for methanol formation [Eqs. (7.18) and (7.19)].

$$CO_2 \rightarrow {}^*COOH \rightarrow {}^*CO \rightarrow {}^*COH \rightarrow {}^*CHOH \rightarrow {}^*CH_2OH \rightarrow CH_3OH \quad (7.18)$$

$$CO_2 \rightarrow {}^*CO_2 \rightarrow {}^*HCOO \rightarrow {}^*HCOOH \rightarrow {}^*CHO \rightarrow {}^*CH_2OH \rightarrow CH_3OH \quad (7.19)$$

Cu SAC on carbon nanofibers (CuSAs/TCNFs) showed good activity for CO_2 electroreduction to methanol (44% FE) because of the high binding energy for *CO of Cu [65]. Cu SAC on MXene produces more methanol (59.1% FE) [66]. In contrast to CuSAs/TCNFs, CO_2 is more suited to continue along the *HCOO pathway, as it is thermodynamically downhill.

7.3.5 SACs for CO_2 to CH_4

Electrochemical reduction of CO_2 to CH_4 proceeds through an $8e^-$ transfer process which is associated with two different types of mechanism. In one way, *COOH formation takes place at the beginning and undergoes a stepwise transformation into CH_4 via *CO and *COH formation. In this mechanism, for the selective production of CH_4, *COH should carry high adsorption energy. For example, a carbon-dot-based SAC having CuN_2O_2 units showed exceptional activity for the selective production of CH_4, believed to follow the above-mentioned mechanism (Fig. 7.7A) [67]. On the other hand, an alternative pathway follows the formation of *HCOO and thereby transforms into CH_4 via *OCH_3 species. The formation of CH_4 using Zn SAC on N-doped carbon follows this mechanism. Theoretical studies indicate that oxygen atom tends to form a bond with Zn, which results in lower free energy *HCOO as compared to *COOH. In this case, CO production is also not feasible as it requires the formation of *COOH (Fig. 7.7B) [68].

7.3.6 SACs for CO_2 to C_{2+} products

There is a consensus view that SACs catalysts cannot produce C_{2+} products as it requires the coupling of two intermediate species such as *CO, *CH_x, *CHO, *COH. etc. This is energetically challenging because in SACs the metal atoms do not exist in close proximity. Despite this fact, several multicarbon products are reported to date [47].

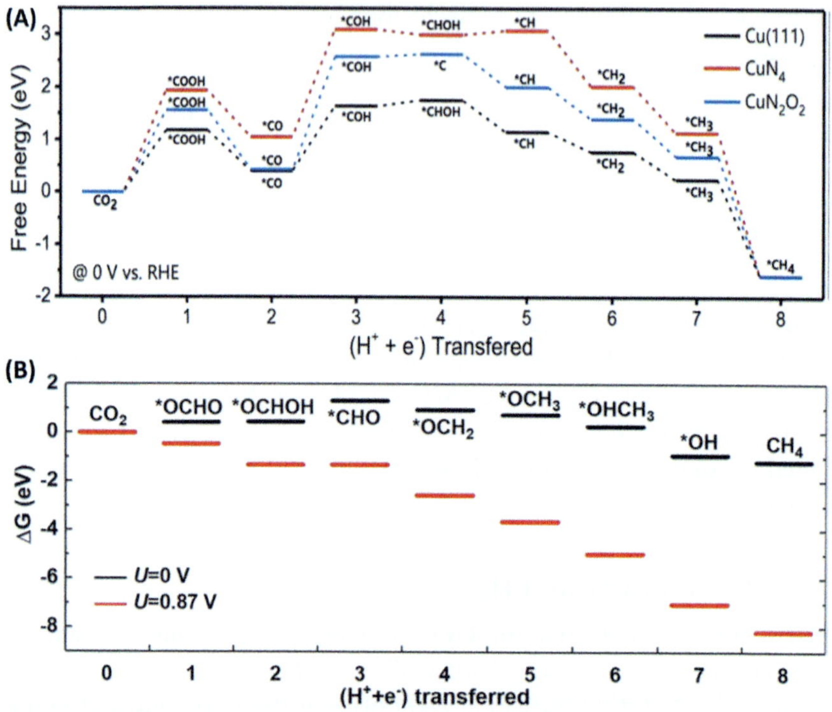

Figure 7.7 DFT simulation of electrocatalytic CO_2 reduction to CH_4 formation on various types of catalysts. (A) Free energy diagram over carbon dot–based SAC margined with unique CuN_2O_2 sites. (B) Free energy diagram over Zn-N_4-graphene. *(A) Reproduced with permission from Ref. [67]. Copyright 2021, Nature. (B) Reproduced with permission from Ref. [68]. Copyright 2020, American Chemical Society.*

Fontecave and coworkers obtained ethanol with a FE of 43% using N-doped carbon-supported Cu SAC [69]. Interestingly, a higher efficiency was achieved when CO was fed. This result indicates that CO* might be an important intermediate for ethanol formation. Insight study through operando XAS revealed that Cu SAC became aggregated to form a Cu cluster, which facilitates the coupling of C-containing intermediates. Interestingly, the in situ aggregation of Cu atoms is reversible as in the post-reaction catalyst only single atoms were observed. In other studies, acetone and acetic acid production are reported from CO_2 with N-doped porous carbon supported Cu SAC and Fe SAC respectively, which is speculated to reduce the energy barrier for C-C coupling [70]. Moreover, the gaseous multicarbon product is also reported using SAC. For example, Zheng et al. reported the production of ethylene using N-doped carbon-supported Cu

SAC, in which coordination sites were tuned to obtain the optimized activity. They claimed that the catalyst with more Cu-N_2 sites showed more activity by lowering the C−C coupling energy barrier as compared to that of Cu-N_4 [71].

7.4 Conclusion

In summary, this chapter covers the fundamental aspects of the electrochemical CO_2RR such as reduction pathway, device assembly including various electrolyzer components, WE features, nature of electrolytes, and effect of other physical parameters, such as pH, temperature, and pressure, on the catalytic activity. This chapter also overviews the recent progress in the development of SACs, their activity, and product selectivity for electrochemical CO_2RR. The catalytic activity and selectivity are largely interconnected with the HER suppression, the nature of the active sites (e.g., metal site), and its local electronic environment (coordination type, metal aggregation, etc.). Although significant progress has been done in this field, there are still many obstacles to overcome. For instance, the production of C_{2+} products is very difficult with SACs, and in this view, the catalyst with di/tri atomic metal site might be helpful. Moreover, the mechanistic study for the CO_2RR on SACs-based electrodes including the catalyst reconstruction process is not well explored. Even though most of the reports supported the activity and product selectivity with simple theoretical calculations, the consideration of the real reaction condition is not implemented. In this regard, time-resolved in situ/*operando* measurements (through vibrational and X-ray spectroscopy) could be helpful and a prudent choice to build a structure−performance relationship.

References

[1] Intergovernmental Panel on Climate Change, Climate Change 2022: Synthesis Report, 2023.

[2] Intergovernmental Panel on Climate Change, IPCC Special Report on the Impacts of Global Warming of 1.5°C—Summary for Policy Makers, Incheon, 2018.

[3] A.M. Appel, J.E. Bercaw, A.B. Bocarsly, H. Dobbek, D.L. DuBois, M. Dupuis, et al., Frontiers, opportunities, and challenges in biochemical and chemical catalysis of CO_2 fixation, Chem. Rev. 113 (2013) 6621−6658.

[4] S. Rönsch, J. Schneider, S. Matthischke, M. Schlüter, M. Götz, J. Lefebvre, et al., Review on methanation—from fundamentals to current projects, Fuel 166 (2016) 276−296.

[5] J. Yang, W. Ma, D. Chen, A. Holmen, B.H. Davis, Fischer−Tropsch synthesis: a review of the effect of CO conversion on methane selectivity, Appl. Catal. A Gen. 470 (2014) 250−260.

[6] S.G. Jadhav, P.D. Vaidya, B.M. Bhanage, J.B. Joshi, Catalytic carbon dioxide hydrogenation to methanol: a review of recent studies, Chem. Eng. Res. Des. 92 (2014) 2557–2567.
[7] E.V. Kondratenko, G. Mul, J. Baltrusaitis, G.O. Larrazábal, J. Pérez-Ramírez, Status and perspectives of CO_2 conversion into fuels and chemicals by catalytic, photocatalytic and electrocatalytic processes, Energy Environ. Sci. 6 (2013) 3112.
[8] M. Götz, J. Lefebvre, F. Mörs, A. McDaniel Koch, F. Graf, S. Bajohr, et al., Renewable power-to-gas: a technological and economic review, Renew. Energy 85 (2016) 1371–1390.
[9] C. Graves, S.D. Ebbesen, M. Mogensen, K.S. Lackner, Sustainable hydrocarbon fuels by recycling CO_2 and H_2O with renewable or nuclear energy, Renew. Sustain. Energy Rev. 15 (2011) 1–23.
[10] G. Wang, J. Chen, Y. Ding, P. Cai, L. Yi, Y. Li, et al., Electrocatalysis for CO_2 conversion: from fundamentals to value-added products, Chem. Soc. Rev. 50 (2021) 4993–5061.
[11] H. Li, K. Jiang, S.-Z. Zou, W.-B. Cai, Fundamental aspects in CO_2 electroreduction reaction and solutions from in situ vibrational spectroscopies, Chin. J. Catal. 43 (2022) 2772–2791.
[12] T.E. Teeter, P. Van Rysselberghe, Reduction of carbon dioxide on mercury cathodes, J. Chem. Phys. 22 (4) (1954) 759–760.
[13] W. Paik, T.N. Andersen, H. Eyring, Kinetic studies of the electrolytic reduction of carbon dioxide on the mercury electrode, Electrochim. Acta 14 (1969) 1217–1232.
[14] P.G. Russell, N. Kovac, S. Srinivasan, M. Steinberg, The electrochemical reduction of carbon dioxide, formic acid, and formaldehyde, J. Electrochem. Soc. 124 (1977) 1329–1338.
[15] K.W. Frese, S. Leach, Electrochemical reduction of carbon dioxide to methane, methanol, and CO on Ru electrodes, J. Electrochem. Soc. 132 (1985) 259–260.
[16] D.P. Summers, S. Leach, K.W. Frese, The electrochemical reduction of aqueous carbon dioxide to methanol at molybdenum electrodes with low overpotentials, J. Electroanal. Chem. Interfacial Electrochem. 205 (1986) 219–232.
[17] S. Nitopi, E. Bertheussen, S.B. Scott, X. Liu, A.K. Engstfeld, S. Horch, et al., Progress and perspectives of electrochemical CO_2 reduction on copper in aqueous electrolyte, Chem. Rev. 119 (2019) 7610–7672.
[18] L. Ma, S. Fan, D. Zhen, X. Wu, S. Liu, J. Lin, et al., Electrochemical reduction of CO_2 in proton exchange membrane reactor: the function of buffer layer, Ind. Eng. Chem. Res. 56 (2017) 10242–10250.
[19] J. Wyndorps, H. Ostovari, N. von der Assen, Is Electrochemical CO_2 reduction the future technology for power-to-chemicals? An environmental comparison with H_2-based pathways, Sustain. Energy Fuels 5 (2021) 5748–5761.
[20] D.M. Weekes, D.A. Salvatore, A. Reyes, A. Huang, C.P. Berlinguette, Electrolytic CO_2 reduction in a flow cell, Acc. Chem. Res. 51 (2018) 910–918.
[21] C.M. Gabardo, C.P. O'Brien, J.P. Edwards, C. McCallum, Y. Xu, C.-T. Dinh, et al., Continuous carbon dioxide electroreduction to concentrated multi-carbon products using a membrane electrode assembly, Joule 3 (2019) 2777–2791.
[22] B. Endrődi, E. Kecsenovity, A. Samu, T. Halmágyi, S. Rojas-Carbonell, L. Wang, et al., High carbonate ion conductance of a robust piperion membrane allows industrial current density and conversion in a zero-gap carbon dioxide electrolyzer cell, Energy Environ. Sci. 13 (2020) 4098–4105.
[23] A. Gawel, T. Jaster, D. Siegmund, J. Holzmann, H. Lohmann, E. Klemm, et al., Electrochemical CO_2 reduction—the macroscopic world of electrode design, reactor concepts & economic aspects, iScience 25 (2022) 104011.

[24] D.A. Salvatore, D.M. Weekes, J. He, K.E. Dettelbach, Y.C. Li, T.E. Mallouk, et al., Electrolysis of gaseous CO2 to CO in a flow cell with a bipolar membrane, ACS Energy Lett. 3 (2018) 149–154.
[25] X. Sun, Q. Zhu, X. Kang, H. Liu, Q. Qian, Z. Zhang, et al., Molybdenum-bismuth bimetallic chalcogenide nanosheets for highly efficient electrocatalytic reduction of carbon dioxide to methanol, Angew. Chem. 128 (2016) 6883–6887.
[26] M. Sassenburg, M. Kelly, S. Subramanian, W.A. Smith, T. Burdyny, Zero-gap electrochemical CO2 reduction cells: challenges and operational strategies for prevention of salt precipitation, ACS Energy Lett. 8 (2023) 321–331.
[27] M. Bevilacqua, J. Filippi, A. Lavacchi, A. Marchionni, H.A. Miller, W. Oberhauser, et al., Energy savings in the conversion of CO_2 to fuels using an electrolytic device, Energy Technol. 2 (2014) 522–525.
[28] C. Xia, P. Zhu, Q. Jiang, Y. Pan, W. Liang, E. Stavitski, et al., Continuous production of pure liquid fuel solutions via electrocatalytic CO_2 reduction using solid-electrolyte devices, Nat. Energy 4 (2019) 776–785.
[29] L. Fan, C. Xia, P. Zhu, Y. Lu, H. Wang, Electrochemical CO_2 reduction to high-concentration pure formic acid solutions in an all-solid-state reactor, Nat. Commun. 11 (2020) 3633.
[30] Y. Jiang, F. Chen, C. Xia, A review on cathode processes and materials for electro-reduction of carbon dioxide in solid oxide electrolysis cells, J. Power Sources 493 (2021) 229713.
[31] R. Kas, K. Yang, D. Bohra, R. Kortlever, T. Burdyny, W.A. Smith, Electrochemical CO_2 reduction on nanostructured metal electrodes: fact or defect? Chem. Sci. 11 (2020) 1738–1749.
[32] M. Moura de Salles Pupo, R. Kortlever, Electrolyte effects on the electrochemical reduction of CO_2, ChemPhysChem 20 (2019) 2926–2935.
[33] N.V. Rees, R.G. Compton, Electrochemical CO_2 sequestration in ionic liquids; a perspective, Energy Environ. Sci. 4 (2011) 403–408.
[34] H.-K. Lim, Y. Kwon, H.S. Kim, J. Jeon, Y.-H. Kim, J.-A. Lim, et al., Insight into the microenvironments of the metal–ionic liquid interface during electrochemical CO_2 reduction, ACS Catal. 8 (2018) 2420–2427.
[35] M. König, J. Vaes, E. Klemm, D. Pant, Solvents and supporting electrolytes in the electrocatalytic reduction of CO_2, iScience 19 (2019) 135–160.
[36] J. Liu, Y. Wang, H. Jiang, H. Jiang, X. Zhou, Y. Li, et al., Ag@Au Core-Shell Nanowires for nearly 100% CO_2 -to-CO electroreduction, Chem.—An. Asian J. 15 (2020) 425–431.
[37] A.S. Varela, W. Ju, T. Reier, P. Strasser, Tuning the catalytic activity and selectivity of Cu for CO_2 electroreduction in the presence of halides, ACS Catal. 6 (2016) 2136–2144.
[38] Y. Hori, Electrochemical CO_2 reduction on metal electrodes, in: Modern Aspects of Electrochemistry, Springer New York, New York, pp. 89–189.
[39] A.S. Varela, M. Kroschel, T. Reier, P. Strasser, Controlling the selectivity of CO_2 electroreduction on copper: the effect of the electrolyte concentration and the importance of the local PH, Catal. Today 260 (2016) 8–13.
[40] M. Ma, K. Djanashvili, W.A. Smith, Controllable hydrocarbon formation from the electrochemical reduction of CO_2 over Cu nanowire arrays, Angew. Chem. Int. Ed. 55 (2016) 6680–6684.
[41] T. Mizuno, K. Ohta, A. Sasaki, T. Akai, M. Hirano, A. Kawabe, Effect of temperature on electrochemical reduction of high-pressure CO_2 with In, Sn, and Pb electrodes, Energy Sources 17 (1995) 503–508.

[42] J. Li, Y. Kuang, Y. Meng, X. Tian, W.-H. Hung, X. Zhang, et al., Electroreduction of CO_2 to formate on a copper-based electrocatalyst at high pressures with high energy conversion efficiency, J. Am. Chem. Soc. 142 (2020) 7276–7282.

[43] S. Nakagawa, A. Kudo, M. Azuma, T. Sakata, Effect of pressure on the electrochemical reduction of CO_2 on group VIII metal electrodes, J. Electroanal. Chem. Interfacial Electrochem. 308 (1991) 339–343.

[44] S. Garg, M. Li, A.Z. Weber, L. Ge, L. Li, V. Rudolph, et al., Advances and challenges in electrochemical CO_2 reduction processes: an engineering and design perspective looking beyond new catalyst materials, J. Mater. Chem. A 8 (2020) 1511–1544.

[45] B. Qiao, A. Wang, X. Yang, L.F. Allard, Z. Jiang, Y. Cui, et al., Single-atom catalysis of CO oxidation using Pt1/FeO_x, Nat. Chem. 3 (2011) 634–641.

[46] J. Zhang, W. Cai, F.X. Hu, H. Yang, B. Liu, Recent advances in single atom catalysts for the electrochemical carbon dioxide reduction reaction, Chem. Sci. 12 (2021) 6800–6819.

[47] T.N. Nguyen, M. Salehi, Q.V. Le, A. Seifitokaldani, C.T. Dinh, Fundamentals of electrochemical CO_2 reduction on single-metal-atom catalysts, ACS Catal. 10 (2020) 10068–10095.

[48] P.C.K. Vesborg, B. Seger, I. Chorkendorff, Recent development in hydrogen evolution reaction catalysts and their practical implementation, J. Phys. Chem. Lett. 6 (2015) 951–957.

[49] Y. Zheng, Y. Jiao, M. Jaroniec, S.Z. Qiao, Advancing the electrochemistry of the hydrogen-evolution reaction through combining experiment and theory, Angew. Chem. Int. Ed. 54 (2015) 52–65.

[50] R.M. Enick, S.M. Klara, CO_2 solubility in water and brine under reservoir conditions, Chem. Eng. Commun. 90 (1990) 23–33.

[51] T. Zhang, X. Han, H. Yang, A. Han, E. Hu, Y. Li, et al., Atomically dispersed Nickel(I) on an alloy-encapsulated nitrogen-doped carbon nanotube array for high-performance electrochemical CO_2 reduction reaction, Angew. Chem. Int. Ed. 59 (2020) 12055–12061.

[52] C. Zhang, S. Yang, J. Wu, M. Liu, S. Yazdi, M. Ren, et al., Electrochemical CO_2 reduction with atomic iron-dispersed on nitrogen-doped graphene, Adv. Energy Mater. 8 (2018) 1703487.

[53] J. Gu, C.-S. Hsu, L. Bai, H.M. Chen, X. Hu, Atomically dispersed Fe^{3+} sites catalyze efficient CO_2 electroreduction to CO, Science 364 (2019) 1091–1094.

[54] L. Gong, D. Zhang, C. Lin, Y. Zhu, Y. Shen, J. Zhang, et al., Catalytic mechanisms and design principles for single-atom catalysts in highly efficient CO_2 conversion, Adv. Energy Mater. 9 (2019) 1902625.

[55] L. Jiao, W. Yang, G. Wan, R. Zhang, X. Zheng, H. Zhou, et al., Single-atom electrocatalysts from multivariate metal–organic frameworks for highly selective reduction of CO_2 at low pressures, Angew. Chem. Int. Ed. 59 (2020) 20589–20595.

[56] X. Wang, Z. Chen, X. Zhao, T. Yao, W. Chen, R. You, et al., Regulation of coordination number over single co sites: triggering the efficient electroreduction of CO_2, Angew. Chem. Int. Ed. 57 (2018) 1944–1948.

[57] C. Zhao, Y. Wang, Z. Li, W. Chen, Q. Xu, D. He, et al., Solid-diffusion synthesis of single-atom catalysts directly from bulk metal for efficient CO_2 reduction, Joule 3 (2019) 584–594.

[58] J. Feng, H. Gao, L. Zheng, Z. Chen, S. Zeng, C. Jiang, et al., A Mn-N_3 single-atom catalyst embedded in graphitic carbon nitride for efficient CO_2 electroreduction, Nat. Commun. 11 (2020) 4341.

[59] L. Sun, Z. Huang, V. Reddu, T. Su, A.C. Fisher, X. Wang, A planar, conjugated N 4 -macrocyclic cobalt complex for heterogeneous electrocatalytic CO_2 reduction with high activity, Angew. Chem. Int. Ed. 59 (2020) 17104–17109.

[60] H.B. Yang, S.-F. Hung, S. Liu, K. Yuan, S. Miao, et al., Atomically dispersed Ni(I) as the active site for electrochemical CO_2 reduction, Nat. Energy 3 (2018) 140–147.

[61] B. Zhang, J. Zhang, J. Shi, D. Tan, L. Liu, F. Zhang, et al., Manganese acting as a high-performance heterogeneous electrocatalyst in carbon dioxide reduction, Nat. Commun. 10 (2019) 2980.

[62] W. Zhu, L. Zhang, S. Liu, A. Li, X. Yuan, C. Hu, et al., Enhanced CO_2 electroreduction on neighboring Zn/Co monomers by electronic effect, Angew. Chem. Int. Ed. 59 (2020) 12664–12668.

[63] X. Ren, S. Liu, H. Li, J. Ding, L. Liu, Z. Kuang, et al., Electron-withdrawing functional ligand promotes CO_2 reduction catalysis in single atom catalyst, Sci. China Chem. 63 (2020) 1727–1733.

[64] W.-Y. Li, X. Zhao, J.-S. Dang, A mechanistic study of B 36 -supported atomic Au promoted CO_2 electroreduction to formic acid, J. Mater. Chem. A 7 (2019) 13935–13940.

[65] H. Yang, Y. Wu, G. Li, Q. Lin, Q. Hu, Q. Zhang, et al., Scalable production of efficient single-atom copper decorated carbon membranes for CO_2 electroreduction to methanol, J. Am. Chem. Soc. 141 (2019) 12717–12723.

[66] Q. Zhao, C. Zhang, R. Hu, Z. Du, J. Gu, Y. Cui, et al., Selective etching quaternary MAX phase toward single atom copper immobilized MXene ($Ti_3C_2Cl_x$) for efficient CO_2 electroreduction to methanol, ACS Nano 15 (2021) 4927–4936.

[67] Y. Cai, J. Fu, Y. Zhou, Y.-C. Chang, Q. Min, J.-J. Zhu, et al., Insights on forming N,O-coordinated Cu single-atom catalysts for electrochemical reduction CO_2 to methane, Nat. Commun. 12 (2021) 586.

[68] L. Han, S. Song, M. Liu, S. Yao, Z. Liang, H. Cheng, et al., Stable and efficient single-atom Zn catalyst for CO_2 reduction to CH_4, J. Am. Chem. Soc. 142 (2020) 12563–12567.

[69] D. Karapinar, N.T. Huan, N. Ranjbar Sahraie, J. Li, D. Wakerley, N. Touati, et al., Electroreduction of CO_2 on single-site copper-nitrogen-doped carbon material: selective formation of ethanol and reversible restructuration of the metal sites, Angew. Chem. Int. Ed. 58 (2019) 15098–15103.

[70] K. Zhao, X. Nie, H. Wang, S. Chen, X. Quan, H. Yu, et al., Selective electroreduction of CO_2 to acetone by single copper atoms anchored on N-doped porous carbon, Nat. Commun. 11 (2020) 2455.

[71] A. Guan, Z. Chen, Y. Quan, C. Peng, Z. Wang, T.-K. Sham, et al., Boosting CO_2 electroreduction to CH_4 via tuning neighboring single-copper sites, ACS Energy Lett. 5 (2020) 1044–1053.

CHAPTER 8

In situ/operando X-ray absorption spectroscopy in small molecule–based electrocatalysis

Rui Ren[1], Qingbao Ni[1], GuiFen Li[1], Dongdong Xue[1], Minbin Gao[2], Yan Zhao[1], Jing Sun[1] and Jiangwei Zhang[1]
[1]Science Center of Energy Material and Chemistry, College of Chemistry and Chemical Engineering, Inner Mongolia University, Hohhot, P.R. China
[2]National Engineering Laboratory for Methanol to Olefins, Dalian Institute of Chemical Physics, Chinese Academy of Sciences, Dalian, P.R. China

8.1 Introduction

Energy is at the heart of a country's economy, and over the past two centuries, a society's energy consumption has grown in tandem with the amount of wealth created. However, traditional fossil fuels are nonrenewable energy sources, which have led to serious energy depletion due to excessive exploitation and utilization in modern society. Global warming caused by energy shortages and CO_2 emissions seriously threatens the sustainable development of human society [1,2]. The development of clean renewable energy conversion technology plays a key role in solving the current energy and environmental crisis. The activation and conversion of small molecules in electrocatalytic energy, including electrocatalytic water splitting and carbon dioxide reduction (CO_2RR) to fuel and chemicals, can convert renewable energy into chemical energy [3–6]. Therefore it has attracted extensive attention from researchers, and this field has become a hot spot in current research at home and abroad.

As clean energy with great development prospects, hydrogen energy plays an important role in replacing traditional fossil fuels in the future renewable energy system [7]. The traditional process is to produce hydrogen through steam reforming of hydrocarbons. However, this method has the disadvantages of high energy consumption, heavy pollution, low purity, and poor efficiency. Compared with the traditional process, hydrogen production by electrolysis of water as a new hydrogen production method has received extensive attention, which better avoids the drawbacks of traditional hydrogen production methods and has broad application prospects [8,9].

Single Atom Catalysts
DOI: https://doi.org/10.1016/B978-0-323-95237-8.00005-7

The electrocatalytic water splitting mainly consists of two semireactions, namely, the hydrogen evolution reaction (HER) of the cathode and the oxygen evolution reaction (OER) of the anode [10,11]. In acidic water splitting, HER occurs with fast electrode reaction kinetics. In contrast, because OER involves the four-electron/proton transfer and coupling process formed by O-O bonds. This reaction is relatively complex, the kinetic behavior is slow and the reaction rate is low, which limits the energy conversion efficiency, so a large overpotential is required to cause it to occur [12]. Therefore OER is often considered a bottleneck in electrocatalytic water splitting. The steps for OER also vary in different media. In an acidic water splitting, water is oxidized to oxygen and hydrogen ions, while in neutral or alkaline electrolytes, hydroxyl ions are oxidized to water and oxygen [13]. At present, noble metal-based materials (RuO_2, IrO_2, and Pt) are still highly efficient catalysts for OER and HER. However, due to their slow kinetic process affect commercial applications. To overcome these difficulties, researchers have designed noble metal catalysts with high activity and stability based on a variety of protocols [14–16]. Braman et al. grew one-dimensional and highly crystallinity RuO_2 nanowires (1D-RuO_2-CNx) on the surface of carbon nitride, which electrochemical experiments showed good activity and stability [17]. Electrochemical N_2 reduction reaction (NRR) has been widely considered as a promising route for green NH_3 synthesis, while the efficient electrocatalysts are required to promote the NRR and retard the HER [18,19]. Recent theoretical predictions have demonstrated that some single-atom possess an outstanding NRR performance with low limiting potential, good thermodynamical stability, and favorable selectivity against HER [20–22].

The conversion of CO_2 into usable low-carbon materials has great practical application value, which can not only solve the problem of greenhouse gases but also provide raw materials for the development of new energy sources. The production of high-value-added products by CO_2RR reaction is an ideal method. However, the biggest challenge in CO_2RR reactions is the slow catalytic reaction rate and low energy efficiency, which is due to the complex kinetic process mechanism involved in the entire catalytic process [23,24]. The reaction consists of a series of multielectron and proton transfer steps with many reaction intermediates and products [25]. Therefore the characteristics of the catalyst determine the efficiency of CO_2RR. Luo et al. adjusted the catalytic performance of CO_2RR by changing the crystal plane exposed by the catalyst, and the experimental results showed that CO_2 electroreduction was sensitive to the surface morphology of Cu electrode. DFT calculations show that Cu(111) promotes the formation of COH* intermediates and Cu(100) facilitates the formation of CHO* intermediates [26]. Wei et al. achieved CO_2RR selectivity for multi-carbon products by constructing composite materials [27]. However, the reaction mechanism of electrocatalytic CO_2RR is still controversial, especially the adsorption mode of intermediates at the active site is unclear.

Most of the current research is only exploring the original structure of the catalyst, that is, the static structure of the active site. Therefore real-time monitoring of adsorption and dissociation at active sites and their influence on local atomic and electronic structure changes plays a decisive role in establishing the structure–activity relationship of catalysts and optimizing catalyst and reaction conditions. X-ray absorption spectroscopy (XAS) is a powerful tool for tracking changes in catalyst structure during reactions and has been used in many electrochemical systems [28,29]. XAS is sensitive to the structure and electronic structure of the central atom and can reveal the properties of the active ingredient under reaction conditions, the structural evolution of the catalyst, and the reaction intermediate in real-time, which makes it a practical and fast analytical method [30]. In this monograph, we present recent progress of in situ/operando XAS in electrocatalytic water splitting and CO_2RR systems, especially in situ XAS, to thoroughly understand the electrocatalytic small molecule transformation process of water splitting and CO_2 reduction.

8.2 X-ray absorption spectroscopy

XAS shows the functional relationship between the X-ray absorption coefficient of a specific element and the photon energy. After X-ray absorption, electrons are excited from the core-occupied state to the empty state. According to the energy difference of the emitted photoelectrons, the spectrum includes two different regions, X-ray absorption near-edge spectroscopy (XANES) and extended X-ray absorption fine structure (EXAFS). XANES and EXAFS can provide complementary information to each other to determine the local coordination environment of each atom in the catalyst and the geometric structure of the metal center and ligand. The physical nature of XANES and EXAFS is photoelectron effect

[31]. The XANES is a part of the spectrum, including the features before the edge and the edge itself, while the position of the distinguishing feature and the edge itself is usually used to identify the oxidation state of the element. XANES is very sensitive to the oxidation state and the coordination chemical form of the absorption atom (for example, octahedral and tetrahedral coordination) [32]. Extending from the absorption edge to 100 eV, XANES can provide information about valence state perturbations as electrons transition from the core energy level and generate information about orbital hybridization, which is determined by selection rules.

8.2.1 X-ray absorption near-edge structure

The XANES performs more detailed feature detection on regions such as weak front edge peaks and shoulders that are more sensitive to electronic structure and geometric symmetry [33]. X-ray absorption fine structure (XAFS) spectroscopy is a powerful tool for tracking changes in atomic and electronic structures during reactions and has been used in multiple electrochemical systems. The valence state of specific elements in the catalyst can be qualitatively verified by comparing the offset of the absorption edge position with the standard sample [34]. Han et al. synthesized Mo single atoms anchored to nitrogen-doped porous carbon (SA-Mo/NPC) as a low-cost and efficient NRR catalyst. The Mo K-edge XANES of SA-Mo/NPC shows a valence state of $> +2$ with Mo foil and Mo_2C as references (Fig. 8.1A) [35]. Chen et al. report a nitrogen-doped γ-Fe_2O_3 ($x$$Fe_2O_3$-N@CN) electrocatalyst, which yield C_2H_6 as the major product in an H-cell. The Fe K-edge XANES curves and the first derivatives of the as-prepared $x$$Fe_2O_3$-N@CN demonstrate that the near-edge absorption energy is between Fe foil and Fe_2O_3, implying that the metal is in the Fe^{3+} state (Fig. 8.1B) [36]. The Pd K-edge XANES spectrum of $PdFe_1$ overlap well with the Pd metallene and Pd foil. And the Fe K-edge XANES. Wang et al. prepared $CoFe_2O_4/Cr^{3+}/S^{2-}$ (CFCS) hollow nanospheres by Cr^{3+} substitution and S^{2-} exchange to achieve outstanding OER activity. As seen from Fig. 8.1A and C–D, XANES of CFC overlaps well with $CoFe_2O_4$, indicating the valence of Co and Fe remain almost unchanged after the addition of Cr^{3+} [37]. Only a low shift can be observed from the absorption edges of CFCS, suggesting that the average valence of Co and Fe in CFCS sample gradually decreased compared with CFC and $CoFe_2O_4$.

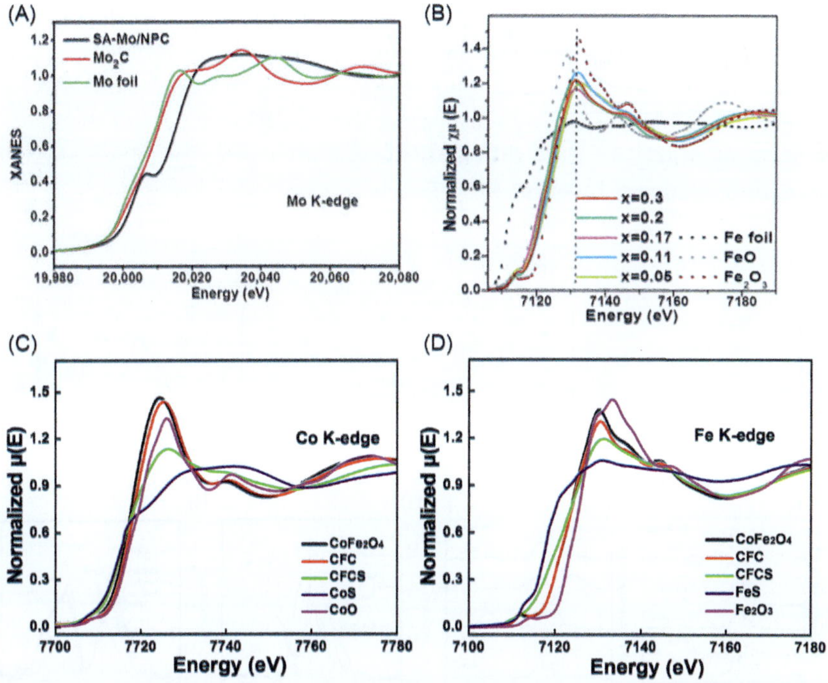

Figure 8.1 (A) XANES measurements of Mo K-edge. (B) Fe K-edge XANES spectra of $x$$Fe_2O_3$-N@CN, Fe_2O_3, and Fe foil (*blank line*). (C–D) Co K-edge and Fe K-edge of different catalysts. *Reproduced with permission (A) from L. Han, X. Liu, J. Chen, R. Lin, H. Liu, F. Lü, et al., Angew. Chem. 131 (2019) 2343–2347, Copyright Angewandte Chemie; (B) from P. Chen, P. Zhang, X. Kang, L. Zheng, G. Mo, R. Wu, et al., J. Am. Chem. Soc. 144 (2022) 14769–14777, Copyright American Chemical Society; (C–D) from J. Sun, H. Xue, Y. Zhang, X.-L. Zhang, N. Guo, T. Song, et al., Nano Lett. 22 (2022) 3503–3511, Copyright Nano Letters.*

8.2.2 Extended X-ray absorption fine structure

On the basis of ANES, more detailed local atomic structure information is obtained through further analysis of EXAFS. It can quickly identify the type of atom nearest to the absorption atom and the atomic separation. The local structure and environmental characteristics of the absorption atom can be basically obtained from the first peak in the Fourier transform EXAFS spectrum. For some metal absorption edges in XAS data collection, the first peak in FT-EXAFS usually corresponds to the low Z neighborhood (O, N, C, etc.), which represents the cation property of the absorption site, as shown in Fig. 8.2A [38].

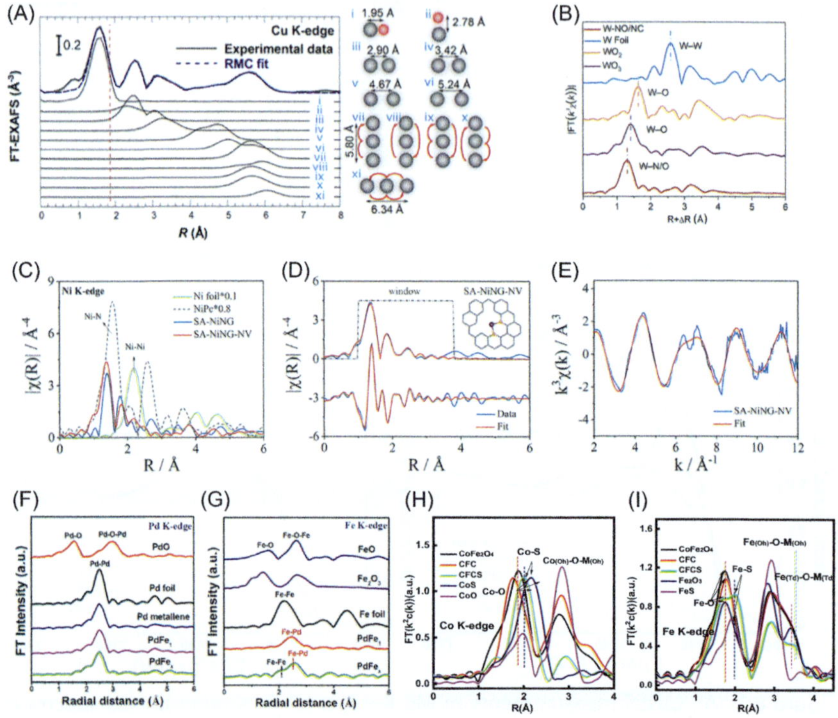

Figure 8.2 (A) Fourier-transformed (FT) Cu K-edge EXAFS spectrum and the contributions of the 11 most important photoelectron scattering paths, whose geometries are shown on the right. (B) *k*3-weighted FT-EXAFS curves of the W-NO/NC at W L3-edge.(C) FT-EXAFS spectra, in which the Ni foil and NiPc spectra have been reduced in size. (D) The magnitude (*top*) and imaginary (*bottom*) parts of R′ space Ni K-edge spectra and fitting curves of SA-NiNG-NV. The inset shows the coordination structure in SA-NiNG-NV; gray, orange, and purple spheres represent C, N, and Ni atoms, respectively. (E) EXAFS fitting on SA-NiNG-NV in K space. (C−E) (F) EXAFS spectra of Pd metallene, $PdFe_1$, PdFex, and reference samples of Pd foil and PdO. (G) EXAFS spectra of $PdFe_1$, PdFex, and reference samples of Fe foil, FeO, and Fe_2O_3. (H) Fe K-edge and (I) FT- EXAFS $k2\chi(k)$ data of Cr K-edge [38−40]. *Reproduced with permission (A) from J. Timoshenko, B. Roldan Cuenya, Chem. Rev. 121 (2020) 882−961, Copyright Chemical Reviews; (B) from Y. Gu, B. Xi, W. Tian, H. Zhang, Q. Fu, S. Xiong, Adv. Mater. 33 (2021) e2100429, Copyright Advanced Materials; (E) from J. Chen, S. Li, Y. Zhao, R.K. Hocking, W. Ren, X. Chen, et al., Adv. Funct. Mater. 31 (2021) 2107072, Copyright Advanced Functional Materials; (F,G) from X. Li, P. Shen, Y. Luo, Y. Li, Y. Guo, H. Zhang, et al., Angew. Chem. 61 (2022) e202205923, Copyright Angewandte Chemie; (I) from J. Sun, H. Xue, Y. Zhang, X.-L. Zhang, N. Guo, T. Song, et al., Nano Lett. 22 (2022) 3503−3511, Copyright Nano Letters.*

The size of the reducing ions is large, and the metal bond is usually long. Therefore such characteristics in FT-EXAFS spectra indicate the metal properties of the absorbing species. Note that the position of FT-EXAFS spectral peak does not correspond to the actual distance between atoms. By comparing the peak positions, we can find that the FT-EXAFS function is located at the R value, which is shorter than the actual length of the corresponding key. Although the visual inspection of FT-EXAFS data only provides information about the distance between atoms, the EXAFS feature itself has much more information and provides the possibility to distinguish different types of neighbors, even if they are located at the same distance between atoms. The type of adjacent atoms determines the correlation scattering function, which is the basis of EXAFS chemical sensitivity. Therefore different neighboring atoms have different effects on the data acquisition of EXAFS. It should be noted that in the periodic table of elements, the contribution of adjacent elements to the data acquisition of EXAFS is small, which is difficult to distinguish, because the contrast between scattering functions is small.

The wavelet transform method of WT-EXAFS spectrum can be used to visually determine the information of designated neighboring atoms [41–43]. Similar to Fourier transform, it is an integral transform used to analyze the frequency content information in the spectrum. The calculation and interpretation of experimental spectra by wavelet transform are very common in recent years and are widely used in the research and design of thermoelectric catalysts. Avoid overinterpretation in the application process. Many factors will change the characterization appearance in WT-EXAFS. In practical applications, only the first peak is dominant, and the first few peaks can be explained by visual inspection. However, the data of the coordination shell farther away often overlap, so modeling and data fitting are required to explain the corresponding characteristics in WTEXAFS and FT-EXAFS spectra. Wavelet analysis is not suitable for distinguishing the contributions of adjacent light elements in EXAFS spectra [44], because the small changes in relative atomic position and disorder have a greater impact on WT-EXAFS characteristics than the subtle differences between the scattering functions of similar elements. Gu et al. have prepared a novel W-SAC with oxygen-nitrogen coordination (W-NO/NC) and a high metal load of over 10 wt.% as an effective and robust NRR electrocatalyst. More detailed local coordination chemistry of the W site in W-NO/NC was obtained by EXAFS analysis. FT-EXAFS spectra at the edge of W L3 (Fig. 8.2B) show that W-NO/NC has only A peak at 1.31 Å, which is assigned to the W-O/N coordination [39]. Chen et al. reported a nitrogen vacancy (NV)-induced coordination reconstruction strategy to construct a highly defective Ni-pyridine nitrogen single-atom catalyst (SA-Ni-NG-NV) for electrochemical CO_2

reduction, with the catalyst exhibiting high activity and stability. To further investigate and uncover the coordination environment around Ni center, the FT-EXAFS of SA-NiNG-NV and SA-NiNG was fitted. Fourier transform EXAFS (FT-EXAFS) analysis shows that SA-NiNG-NV has no Ni-Ni peak at 2.1 Å, indicating that Ni atoms are well dispersed on the graphene framework (Fig. 8.2C). The main peak at R' = 1.4 Å (Ni-N) is due to the first-shell scattering interaction in the catalyst SA-NiNG-NV. As shown in Fig. 8.2D and E, it is worth noting that each Ni atom coordinates with two N atoms on average to form a Ni-pyridine N_2V_2 unsaturated structure, and the defective coordination structure of pyrrole nitrogen species is rapidly formed. As the defects in the carbon frame increase, these pyrrolidine N atoms create vacancies, which leads to the local reconstitution of the unstable pentagonal defects into more stable pyridine nitrogen Ni-N species with NVs, and these vacancies further alter the coordination environment of the central Ni atoms, resulting in inconsistent Ni-N bond lengths [40].

Li et al. prepared $PdFe_1$ single-atom alloy metalene (in which Fe monoatoms are confined to Pd metalene support) catalyst for the first time, which has excellent NRR performance. The coordination environment of $PdFe_1$ was studied by EXAFS spectroscopy. The findings showed that the introduction of Fe atoms in $PdFe_1$ does not affect the coordination structure of the Pd host (Fig. 8.2F). Fig. 8.2G demonstrates that $PdFe_1$ only has the main peak of the Fe-Pd coordination bond, there are no Fe-Fe and Fe-O bonds [45]. EXAFS spectroscopy is applied to further investigate the influence of Cr^{3+} and S^{2-} species on the electronic structure of $CoFe_2O_4$. The Fourier-transformed (FT) Co and Fe K-edge EXAFS shows a prominent peak at ~1.5 assigned to the M-O, and another main peak reveals the distance between metal atoms and their neighbors at octahedral or tetrahedral positions (Co(Td)-O-M(Td)/Fe (Td)-O-M(Td)). Fig. 8.2H also shows that the second-shell Co-O-Co features of CFCS split into two FT peaks at 2.9 and 3.3 Å, indicating the coexistence of Co centers in the Oh and Td sites. Cr atoms have 2.5 distance away from their neighboring metal atoms, implying that Cr cations only occupy the Oh position (Fig. 8.2I) [37].

8.3 In situ X-ray absorption spectroscopy

In situ XAS is sensitive to the local atomic structure and electronic structure of central absorbing atoms and is a powerful detection technique with essential understanding of the basic principles of fine structure. In addition, with the construction and application of synchrotron radiation devices, XAS has become a practical and fast analytical method. In situ XAS, mainly composed of X-ray spectrometers and unique electrochemical cells, has been used for more than a decade to systematically study the mechanism of electrocatalytic CO_2RR. At the same time, in situ XAS instruments for electrochemical testing are also developing rapidly [46]. Synchrotron radiation X-rays have high-energy and penetrating properties, and in situ techniques can provide insight into electrochemical reactions at the atomic level [47–49]. In addition, soft X-ray based in situ XAS technology has been widely used for the state-of-the-art transition metal monatomic catalysts for nitrogen and carbon fixation [47,50].

8.3.1 In situ X-ray absorption spectroscopy instruments for electrocatalysis

Recent developments in situ XAS instruments for electrocatalysis have focused on the rational design of electrochemical cells and their related components. The cell consists of five sections: body, working electrode (WE), counter electrode (CE), reference electrode (RE), windows (Fig. 8.3A). The basic principle of XAS can be explained by Lambert-Beer's law. When the X-ray energy rises to a certain level, a large change in the value of μ can be seen. This is because, at this time, the X-ray bombards the electrons around the atom into the higher energy level of the empty space, causing the X-ray to be strongly absorbed. As the X-ray energy continues to rise, the core electrons around the atom are blasted out of orbit, giving off light. Due to the wave-particle duality of light waves, the light waves generated between atoms will interfere [38]. The interference enhancement is reflected in the convex part, and the interference cancellation is reflected in the concave part as shown in Fig. 8.3B. At present, the electrochemical cell can be conducted in two patterns according to the detection modes: fluorescence and transmission, as shown in Fig. 8.3C. The left mode is fluorescence mode, which is often used in single-atom systems because the X-ray absorption capacity of single atoms is limited and needs to be measured by fluorescence. In fluorescence mode, the fluorescence intensity is proportional to the intensity of the incident light. The other is the transmission mode, which requires high concentration of samples to ensure full absorption of X-ray; otherwise, it will not get a good spectrum. Transmission mode X-rays can be quickly collected during the detection process [51]. However, the thickness of the

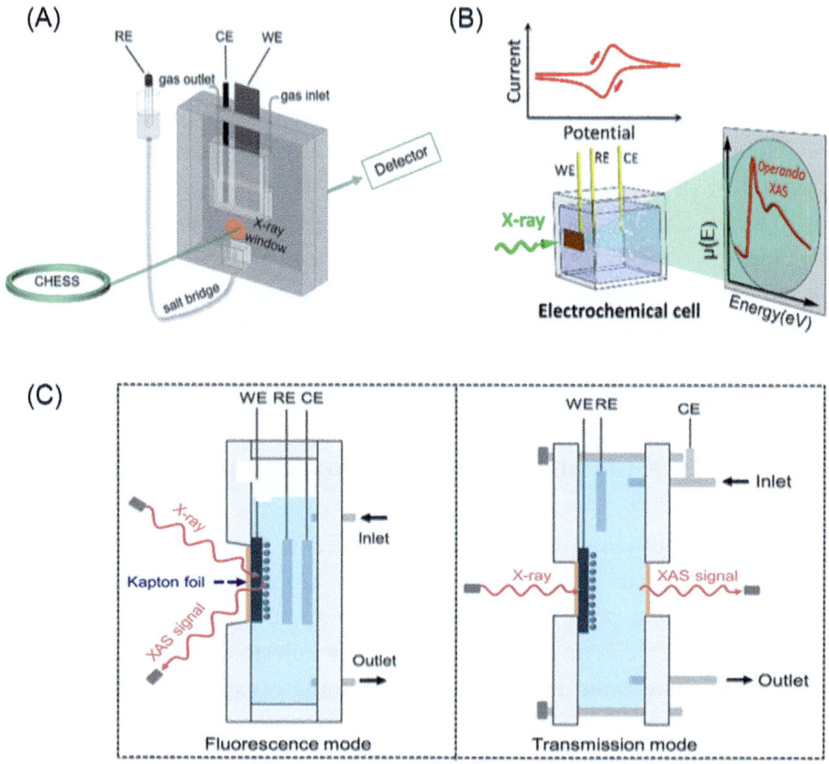

Figure 8.3 (A) Schematic of the in situ XAS electrochemical cell. (B) Schematic of how XAS works. (C) Illustration of in situ XAS electrochemical cells for (*left*) fluorescence mode and (*right*) transmission mode. WE, RE, and CE represent the working electrode, reference electrode, and counter electrode, respectively [52]. *Reproduced with permission (A) from Y. Yang, Y. Wang, Y. Xiong, X. Huang, L. Shen, R. Huang, et al., J. Am. Chem. Soc. 141 (2019) 1463–1466, Copyright Journal of the American Chemical Society; (B) from J. Timoshenko, B. Roldan Cuenya, Chem. Rev. 121 (2020) 882–961, Copyright Chemical Reviews; (C) from L. Cao, X. Liu, X. Shen, D. Wu, T. Yao, Acc. Chem. Res. 55 (2022) 2594–2603, Copyright Accounts of Chemical Research.*

sample must be thin enough to allow transmission. The in situ XAS device can accurately track the dynamic changes of the catalyst surface during the whole electrocatalytic process.

8.3.2 In situ X-ray absorption spectroscopy for nitrogen reduction reaction and CO_2RR

In general, the active atom in an electrocatalyst acts as the catalytic center, and other ligand atoms play a supporting role in changing the properties of the associated atoms or providing a suitable environment for adsorption. In situ XAS usually identifies the active site by analyzing the near-edge position changes, which indicate the valence changes and bond length deviations between the metal center and the ligand atoms during electrocatalysis. NRR and CO_2RR require high applied bias voltage decomposition and involve a multielectron transfer process.

In situ soft XAS was used to track the changes in copper catalyst valence during CO_2 reduction. Mehta et al. combined in situ grazing incident X-ray absorption spectroscopy (GIXAS) with X-ray diffraction (GIXRD) to reveal the structural changes on the surface of polycrystalline copper electrocatalysts under CO_2 reduction conditions [53]. In situ GIXAS showed that surface reconstruction of polycrystalline Cu was preferred to Cu(100) surface at CO_2 reduction potential, and the degree of reconstruction increased with the increase in applied negative potential (Fig. 8.4A and B). In situ GIXRD results show that the Cu surface can be reoxidized to Cu_2O by a reversible process after removing the CO_2 reduction potential.

Liu et al. prepared an inexpensive metal oxide Bi_2CuO_4 with molecular-level structural distortions for efficient CO_2 electroreduction to formate [54]. Determining the active structure during the reaction is an important guide to design the catalyst. The Cu K-edge XANES showed that the valence state of Cu remained constant at +1 with increasing reaction time, the table was not reduced to metallic Cu, and O was always present in the structure of BiOCu (Bi_2CuO_4-D). However, the valence state of Cu in BiCu (Bi_2CuO_4-S) was 0 after the reaction (Fig. 8.4C). It is similar to the former in the XANES spectrum of the Bi L_3-edge (Fig. 8.4D). It can be seen that the valence state of Bi in BiOCu cannot reach 0, while Bi in BiCu can be reduced to the metallic state. The changes in the valence state of the metal sites and their coordination environment during the CO_2RR process were detected online with the help of Operando XAS, which confirmed that the structural distortion led to structural stabilization after the CO_2 electrochemical reduction reaction, which facilitated the retention of many lattice oxygens during the reaction process and finally enhanced the CO_2RR performance of the catalyst. Li et al. studied the dynamic evolution of $PdFe_1$ active sites during NRR by In Situ XAS. The Fe K- edge XANES spectra (Fig. 8.4E) show that the absorption edge exhibits a progressive low-E shift as the potential increases, indicating that the valence state of the Fe atom decreases during NRR. The electronic structure and coordination environment of $PdFe_1$ were investigated. The results show that the addition of Fe atoms in $PdFe_1$

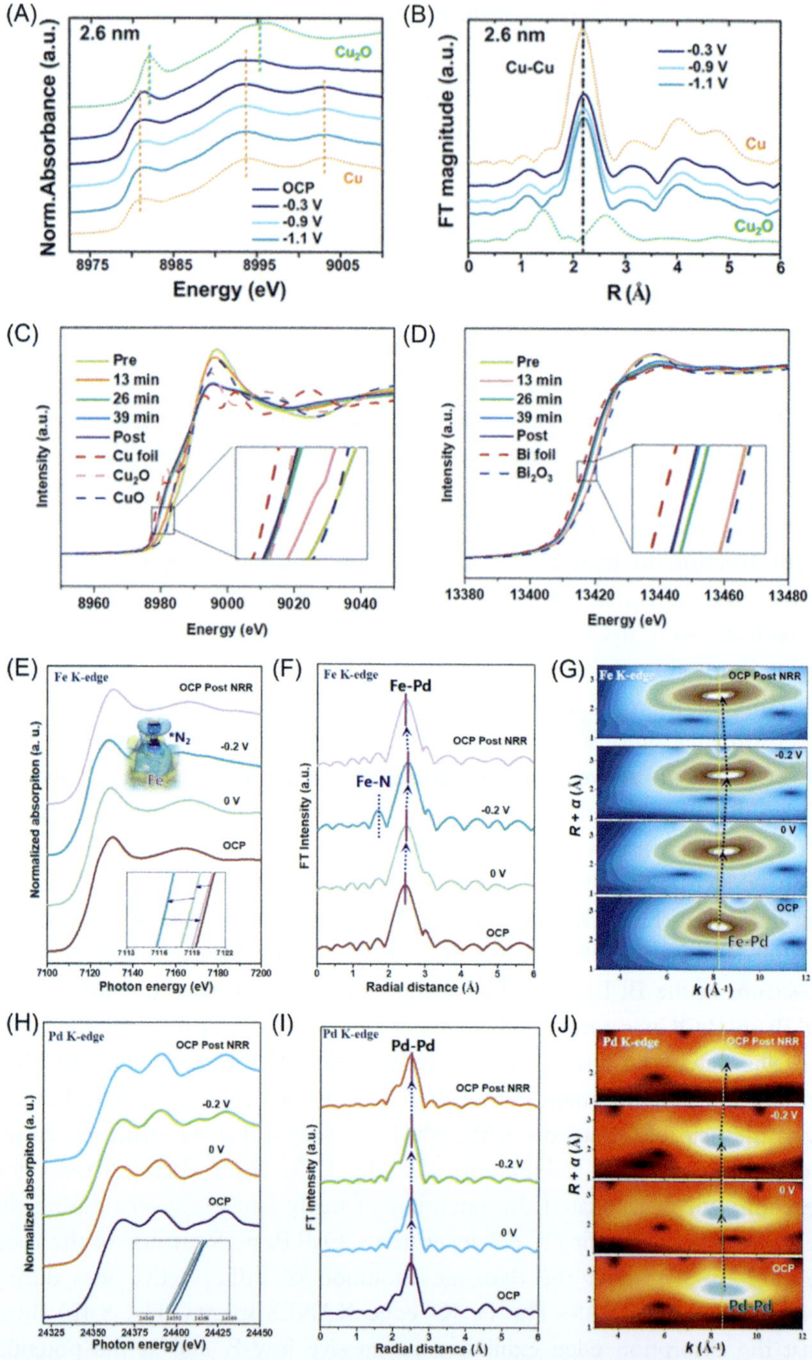

Figure 8.4 (A) and (B) In situ GIXAS of Cu(pc) catalyst, spectra of XANES and EXAFS of Cu(pc) catalyst at 2.6 nm probe depth at different applied potentials. (C) Cu K-edge and (D) Bi L3-edge XANES spectra of Bi_2CuO_4-D. (E—J) Operando XAS measurements of $PdFe_1$ during the NRR electrolysis at the potentials from open-circuit potential (OCP) to 0 V, 0.2 V (RHE) and then back to OCP. (E) Fe K-edge XANES, (F) Fe K-edge

does not significantly affect the valence and coordination structure of Pd host; however, more Fe incorporation (7.07 at%) can slightly change the electronic structure of Pd host. It is proved that single-atom Fe confinement on Pd metal support ($PdFe_1$) is an effective and robust catalyst with excellent NRR performance. The peak of the Fe-Pd coordination reveals a high-R shift at the operating potential (Fig. 8.4F and G), which returns to its original state when the potential is set back to OCP. The reversibility of the Fe active site in the NRR cycle is indicated. In contrast, no significant change was found in the Pd K-edge XANES (Fig. 8.4H), EXAFS (Fig. 8.4I), and WT profiles (Fig. 8.4J), demonstrating that the Pd support is almost catalytically inert to NRR [45].

8.4 Summary

We have described and analyzed the latest characterization of XAS, especially in situ XAS, to thoroughly understand the electrocatalytic small molecule transformation process of water splitting and CO_2 reduction. In particular, small molecule reaction processes (HER, OER, NRR, and CO_2RR) are critical for energy conversion and practical applications. However, most of the current research is only exploring the original structure of the catalyst, that is, the static structure of the active site. In situ XAS provides powerful evidence for probing the in-depth mechanism of electrocatalytic water splitting, NRR, and CO_2RR, including the recognition of the active component, tracking the dynamic structural evolution of the electrocatalysts, and observation of steady reaction intermediates in nitrogen/carbon-immobilized transition metal single-atom catalysts, transition metal oxides, and metal polycrystals among others. Although encouraging progress has been made on the influence of XAS technique on monitoring the structural changes, many scientific issues remain to be further improved. Due to the similar X-ray scattering ability of light elements, adsorbed substances can only be detected indirectly by monitoring the

EXAFS spectra, and corresponding (G) WT contour profiles, (H) Pd K-edge XANES, (I) Pd K-edge EXAFS spectra, and corresponding (J) WT contour profiles. *Reproduced with permission (A,B) from S.H. Lee, J.C. Lin, M. Farmand, A.T. Landers, J.T. Feaster, J.E. Avilés Acosta, et al., J. Am. Chem. Soc. 143 (2020) 588–592, Copyright Journal of the American Chemical Society, (C,D) from Y. Liu, Z.X. Lou, X. Wu, B. Mei, J. Chen, J.Y. Zhao, et al., Adv. Mater. 34 (2022) 2202568, Copyright Advanced Materials; (J) from X. Li, P. Shen, Y. Luo, Y. Li, Y. Guo, H. Zhang, et al., Angew. Chem. 61 (2022) e202205923, Copyright Angewandte Chemie.*

subtle changes in the coordination environment of the catalyst, and the local geometric structure of elements such as C/H/O cannot be directly detected. The conventional XAS technique hitherto requires minute-level collection of a set of XAS spectra. In addition, this relatively low-speed acquisition of data cannot allow for real-time analysis of the kinetic process. Therefore a single technique cannot provide comprehensive details of the processes of complex electrocatalytic systems, and these technical shortcomings need to be overcome in combination with other in situ characterization and theoretical calculations. We believe that the further development of XAS and other operando technologies, such as Raman spectroscopy and Mössbauer, combined with theoretical calculations, will lead to a further in-depth understanding of the fundamental processes of electrocatalytic reactions.

Acknowledgments

This work was financially supported by the "High level talent steed plan" of Inner Mongolia University. "Grassland Talents" of Inner Mongolia Autonomous Region; Young Talents of Science and Technology in Universities of Inner Mongolia Autonomous Region (NJYT23030); Carbon neutralization research project (STZX202218); National Natural Science Foundation of China (U22A20107); Inner Mongolia Autonomous Region Natural Science Foundation (2023MS02002); National Key R&D Program of China (2022YFA1205201).

References

[1] C. Li, X. Tong, P. Yu, W. Du, J. Wu, H. Rao, et al., J. Mater. Chem. A 7 (2019) 16622–16642.
[2] W. Zhang, Z. Jin, Z. Chen, Adv. Sci. 9 (2022) 2105204.
[3] X. Li, H. Wang, Z. Cui, Y. Li, S. Xin, J. Zhou, et al., Sci. Adv. 5 (2019). eaav6262.
[4] X.F. Lu, L. Yu, X.W. Lou, Sci. Adv. 5 (2019). eaav6009.
[5] E. Boutin, L. Merakeb, B. Ma, B. Boudy, M. Wang, J. Bonin, et al., Chem. Soc. Rev. 49 (2020) 5772–5809.
[6] D.C. Liu, D.C. Zhong, T.B. Lu, EnergyChem 2 (2020) 100034.
[7] Z. Chen, X. Duan, W. Wei, S. Wang, B.J. Ni, J. Mater. Chem. A 7 (2019) 14971–15005.
[8] X. Du, G. Ma, X. Zhang, ACS Sustain. Chem. Eng. 7 (2019) 19257–19267.
[9] F. Podjaski, D. Weber, S. Zhang, L. Diehl, R. Eger, V. Duppel, et al., Nat. Catal. 3 (2020) 55–63.
[10] A.B. Laursen, S. Kegnæs, S. Dahl, I. Chorkendorff, Energy Environ. Sci. 5 (2012) 5577–5591.
[11] J. Zhu, L. Hu, P. Zhao, L.Y.S. Lee, K.-Y. Wong, Chem. Rev. 120 (2019) 851–918.
[12] N.T. Suen, S.F. Hung, Q. Quan, N. Zhang, Y.J. Xu, H.M. Chen, Chem. Soc. Rev. 46 (2017) 337–365.

[13] Y. Guo, T. Park, J.W. Yi, J. Henzie, J. Kim, Z. Wang, et al., Adv. Mater. 31 (2019) 1807134.
[14] Y. Pi, Q. Shao, P. Wang, J. Guo, X. Huang, Adv. Funct. Mater. 27 (2017) 1700886.
[15] J. Feng, F. Lv, W. Zhang, P. Li, K. Wang, C. Yang, et al., Adv. Mater. 29 (2017) 1703798.
[16] T. Reier, Z. Pawolek, S. Cherevko, M. Bruns, T. Jones, D. Teschner, et al., J. Am. Chem. Soc. 137 (2015) 13031–13040.
[17] T. Bhowmik, M.K. Kundu, S. Barman, ACS Appl. Mater. Interfaces 8 (2016) 28678–28688.
[18] Y. Pi, N. Zhang, S. Guo, J. Guo, X. Huang, Nano Lett. 16 (2016) 4424–4430.
[19] P. Shen, X. Li, Y. Luo, N. Zhang, X. Zhao, K. Chu, Appl. Catal. B: Environ. 316 (2022) 121651.
[20] X. Li, G. Hai, J. Liu, F. Zhao, Z. Peng, H. Liu, et al., Appl. Catal. B: Environ. 314 (2022) 121531.
[21] C. Ma, Y. Zhang, S. Yan, B. Liu, Appl. Catal. B: Environ. 315 (2022) 121574.
[22] L. Zhang, G.F. Chen, L.X. Ding, H. Wang, Chemistry 25 (2019) 12464–12485.
[23] S. Ma, P.J. Kenis, Curr. Opin. Chem. Eng. 2 (2013) 191–199.
[24] R.J. Bellows, E.P. Marucchi-Soos, D.T. Buckley, Ind. Eng. Chem. Res. 35 (1996) 1235–1242.
[25] M.G. Kibria, J.P. Edwards, C.M. Gabardo, C.T. Dinh, A. Seifitokaldani, D. Sinton, et al., Adv. Mater. 31 (2019) 1807166.
[26] W. Luo, X. Nie, M.J. Janik, A. Asthagiri, ACS Catal. 6 (2016) 219–229.
[27] J. Wei, Q. Ge, R. Yao, Z. Wen, C. Fang, L. Guo, et al., Nat. Commun. 8 (2017) 1–9.
[28] L. Ma, W. Hu, B. Mei, H. Liu, B. Yuan, J. Zang, et al., ACS Catal. 10 (2020) 4534–4542.
[29] L. Cao, Q. Luo, W. Liu, Y. Lin, X. Liu, Y. Cao, et al., Nat. Catal. 2 (2019) 134–141.
[30] B. Wang, S. Chu, L. Zheng, X. Li, J. Zhang, F. Zhang, Small Sci. 1 (2021) 2100023.
[31] J. Yang, W. Li, D. Wang, Y. Li, Small Struct. 2 (2021) 2000051.
[32] H. Huang, A.E. Russell, Curr. Opin. Electrochem. 27 (2021) 100681.
[33] J. Park, J. Cho, Angew. Chem. 132 (2020) 15427–15437.
[34] M. Filez, E.A. Redekop, J. Dendooven, R.K. Ramachandran, E. Solano, U. Olsbye, et al., Angew. Chem. Int. Ed. 58 (2019) 13220–13230.
[35] L. Han, X. Liu, J. Chen, R. Lin, H. Liu, F. Lü, et al., Angew. Chem. 131 (2019) 2343–2347.
[36] P. Chen, P. Zhang, X. Kang, L. Zheng, G. Mo, R. Wu, et al., J. Am. Chem. Soc. 144 (2022) 14769–14777.
[37] J. Sun, H. Xue, Y. Zhang, X.-L. Zhang, N. Guo, T. Song, et al., Nano Lett. 22 (2022) 3503–3511.
[38] J. Timoshenko, B. Roldan Cuenya, Chem. Rev. 121 (2020) 882–961.
[39] Y. Gu, B. Xi, W. Tian, H. Zhang, Q. Fu, S. Xiong, Adv. Mater. 33 (2021) e2100429.
[40] J. Chen, S. Li, Y. Zhao, R.K. Hocking, W. Ren, X. Chen, et al., Adv. Funct. Mater. 31 (2021) 2107072.
[41] H. Funke, A. Scheinost, M. Chukalina, Phys. Rev. B 71 (2005) 094110.
[42] J. Timoshenko, A. Kuzmin, Comput. Phys. Commun. 180 (2009) 920–925.
[43] Z. Xia, H. Zhang, K. Shen, Y. Qu, Z. Jiang, Phys. B: Condens. Matter 542 (2018) 12–19.
[44] H. Fei, J. Dong, M.J. Arellano-Jiménez, G. Ye, N. Dong Kim, E.L. Samuel, et al., Nat. Commun. 6 (2015) 1–8.

[45] X. Li, P. Shen, Y. Luo, Y. Li, Y. Guo, H. Zhang, et al., Angew. Chem. 61 (2022) e202205923.
[46] W. Cheng, H. Su, Q. Liu, Acc. Chem. Res. 55 (2022) 1949−1959.
[47] A. Kumar, V.K. Vashistha, D.K. Das, S. Ibraheem, G. Yasin, R. Iqbal, et al., Fuel 304 (2021) 121420.
[48] M.A. Soldatov, P.V. Medvedev, V. Roldugin, I.N. Novomlinskiy, I. Pankin, H. Su, et al., Nanomaterials 12 (2022) 839.
[49] J. Li, M.N. Banis, Z. Ren, K.R. Adair, K. Doyle-Davis, D.M. Meira, et al., Small 17 (2021) 2007245.
[50] X. Su, Z. Jiang, J. Zhou, H. Liu, D. Zhou, H. Shang, et al., Nat. Commun. 13 (2022) 1−11.
[51] L. Cao, X. Liu, X. Shen, D. Wu, T. Yao, Acc. Chem. Res. 55 (2022) 2594−2603.
[52] Y. Yang, Y. Wang, Y. Xiong, X. Huang, L. Shen, R. Huang, et al., J. Am. Chem. Soc. 141 (2019) 1463−1466.
[53] S.H. Lee, J.C. Lin, M. Farmand, A.T. Landers, J.T. Feaster, J.E. Avilés Acosta, et al., J. Am. Chem. Soc. 143 (2020) 588−592.
[54] Y. Liu, Z.X. Lou, X. Wu, B. Mei, J. Chen, J.Y. Zhao, et al., Adv. Mater. 34 (2022) 2202568.

CHAPTER 9

Catalysts for Li-S batteries

Shahzad Ahmed[1], Arshiya Ansari[1], Moin Ali Siddiqui[1], Pranay Ranjan[1] and Prashant Kumar[2]

[1]Department of Metallurgical and Materials Engineering, Indian Institute of Technology Jodhpur, Jodhpur, Rajasthan, India

[2]Global Innovative Centre for Advanced Nanomaterials (GICAN), College of Engineering, Science and Environment (CESE), School of Engineering, The University of Newcastle, Callaghan, NSW, Australia

9.1 Introduction

Due to the burden of increased energy needs and growing population expansion, attaining a sustainable energy future without damaging the environment becomes essential for human progress [1,2]. Fossil fuels (crude oil, natural gas, petroleum, gasoline, and coke) are essential to the advancement of the global economy. By relying too heavily on fossil fuels, we are putting the world at risk of oncoming energy problems and ecological crises, such as depletion of natural resources, climate emergency, and pollution. To that aim, we should diminish our dependence on traditional fossil energy sources and instead embrace clean, renewable energy sources including solar, geothermal energy, wind, and tidal.

Moreover, there is indeed a rising global drive to produce green and sustainable energy to put back fossil fuels, even though they still supply more than 80% of the globe's energy needs [3]. Significant opportunities for the everlasting evolution of civilization have been shown by the advancement of renewable and ecologically friendly power sources [4]. The electrochemical units that make up a battery are used to transform chemical energy into electrical energy. Batteries are a type of electric power source and also a frequent source of energy for a variety of commercial and domestic purposes. Since the early 1900s, batteries have quickly changed the world's profit market and social structure [5]. They are presently being manufactured for a wider variety of applications, including microchips and automobiles.

The need for developing environmentally more friendly, renewable, green electrochemical energy storage, and transformation technologies, which include metal-air-based batteries, lithium-sulfur (Li-S) batteries, and, lithium-ion batteries (LIBs) has increased [6]. The advancement of energy

Single Atom Catalysts
DOI: https://doi.org/10.1016/B978-0-323-95237-8.00001-X

storage technologies is yet hampered by low energy density, weak cycle stability, and exorbitant prices. To achieve high energy density and quick rate performances, it is desirable to create new catalytic electrode materials that would encourage electrochemical processes throughout the charging and discharging operation. In addition, over the past ten years, the scientific community has been much more interested in next-generation batteries. For that aim, Li-S batteries are seen as a viable alternative for energy storage owing to their great energy density, inexpensiveness, and environmental stewardship [7]. In contrast to Li-ion batteries, Li-S batteries have the potential for higher theoretical specific energy (~2500 W h kg^{-1}) [8], enhanced safety [9], and are taken into consideration as a possible grid energy storage and mobile rechargeable battery technology. To accomplish quick electrochemical processes by lowering the energy barricade and increasing the kinetics of the electrochemical process, high-performance electrocatalyst configuration and manufacturing are of utmost importance. Up till now, a lot of work has gone into creating new electrocatalysts with outstanding execution by diminishing the electrocatalyst particles. Most of the time, better efficacy is related to the ability to enhance the number of catalytic active regions. For instance, to increase the catalysis activation for diverse electrocatalytic purposes, metal-based nanoclusters and nanoparticles have been effectively produced on matrix materials with substantial surface regions, for example, metal oxides, g-C_3N_4, carbon, and metal sulfides [10]. Large surface energy, nevertheless, makes the tiny and fine metal particles clump into clusters, which results in reduced catalytic activity [11]. Additionally, the unequal size dissemination of the catalysts might have an impact on the electrocatalytic procedure' intermediate conversion, leading to subpar electrochemical properties. Consequently, the fabrication of innovative metal electrocatalysts featuring periodic geometries and homogenous size dissemination is crucial.

In the fields of electrochemical and catalysis frameworks, single-atom catalysts (SACs), which are mainly composed of solitary metal atoms distributed or attached to matrix materials, have drawn a lot of interest [12–14]. Employing SACs with homogeneous and well-distinct atomic dispersion, excellent activity and conversion efficacy for chemical processes may be produced. SACs have a number of advantages over conventional catalysts in the storage and conversion of electrochemical energy, resulting in outstanding electrochemical outcomes for battery arrangements with great energy density and high-rate characteristics. This chapter intends to present a thorough outline of the synthesis methods and use of highly active SACs in durable, highly efficient Li-S batteries. The Li-S battery's mechanism and challenges are also covered. Finally, the summary and conclusion are addressed.

9.2 Lithium-sulfur batteries

Li-S batteries are a potential energy storage device that has received a lot of interest due to their greater energy density than current LIBs. The primary distinction between these two kinds of battery systems is their energy storing strategy. LIBs work by allowing lithium ions to intercalate into stacked electrode materials such as lithium metal oxide cathodes and graphite anodes. The theoretical energy density of LIBs is often restricted to ~ 420 or 1400 W h L^{-1} because lithium ions can only be topotactically intercalated into some of these precise locations [15–17]. The energy density of traditional LIBs featuring graphite anodes and insertion-compound cathodes (such as $LiCoO_2$, $LiMn_2O_4$, and $LiFePO_4$) is insufficient to meet the needs of electric vehicles and smart grids. As a consequence, research into novel battery chemistries is intensive [18,19].

Lithium-air (Li-air) and Li-S batteries, which have potential energy densities of 3500 and 2500 W h kg^{-1}, respectively, have lately received a lot of interest. Nevertheless, due to a number of inherent issues and a poor operational energy density, Li-air batteries face insurmountable difficulties. The obstacles with Li-S batteries are thought to be less challenging than those with Li-air batteries, making them more practical and useful [18]. The operation of Li-S batteries relies on metal plating and stripping on the lithium anode side and a conversion reaction on the sulfur cathode side, as opposed to intercalation. Lithium anodes and sulfur cathodes have relatively high specific capacities of 3860 and 1673 mA h g^{-1}, respectively, due to the nontopotactic behavior of these reactions [20]. Additionally, sulfur is plentiful and safe for the environment. Li^+ ions are created at the lithium-metal anode during discharge and flow through the electrolyte to the sulfur cathode, whereas the electrons move to the external circuit to form Li_2S, which is the ultimate discharge product produced at the cathode [18].

Given their superior performance compared to current Li-ion technology, Li-S batteries may be a viable replacement. The earliest mention of Li-S batteries is from the 1960s when sulfur was employed as the cathode and lithium metal as the anode. Due to the high availability of sulfur in the Earth's crust and the fact that the chemistry of Li-S batteries relies on the reversible conversion between sulfur and lithium sulfides, Li-S batteries are also more affordable than conventional Li-ion batteries with a Li electrode. When sulfur is combined with a lithium metal anode, Li-S batteries may provide an effective energy density of 400–600 W h kg^{-1}, which is about three times more than that of existing LIBs [21,22]. Li-S batteries' electrochemical characteristics while using different RGO interlayers are tabulated in Table 9.1.

Table 9.1 Electrochemical properties of Li-S batteries with various RGO interlayers.

Description	Current density	Initial capacities (mA h g^{-1})	Cycling stability	References
RGO	0.2 C	1322	646 mA h g^{-1} after 200 cycles	[23]
RGO/carbon black	167.5 mA g^{-1}	1260	895 mA h g^{-1} after 100 cycles	[24]
MoS_2/RGO nanoflowers	0.2 A g^{-1}	1642	720 mA h g^{-1} after 100 cycles	[25]
Active carbon/RGO	0.1 C	1078	655 mA h g^{-1} after 100 cycles	[26]
MoS_2/RGO	0.2 C	1122	0.116% fading rate for 500 cycles at 1 C	[27]
MgO/RGO	0.2 C	923	767 mA h g^{-1} after 100 cycles	[28]
PEDOT-PSS/RGO	0.1 C	1249.4	812.8 mA h g^{-1} after 100 cycles	[29]
N-doped RGO	0.2 C	856	495 mA h g^{-1} after 200 cycles	[30]
N-doped 3D RGO	0.2 C	1150.3	852 mA h g^{-1} after 200 cycles	[31]
Fe_3O_4/RGO	1 C	~820	732 mA h g^{-1} after 500 cycles	[32]
SnO_2/RGO	1 C	~1100	734 mA h g^{-1} after 200 cycles at 1 C	[33]
B-doped RGO	0.2 C	902	435 mA h g^{-1} after 200 cycles	[30]
Amino-BN/RGO	3 C	1100	0.0037% fading rate at 1000 cycles	[34]
Sulfonated RGO	0.5 C	1240	802 mA h g^{-1} after 250 cycles at 0.5 C	[35]
ZnO/RGO foam	0.5 C	1051	672 mA h g^{-1} after 100 cycles	[36]

9.3 Fabrication of single-atom catalysts

SACs are solid-substrate-anchored single atoms with constrained size distribution and an anticipated atom resource utilization effectiveness of 100%. The development of SACs with excellent catalytic properties, stability, and specificity has opened up novel possibilities for catalysts [37]. Owing to the large surface energy of catalysts, single atoms may readily cluster into massive nanoparticles, which results in decreased catalytic activity [11]. They incorporate the advantages of heterogeneous and homogeneous catalysts [38]. The conditions that must be gathered to produce greatly effectual and stable SACs are as follows: (1) effective mass and electron transportation; (2) readily approachable active sites; (3) homogeneous spread of single metallic atoms onto the surface of the substrates; and (4) robust fastening between the support matrix and the single-atom metals [39]. As a result, the synthesis techniques for SACs are crucial in meeting the aforementioned significant requirements. Moreover, SACs have attracted the attention of the scientific community for a wide range of chemical processes and energy conversions owing to their unique catalytic features and optimal metal usage efficacy [40]. SACs will impulsively scatter metal atoms inserted on suitable scaffolds. SACs, which only contain isolated single atoms distributed over the substrate, have lately drawn plenty of attention in the catalysis field due to their extraordinary atom-utilization efficacy, high selectivity, and unique features [41,42]. In comparison with other composite substances, SACs not only possess homogenous active species for chemical reactions, as do other homogeneous materials but they also give the further advantages of better reusability and extended lifespan just like heterogeneous materials do. SACs were first documented in 2011, when platinum SACs were produced via a coprecipitation method on an iron oxide (FeO_x) surface, showing excellent durability and performance in the oxidation of cobalt. SACs thus provide substantial advantages [40–43]. SACs have drawn a significant amount of attention in a range of catalytic implementations due to this distinctive property. Single atoms could be effectively contained in the pore features in or on the matrices to prevent them from aggregating as big nanoparticles. The processes in this fabrication of spatial confinement approach are as follows: (1) preparing porous substances, such as covalent organic frameworks, metal-organic frameworks (MOFs), and zeolites to serve as platforms; (2) incorporating the metallic precursors and the matrices; and (3) carbonizing the finished products in an inert environment. Due to their vast surface area, homogeneous pore dimension, and high chemical stability, MOFs are being employed as heterogeneous catalysts extensively [44]. By the way of the pore confinement effect, adequate mononuclear metal precursors may be contained inside the porous geometry. Additionally, MOFs may be altered to produce a lot of pivoting sites, which promotes high atom distribution and stability [45]. Zeolite imidazolate frameworks, one of the typical groups of MOFs, have received a lot of interest because of their well-controlled shape and porosity, huge surface region, texture adjustability, and excellent chemical and thermal stability. The following factors make SACs with well-distributed metal atoms on constructed foundation extremely beneficial in Li-S batteries: (1) their active regions, which include secluded metal atoms and the local linkage surroundings, could be examined by cutting-edge techniques that is, microscopy and spectroscopy, which assist in comprehending the structure; and (2) they could be removed easily from the catalytic system and exhibit strong reusability [46]. Moreover, direct pyrolysis, atomic layer deposition (ALD), and ball milling receive special emphasis as methods employed for the production of SACs.

9.3.1 Direct pyrolysis method

For the pyrolysis-based production of SACs, MOFs with particular pore architectures and metal ion agglomerations coupled to ordered organic ligands are potential precursors [47]. Under heat processing, the evenly spread metal domains of pure MOFs are easily transformed into atomically scattered single metal atoms. It is possible to create desired MOFs by using functional organic ligands to coordinate various functional groups and atoms. Due to the ability of zinc to be extracted specifically upon pyrolysis, zinc-based bimetallic MOFs are typically utilized as precursors.

9.3.2 Atomic layer deposition

Direct pyrolysis at high-temperature has been utilized frequently to make SACs, however, because of the intricacy of the reactions that take place at these temperatures, it is difficult to precisely control the architecture of the desired products. To develop and create SACs with carefully regulated contents and thicknesses, ALD has been used. ALD is a process that uses gaseous state precursors and substrates to produce monoatomic coatings over them. This process depends on self-limited surface links. A typical ALD process entails four basic phases: the insertion of the primary precursor, reaction compartment purging, the insertion of the second precursor, and yet another reaction compartment purgation. By changing the number of ALD cycles, one may get the required level of single atoms. By selecting the required metal precursors and pumping them into the reactor vessel sequentially, the target SACs may be created.

9.3.3 Ball milling technique

While ALD has the potential to create stable SACs, its expensiveness and harsh circumstances prevent it from being employed in the industry. On the contrary, the creation of SACs using ball milling is quick, affordable, and configurable. Ball milling may create intense shear pressures and temperatures of up to 1000°C, which encourages the rebuilding of organic group chemical bonding and the breaking of raw substance particulates to produce new surfaces. Additionally, the newly manifested surfaces' dangling bonds are frequently chemically reactive and prone to interact with one another, thus facilitating the creation of the optimal cooperative arrangement between metal atoms and matrix substances.

9.4 Li-S batteries: the mechanism

Due to the electrochemical reaction $S_8 + 16Li_2 \rightarrow 8Li_2S$, Li-S batteries have a high theoretical specific capacity ($1673\ mA\ h\ g^{-1}$). The lithium metal anode (negative electrode) is likely to oxidize during the discharge process to produce electrons and lithium ions, which move via the external circuit and the electrolyte, respectively, to reach the sulfur cathode (positive electrode). At the cathode, sulfur interacts with the lithium ions and electrons to generate lithium sulfide by reduction. During the charge response, the reverse happens [15,48]. A typical schematic representation of the electrochemistry is shown in Fig. 9.1.

Charging: positive electrode: $8Li_2S \rightarrow S_8 + 16Li^+ + 16e^-$
Negative electrode: $Li^+ + e^- \rightarrow Li$
Discharging: positive electrode: $S_8 + 16Li^+ + 16e^- \rightarrow 8Li_2S$
Negative electrode: $Li \rightarrow Li^+ + e^-$

9.5 Challenges of lithium-sulfur batteries

It is difficult to realize the practical application of Li-S batteries, despite their benefits of high energy density and low cost. Despite promising developments in this area, significant obstacles such as low sulfur usage and uneven Li layering prevent the widespread use of Li-S batteries.

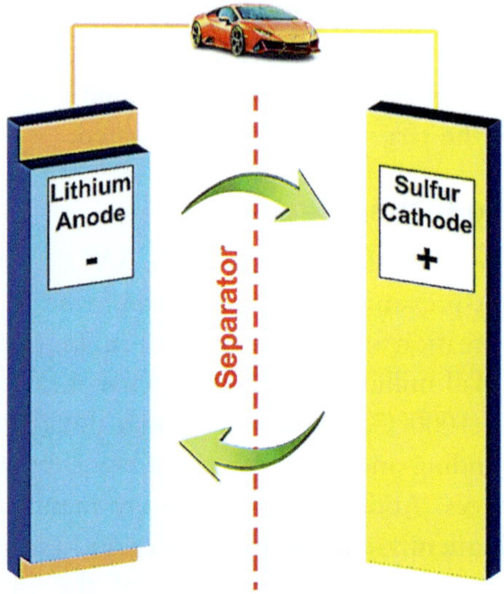

Figure 9.1 Schematic representation of the electrochemistry.

Furthermore, to the anode being corroded by polysulfides, lithium dendrites that developed during the breakdown of solid-electrolyte interphase (SEI) coatings might harm the separator, leading to interior short circuiting with a significant risk of process. Quite substantially, during the discharge operation, polysulfides provide a propensity to solubilize in the ether electrolyte and separate from the cathode, resulting in the unfavorable "shuttle effect" (Fig. 9.2). The poor electrochemical consumption of sulfur and rapid capacity fading is the key barriers preventing the commercialization of Li-S batteries. In the beginning, both sulfur and the discharge product Li_2S are ionically and electronically insulating. This lowers the energy density since it generally necessitates the integration of sulfur into a conductive matrix (such as carbon, a polymer, or metal). Moreover, in contrast to insertion-compound cathodes, sulfur experiences a number of complex structural and compositional changes as a result of the creation of soluble polysulfide intermediates, leading to poor mechanical stability and significant capacity loss. Furthermore, the dissolved entities of polysulfide transit between the cathode and anode, resulting in active loss of material from the cathode, a poor Coulombic efficiency, and passivation of the lithium-metal surface with insoluble products, such as Li_2S/Li_2S_2. Lastly, the lithium-metal anode deteriorates because of surface passivation and the formation of an unstable SEI with the organic electrolyte, preventing Li-S batteries from achieving long-term cycle durability [18]. A very promising solution to the current problems, such as the poor

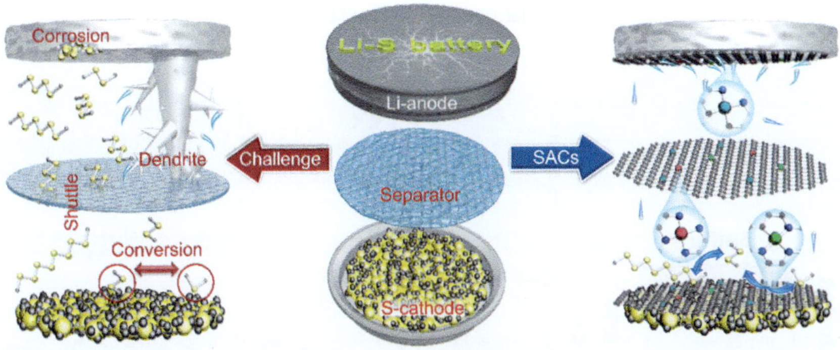

Figure 9.2 Schematic representation of the main issues with Li-S batteries and the significance of electrode materials altered by SACs [49]. *Reproduced with permission from F. Wang, J. Li, J. Zhao, Y. Yang, C. Su, Y.L. Zhong, et al., Single-atom electrocatalysts for lithium sulfur batteries: progress, opportunities, and challenges, ACS Mater. Lett. 2 (11) (2020) 1450–1463, Copyright 2020, ACS Materials Letters.*

utilization efficacy of active sulfur, the serious shuttle impact of polysulfides, and the aging of Li metal, appears to be the employment of SACs as battery parts or to alter battery parts.

9.6 Utilization of single-atom catalysts in other battery application domains

SACs' distinct coordinating properties and active regions make them significantly advantageous for many rechargeable battery technologies. It also has the ability to offer distinct lithium/sodium (Li/Na) nucleation regions, prevent dendrite production, and improve metal affinity in Li/Na-metal battery systems, resulting in exorbitant coulombic efficacy and prolonged cycle life. Additionally, because of their high affinity for soluble polysulfides and catalytic amplification of polysulfide transformation to the end product, SACs might reduce polysulfide mobility and boost the redox reaction kinetics in Li/Na-sulfur battery systems. Additionally, SACs can facilitate electrocatalytic processes including the oxygen evolution reaction (OER) and oxygen reduction reaction (ORR), which enhance the electrochemical stability and activity of zinc-air (Zn-air) and lithium-oxygen ($Li-O_2$) battery systems. This section will provide a very brief overview of SACs' reasonable configuration, their applicability in many kinds of high-tech energy storage systems, and their ideal architecture and chemical composition. Techniques for (SACs) synthesis and battery implementations are shown schematically in Fig. 9.3.

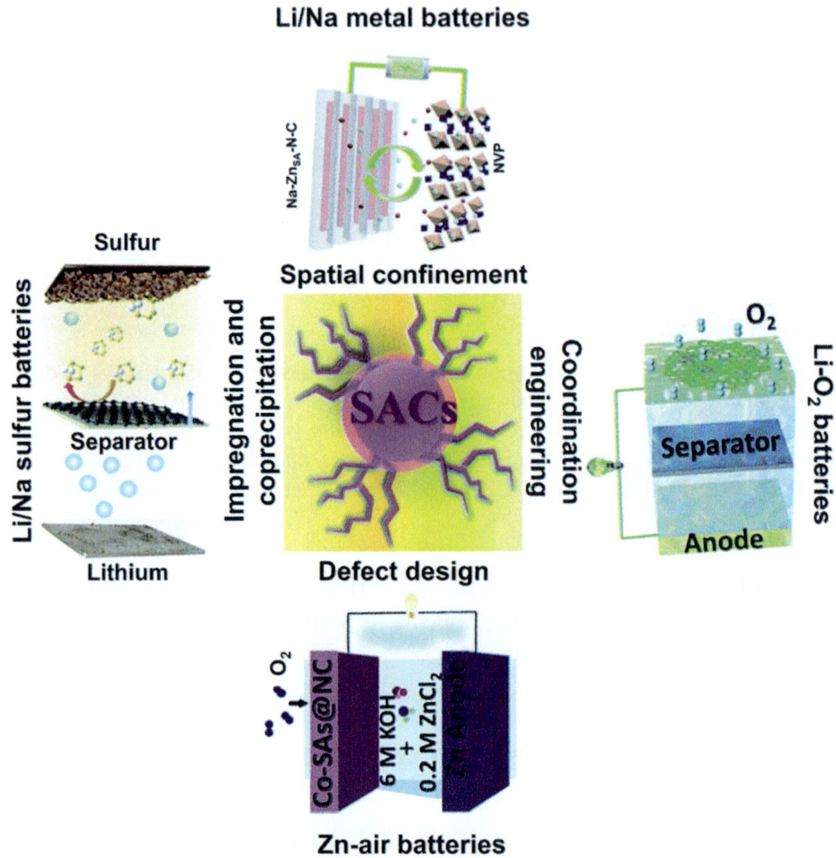

Figure 9.3 Schematic representation of (SACs) synthesis techniques and battery implementations [50]. *Reproduced with permission from H. Tian, A. Song, H. Tian, J. Liu, G. Shao, H. Liu, et al,. Single-atom catalysts for high-energy rechargeable batteries, Chem. Sci. 12 (22) (2021) 7656–7676, Copyright 2021, RSC Chemical Science.*

9.6.1 Li/Na metal batteries

Li-metal-based batteries have been intensively researched in recent decades due to their exorbitant theoretical specific capacity (3860 mA h g^{-1}) and the minimum electrochemical potential value (3.04 V vs SHE) of Li-metal anodes. However, using untreated Li-metal electrodes makes them vulnerable to the development of Li dendrites, substantial electrode volume capacity changes, and the formation of an unstable junction during discharging–charging cycles. The high activity of Li-metal causes these impacts, which can manifest in internal short circuits, poor coulombic performances, and shorter cycle times. Great attempts have been made to address these problems, including the use of SEI films, electrolyte surfactants, and electrode structuring for dendritic reduction and impediment [51].

9.6.2 Li/Na-S batteries

In recent years, surrogate rechargeable batteries have been designed and extensively studied to obtain high energy density. Li-S and Na-S batteries systems have earned the most emphasis of these due to a number of benefits, including the abundant sulfur found naturally, an exorbitant specific energy density (2600 $W\ h\ kg^{-1}$), and a great theoretical capacity (1675 $mA\ h\ g^{-1}$).

9.6.3 Li-O_2 batteries

Li-O_2 batteries have been regarded as one of the most important energy storage technologies due to their potentially great gravimetric energy density ($\sim$3505 $W\ h\ kg^{-1}$, following the reaction $O_2 + 2Li^+ + 2e^- \leftrightarrow Li_2O_2$), steady discharge voltage, and prolonged durability. When the Li-O_2 battery systems are being discharged, O_2 and Li^+ within the porous cathode medium could react to generate Li_2O or Li_2O_2 via ORR. Li_2O or Li_2O_2 can decompose and produce oxygen by OER during the transition procedure in Li-O_2 batteries. Due to the insulating and insoluble Li_2O_2, the development of Li-O_2 batteries has been hampered by difficulties with lower coulombic efficacy, poor rate capability, and poorer cycle life. Numerous electrocatalytic substances, including sulfides, metal oxides, and noble metals, have been utilized to stimulate electrochemical processes during charging and discharging to solve these issues, resulting in a reduction in the overpotential between ORR and OER. SACs have been characterized as efficient catalysts for Li-O_2 battery systems in recent years, producing batteries with lower charge overpotentials, large reversible capacities, and good cycle stability [52–60].

9.6.4 Zinc-air batteries

Zn-air batteries with rechargeable capabilities have higher theoretical specific energy density (1350 $W\ h\ kg^{-1}$, without O_2), intrinsic safeness, ecological friendliness, and economic feasibility than the previously stated Li/Na-metal batteries and Li/Na-S batteries have a wider range of applicability prospects for the consumer electronics trade and worthwhile devices. Zn-air battery systems, in contrast to Li-O_2 batteries, have consistent stability to shield Zn anodes from moisture. Nevertheless, significant challenges remain for the production of Zn-air batteries, such as reduced energy effectiveness (65%) driven by the slow kinetic model of ORR and OER at air cathode materials and zinc dendrite growth on Zn anodes. SACs were lately introduced as bifunctional catalysts for the air electrodes system of Zn-air batteries, which has had a significant impact on the research in this area. SACs have potential charging and discharging properties when constructed as a cathode with recharging capabilities and solid-state Zn-air batteries owing to their homogenous active regions, distinctive electrical architectures, and strong catalytic properties.

9.7 Conclusions and outlook

The last decade has seen a rise in curiosity in the production of high-energy, long-lasting Li-S batteries. According to contemporary developments, the future of Li-S batteries used for grid energy storage and mobile electric cars appears to be particularly promising. SACs with adjustable atomic frameworks and unique catalytic properties have created a wide range of opportunities for solving current problems and enhancing Li-S catalysis effectiveness for realistic applicability in the future. SACs could be synthesized in a variety of ways, including by direct pyrolysis, ALD, and ball milling. Direct high-temperature pyrolysis could be employed frequently to manufacture SACs, because of the intricacy of the reactions that take place at these temperatures, it is really challenging to precisely control the architecture of the desired products. It has been demonstrated that producing SACs via ALD is a viable approach for doing so on a large scale; however, the technology's high cost and challenging operating environment prevent it from finding a widespread commercial use. On the other hand, the creation of SACs using ball milling is simple, scalable, and affordable.

Meticulous artificial techniques, such as the spatial confining approach, flaw development methodology, impregnation, and coprecipitation approach, coordinating engineering approach, and other innovative techniques, can lead to SACs with well-defined architectures and unsaturated coordination. Nevertheless, due to the complex coordination settings and electronic architecture of their active components, SACs typically exhibit a variety of distinct catalytic actions. As a result, it is extremely challenging to precisely and consistently fabricate SACs on substrates with controlled coordination number and atomic dispersion. Additionally, itsis critical to use fabrication techniques that can yield SACs extensively. The preparation of precursors, such as MOFs, molecular sieves, organics, or polymers, followed by a calcination procedure is typically required for the manufacturing of SACs. This procedure could be expensive to produce, incorporate violent reaction conditions, and produce potentially dangerous byproducts (like vapors when converting precursors into the final form). SACs via "green" methods for extensive and mass-production manufacturing are widely demanded. Although this area has advanced quickly, certain significant obstacles still need to be overcome to develop the use of SACs in Li-S batteries (Fig. 9.4). Representation of the primary potential developments for 2D hosts in Li-S batteries and the fundamental needs of 2D host materials is shown in Fig. 9.5.

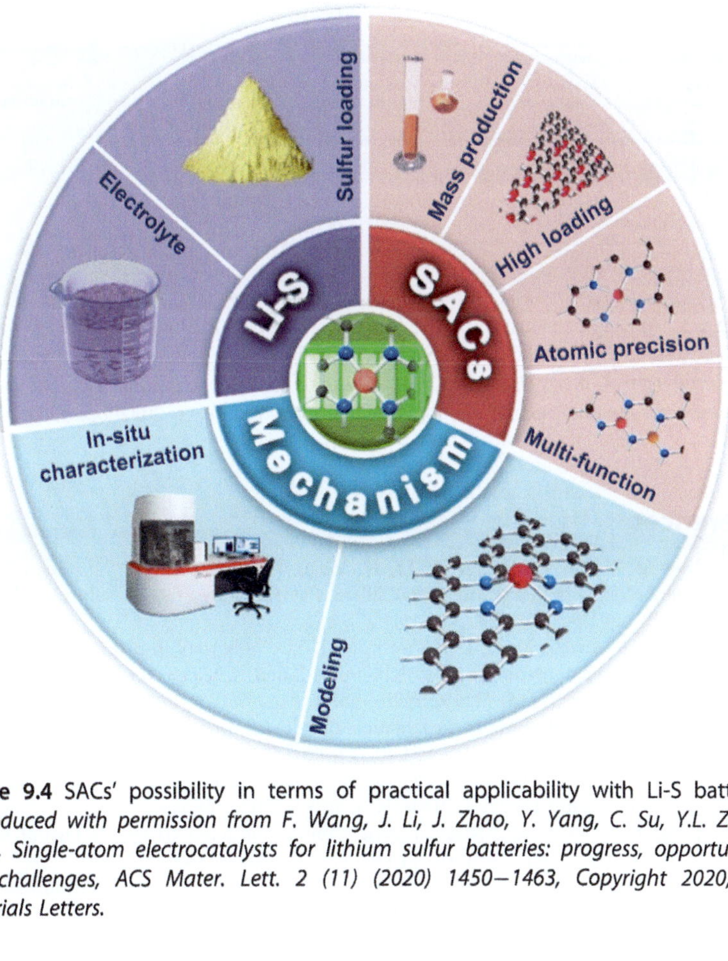

Figure 9.4 SACs' possibility in terms of practical applicability with Li-S batteries. *Reproduced with permission from F. Wang, J. Li, J. Zhao, Y. Yang, C. Su, Y.L. Zhong, et al., Single-atom electrocatalysts for lithium sulfur batteries: progress, opportunities, and challenges, ACS Mater. Lett. 2 (11) (2020) 1450–1463, Copyright 2020, ACS Materials Letters.*

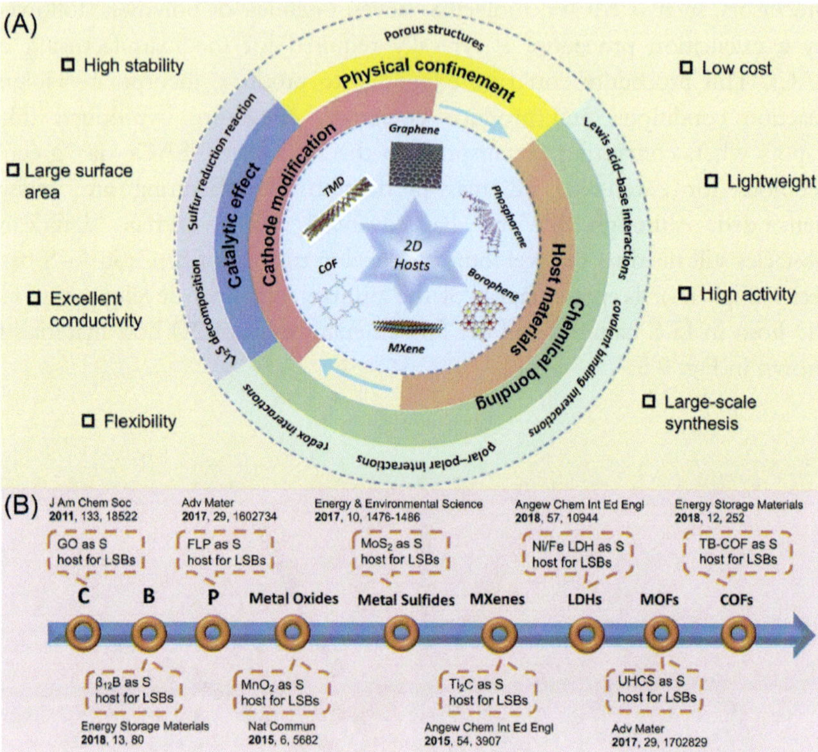

Figure 9.5 (A) Diagram illustrating the fundamental specifications for 2D host materials in Li-S batteries; (B) main 2D host advancements in Li-S batteries [61]. *Reproduced with permission from K. Fan, H. Huang, Two-dimensional host materials for lithium-sulfur batteries: a review and perspective, in: Energy Storage Materials, 2022, pp. 696–717.*

References

[1] Z. Zhuang, Y. Li, J. Huang, Z. Li, K. Zhao, Y. Zhao, et al., Sisyphus effects in hydrogen electrochemistry on metal silicides enabled by silicene subunit edge, Sci. Bull. 64 (9) (2019) 617–624.

[2] Z. Zhuang, J. Huang, Y. Li, L. Zhou, L. Mai, The holy grail in platinum-free electrocatalytic hydrogen evolution: molybdenum-based catalysts and recent advances, ChemElectroChem 6 (14) (2019) 3570–3589.

[3] I. Dincer, C. Acar, Review and evaluation of hydrogen production methods for better sustainability, Int. J. Hydrog. Energy 40 (34) (2015) 11094–11111.

[4] Z. Zhuang, Q. Kang, D. Wang, Y. Li, Single-atom catalysis enables long-life, high-energy lithium-sulfur batteries, Nano Res. 13 (7) (2020) 1856–1866.

[5] X. Zeng, J. Li, N. Singh, Recycling of spent lithium-ion battery: a critical review, Crit. Rev. Environ. Sci. Technol. 44 (10) (2014) 1129–1165.

[6] J.M. Tarascon, M. Armand, Issues and challenges facing rechargeable lithium batteries, Materials For Sustainable Energy, Nature Publishing Group, 2011, pp. 171–179.

[7] H. Yu, P. Zeng, X. Zhou, C. Guo, X. Liu, K. Wang, et al., Atomically dispersed and O, N-Coordinated Mn-Based catalyst for promoting the conversion of polysulfides in li2s-based Li–S battery, ACS Appl. Mater. Interfaces 13 (45) (2021) 54113–54123.
[8] X. Ji, L.F. Nazar, Advances in Li–S batteries, J. Mater. Chem. 20 (44) (2010) 9821–9826.
[9] K. Zhu, C. Wang, Z. Chi, F. Ke, Y. Yang, A. Wang, et al., How far away are lithium-sulfur batteries from commercialization? Front. Energy Res. 7 (2019) 123.
[10] J. Mahmood, F. Li, S.M. Jung, M.S. Okyay, I. Ahmad, S.J. Kim, et al., An efficient and pH-universal ruthenium-based catalyst for the hydrogen evolution reaction, Nat. Nanotechnol. 12 (5) (2017) 441–446.
[11] J. Jones, H. Xiong, A.T. DeLaRiva, E.J. Peterson, H. Pham, S.R. Challa, et al., Thermally stable single-atom platinum-on-ceria catalysts via atom trapping, Science 353 (6295) (2016) 150–154.
[12] L. Lin, W. Zhou, R. Gao, S. Yao, X. Zhang, W. Xu, et al., Low-temperature hydrogen production from water and methanol using Pt/α-MoC catalysts, Nature 544 (7648) (2017) 80–83.
[13] P. Liu, Y. Zhao, R. Qin, S. Mo, G. Chen, L. Gu, et al., Photochemical route for synthesizing atomically dispersed palladium catalysts, Science 352 (6287) (2016) 797–800.
[14] G. Malta, S.A. Kondrat, S.J. Freakley, C.J. Davies, L. Lu, S. Dawson, et al., Identification of single-site gold catalysis in acetylene hydrochlorination, Science 355 (6332) (2017) 1399–1403.
[15] Z.W. Seh, Y. Sun, Q. Zhang, Y. Cui, Designing high-energy lithium–sulfur batteries, Chem. Soc. Rev. 45 (20) (2016) 5605–5634.
[16] J.B. Goodenough, K.S. Park, The Li-ion rechargeable battery: a perspective, J. Am. Chem. Soc. 135 (4) (2013) 1167–1176.
[17] M.S. Whittingham, Ultimate limits to intercalation reactions for lithium batteries, Chem. Rev. 114 (23) (2014) 11414–11443.
[18] A. Manthiram, S.H. Chung, C. Zu, Lithium–sulfur batteries: progress and prospects, Adv. Mater. 27 (12) (2015) 1980–2006.
[19] J.B. Goodenough, Y. Kim, Challenges for rechargeable Li batteries, Chem. Mater. 22 (3) (2010) 587–603.
[20] P. Adelhelm, P. Hartmann, C.L. Bender, M. Busche, C. Eufinger, J. Janek, From lithium to sodium: cell chemistry of room temperature sodium–air and sodium–sulfur batteries, Beilstein J. Nanotechnol. 6 (1) (2015) 1016–1055.
[21] T. Li, X. Bai, U. Gulzar, Y.J. Bai, C. Capiglia, W. Deng, et al., A comprehensive understanding of lithium–sulfur battery technology, Adv. Funct. Mater. 29 (32) (2019) 1901730.
[22] Y.S. Su, A. Manthiram, Sulfur/lithium-insertion compound composite cathodes for Li–S batteries, J. Power Sources 270 (2014) 101–105.
[23] P. Zhu, J. Zhu, J. Zang, C. Chen, Y. Lu, M. Jiang, et al., A novel bi-functional double-layer rGO–PVDF/PVDF composite nanofiber membrane separator with enhanced thermal stability and effective polysulfide inhibition for high-performance lithium–sulfur batteries, J. Mater. Chem. A 5 (29) (2017) 15096–15104.
[24] X. Wang, Z. Wang, L. Chen, Reduced graphene oxide film as a shuttle-inhibiting interlayer in a lithium–sulfur battery, J. Power Sources 242 (2013) 65–69.
[25] P. Guo, D. Liu, Z. Liu, X. Shang, Q. Liu, D. He, Dual functional MoS2/graphene interlayer as an efficient polysulfide barrier for advanced lithium-sulfur batteries, Electrochim. Acta 256 (2017) 28–36.
[26] H. Li, L. Sun, Y. Zhang, T. Tan, G. Wang, Z. Bakenov, Enhanced cycle performance of Li/S battery with the reduced graphene oxide/activated carbon functional interlayer, J. Energy Chem. 26 (6) (2017) 1276–1281.

[27] L. Tan, X. Li, Z. Wang, H. Guo, J. Wang, Lightweight reduced graphene oxide@ MoS_2 interlayer as polysulfide barrier for high-performance lithium−sulfur batteries, ACS Appl. Mater. Interfaces 10 (4) (2018) 3707−3713.
[28] M. Yu, A. Wang, F. Tian, H. Song, Y. Wang, C. Li, et al., Dual-protection of a graphene-sulfur composite by a compact graphene skin and an atomic layer deposited oxide coating for a lithium-sulfur battery, Nanoscale 7 (12) (2015) 5292−5298.
[29] J.H. Lee, J. Kang, S.W. Kim, W. Halim, M.W. Frey, Y.L. Joo, Effective suppression of the polysulfide shuttle effect in lithium−sulfur batteries by implementing rGO−PEDOT: PSS-coated separators via air-controlled electrospray, ACS Omega 3 (12) (2018) 16465−16471.
[30] P. Han, A. Manthiram, Boron-and nitrogen-doped reduced graphene oxide coated separators for high-performance Li-S batteries, J. Power Sources 369 (2017) 87−94.
[31] H. Wu, Y. Huang, S. Xu, W. Zhang, K. Wang, M. Zong, Fabricating three-dimensional hierarchical porous N-doped graphene by a tunable assembly method for interlayer assisted lithium-sulfur batteries, Chem. Eng. J. 327 (2017) 855−867.
[32] Y. Liu, X. Qin, S. Zhang, G. Liang, F. Kang, G. Chen, et al., Fe3O4-decorated porous graphene interlayer for high-performance lithium−sulfur batteries, ACS Appl. Mater. Interfaces 10 (31) (2018) 26264−26273.
[33] N. Hu, X. Lv, Y. Dai, L. Fan, D. Xiong, X. Li, SnO_2/reduced graphene oxide interlayer mitigating the shuttle effect of Li−S batteries, ACS Appl. Mater. Interfaces 10 (22) (2018) 18665−18674.
[34] Y. Fan, Z. Yang, W. Hua, D. Liu, T. Tao, M.M. Rahman, et al., Functionalized boron nitride nanosheets/graphene interlayer for fast and long-life lithium−sulfur batteries, Adv. Energy Mater. 7 (13) (2017) 1602380.
[35] Y. Lu, S. Gu, J. Guo, K. Rui, C. Chen, S. Zhang, et al., Sulfonic groups originated dual-functional interlayer for high performance lithium−sulfur battery, ACS Appl. Mater. Interfaces 9 (17) (2017) 14878−14888.
[36] R. Yi, C. Liu, Y. Zhao, L.J. Hardwick, Y. Li, X. Geng, et al., A light-weight free-standing graphene foam-based interlayer towards improved Li-S cells, Electrochim. Acta 299 (2019) 479−488.
[37] J. Guo, J. Huo, Y. Liu, W. Wu, Y. Wang, M. Wu, et al., Nitrogen-doped porous carbon supported nonprecious metal single-atom electrocatalysts: from synthesis to application, Small Methods 3 (9) (2019) 1900159.
[38] B. Qiao, A. Wang, X. Yang, L.F. Allard, Z. Jiang, Y. Cui, et al., Single-atom catalysis of CO oxidation using Pt1/FeOx, Nat. Chem. 3 (8) (2011) 634−641.
[39] S. Ji, Y. Chen, X. Wang, Z. Zhang, D. Wang, Y. Li, Chemical synthesis of single atomic site catalysts, Chem. Rev. 120 (21) (2020) 11900−11955.
[40] H. Xu, D. Cheng, D. Cao, X.C. Zeng, A universal principle for a rational design of single-atom electrocatalysts, Nat. Catal. 1 (5) (2018) 339−348.
[41] A. Wang, J. Li, T. Zhang, Heterogeneous single-atom catalysis, Nat. Rev. Chem. 2 (6) (2018) 65−81.
[42] L. Liu, A. Corma, Metal catalysts for heterogeneous catalysis: from single atoms to nanoclusters and nanoparticles, Chem. Rev. 118 (10) (2018) 4981−5079.
[43] H. Yan, H. Cheng, H. Yi, Y. Lin, T. Yao, C. Wang, et al., Single-atom Pd1/graphene catalyst achieved by atomic layer deposition: remarkable performance in selective hydrogenation of 1, 3-butadiene, J. Am. Chem. Soc. 137 (33) (2015) 10484−10487.
[44] R. Banerjee, A. Phan, B. Wang, C. Knobler, H. Furukawa, M. O'Keeffe, et al., High-throughput synthesis of zeolitic imidazolate frameworks and application to CO_2 capture, Science 319 (5865) (2008) 939−943.
[45] A.M. Abdel-Mageed, B. Rungtaweevoranit, M. Parlinska-Wojtan, X. Pei, O.M. Yaghi, R.J. Behm, Highly active and stable single-atom Cu catalysts supported by a metal−organic framework, J. Am. Chem. Soc. 141 (13) (2019) 5201−5210.

[46] H.J. Peng, J.Q. Huang, Q. Zhang, A review of flexible lithium–sulfur and analogous alkali metal–chalcogen rechargeable batteries, Chem. Soc. Rev. 46 (17) (2017) 5237–5288.
[47] Z. Liang, C. Qu, D. Xia, R. Zou, Q. Xu, Atomically dispersed metal sites in MOF-based materials for electrocatalytic and photocatalytic energy conversion, Angew. Chem. Int. Ed. 57 (31) (2018) 9604–9633.
[48] L. Ma, K.E. Hendrickson, S. Wei, L.A. Archer, Nanomaterials: science and applications in the lithium–sulfur battery, Nano Today 10 (3) (2015) 315–338.
[49] F. Wang, J. Li, J. Zhao, Y. Yang, C. Su, Y.L. Zhong, et al., Single-atom electrocatalysts for lithium sulfur batteries: progress, opportunities, and challenges, ACS Mater. Lett. 2 (11) (2020) 1450–1463.
[50] H. Tian, A. Song, H. Tian, J. Liu, G. Shao, H. Liu, et al., Single-atom catalysts for high-energy rechargeable batteries, Chem. Sci. 12 (22) (2021) 7656–7676.
[51] C. Lu, R. Fang, X. Chen, Single-atom catalytic materials for advanced battery systems, Adv. Mater. 32 (16) (2020) 1906548.
[52] L.N. Song, W. Zhang, Y. Wang, X. Ge, L.C. Zou, H.F. Wang, et al., Tuning lithium-peroxide formation and decomposition routes with single-atom catalysts for lithium–oxygen batteries, Nat. Commun. 11 (1) (2020) 1–11.
[53] Z. Peng, S.A. Freunberger, Y. Chen, P.G. Bruce, A reversible and higher-rate Li-O_2 battery, Science 337 (6094) (2012) 563–566.
[54] G. Huang, J. Wang, X. Zhang, Electrode Protection in High-Efficiency Li–O_2 Batteries, ACS Cent. Sci. 6 (12) (2020) 2136–2148.
[55] P.G. Bruce, S.A. Freunberger, L.J. Hardwick, J.M. Tarascon, Li–O_2 and Li–S batteries with high energy storage, Nat. Mater. 11 (1) (2012) 19–29.
[56] S. Kang, Y. Mo, S.P. Ong, G. Ceder, A facile mechanism for recharging Li_2O_2 in Li–O_2 batteries, Chem. Mater. 25 (16) (2013) 3328–3336.
[57] H.T. Le, D.T. Ngo, P.N. Didwal, J.G. Fisher, C.N. Park, I.D. Kim, et al., Highly efficient and stable solid-state Li–O 2 batteries using a perovskite solid electrolyte, J. Mater. Chem. A 7 (7) (2019) 3150–3160.
[58] W.J. Kwak, Rosy, D. Sharon, C. Xia, H. Kim, L.R. Johnson, et al., Lithium–oxygen batteries and related systems: potential, status, and future, Chem. Rev. 120 (14) (2020) 6626–6683.
[59] Y.S. Hong, C.Z. Zhao, Y. Xiao, R. Xu, J.J. Xu, J.Q. Huang, et al., Safe lithium-metal anodes for Li-O_2 batteries: from fundamental chemistry to advanced characterization and effective protection, Batter. Supercaps 2 (7) (2019) 638–658.
[60] Y. Qiao, Q. Wang, X. Mu, H. Deng, P. He, J. Yu, et al., Advanced hybrid electrolyte Li-O_2 battery realized by dual superlyophobic membrane, Joule 3 (12) (2019) 2986–3001.
[61] K. Fan, H. Huang, Two-dimensional host materials for lithium-sulfur batteries: a review and perspective, Energy Stor. Mater. 50 (2022) 696–717.

CHAPTER 10

Conclusions and outlooks

Priyanka Aggarwal[1], Kamlendra Awasthi[1], Debasish Sarkar[1] and Prashanth W. Menezes[2,3]
[1]Department of Physics, Malaviya National Institute of Technology Jaipur, Jaipur, Rajasthan, India
[2]Materials Chemistry Group for Thin Film Catalysis—CatLab, Helmholtz-Zentrum Berlin für Materialien und Energie, Berlin, Germany
[3]Department of Chemistry, Technische Universität Berlin, Berlin, Germany

10.1 Conclusions

Single-atom catalysts (SACs) consist of homogeneous atomic reactive centers on heterogeneous supports, and they play a crucial role in offering an elegant platform to bridge the gap between homogeneous and heterogeneous catalysis [1]. In the last few years, SACs have emerged as a class of powerful catalysts that continue to garner significant scientific interest because of their competitive cost, high atomic utilization efficiency, homogeneous catalytic sites, unique electronic structure, high activity, and selectivity for various electrocatalytic reactions and charge storage devices as discussed throughout various chapters of this book. However, the stability of currently available SACs is questionable due to the highly reactive nature of single atoms (SAs). These SAs, owing to their high surface free energy, are prone to sintering and agglomeration, eventually forming thermodynamically stable clusters/nanoparticles or they can even react with nearby reactants to produce other undesirable species during synthesis or reaction processes. Consequently, catalyst deactivation occurs, drastically reducing their activity, selectivity, and stability. To mitigate these issues, SAs must be dispersed on suitable support that can efficiently anchor SAs, preventing them from aggregation to produce highly dispersed SACs. For this, it is necessary to diminish the interactions between individual metal atoms while strengthening the interactions with the support by regulating the coordination environment around SAs. In this direction, numerous strategies, including doping with heteroatoms (N, S, O, P, etc.), creation of defects or vacancies, and attaching functional groups or ligands, have been employed so as to strongly bind the SAs with the support and prevent them from aggregation. Up to now,

Single Atom Catalysts
DOI: https://doi.org/10.1016/B978-0-323-95237-8.00009-4

coordination environments of active metal atoms often include M-N_x, M-O_x, M-C_x, M-S_x, M-Cl_x, M-N_x-C_y, M-N_x-O_y, M-C_x-N_y, and M-C_x-O_y. Commonly, active sites in SACs are SAs and nearby atoms of the support. In general, SAs comprised of transition metal elements, such as Fe, Co, Ni, Cu, Mo, W, Mn, Zn, Cr, Ru, Pt, Pd, Rh, Ir, Ag, Au, V, Sn, etc., and recently nonmetals like P, I, and O are also included in the list of active SAs. Apart from a single atom serving as an active site, the neighboring atoms, such as N, S, F, C, and O around the SAs can also provide active sites. Since these neighboring atoms are associated with the support only, even supports start to play a key role in influencing the overall electrocatalytic performance of SACs. The support regulates the local coordination environments, optimizing the electronic structure to enhance the adsorption and desorption energies of reaction intermediates. Hence, supports govern the catalytic activity and stability of SACs. Consequently, extensive research has been conducted on support, and finding the appropriate support materials has garnered significant attention in the field of SACs. Different supports have their own special features and offer various sites for SAs to be anchored. Supports that are employed for supporting SACs include metal oxides, MXenes, metals, graphene, g-C_3N_4, transition metal dichalcogenides (TMDs), heteroatom-doped carbons, transition metal phosphides, graphdiyne, zeolites, carbon nanotubes, metal-organic frameworks (MOFs), covalent organic frameworks, and other promising supports (hexagonal boron nitride, phosphomolybdic acid, black phosphorus, perovskite, polyoxometalates, and conductive polymer materials).

In Chapter 2, we have closely examined various synthetic techniques utilized to prepare SACs like high-temperature pyrolysis, wet chemical, electrochemical route, atomic layer deposition (ALD), photochemical reduction, and other effective methods (atom trapping, ball milling, hydrothermal, metal leaching, flame spray pyrolysis, and microwave reduction). We learnt that the pyrolysis method is frequently used to fabricate SACs based on MOFs and bears the advantage that it is simple and does not need specialized instruments. But it generally requires ultrahigh temperature (usually $>800°C$) and prolonged heating time, which is not only time and energy-intensive but also significantly raises the risk of aggregation of SAs. Moreover, this method is fairly complex to comprehend in detail, making it challenging to properly control and optimize the central metal coordination to maximize catalytic performance. Additionally, the wastage of target atoms during acid etching process is another drawback of this method.

Similarly, the conventional wet impregnation technique is straightforward and suitable for mass manufacturing. However, it is challenging to obtain high metal loadings due to the restricted capacity of support to bind significant amounts of the metal precursor. Moreover, it is still elusive in academia to fully understand this method due to its complex procedure and nonuniform mechanism, which requires a trial-and-error process for its optimization. On the other hand, ALD provides unrivaled benefits by offering precise control over size, morphology, structure, and composition during the synthesis of SACs, but because of its high cost and low yield, mass production has been severely hampered. The mechanical ball milling method is straightforward to use and appropriate for large-scale production, its commercial uses are limited by the severe experimental conditions and need for specialized precursors. Therefore, it is desirable to look for suitable synthetic and immobilization techniques that can produce perfect SACs. Thus, the discovery of low-cost, scale-up-ready, environmentally friendly preparation techniques that can yield SACs with high activity represents a further development in the SACs synthesis.

In Chapter 3, we introduced in detail different electron microscopic characterization techniques, including aberration-corrected high-angle annular dark-field scanning transmission electron microscopy (AC-HAADF-STEM), environmental scanning transmission electron microscopy (E(S)TEM), field emission scanning electron microscopy (FESEM), and atom probe tomography (APT). We also discussed various spectroscopic methods such as X-ray absorption spectroscopy (XAS), X-ray photoelectron spectroscopy (XPS), Fourier-transform infrared spectroscopy (FTIR), electron energy loss spectroscopy (EELS), as well as theoretical and analytical approaches. We delved into the applications of AC-HAADF-STEM, which is primarily used for direct observation of individual atoms. E(S)TEM emerged as a powerful tool for studying dynamic reaction mechanisms in real-world synthesis and catalytic conditions. FESEM proved useful for investigating surface morphology, while APT enabled atomic-scale 3D imaging and elemental composition analyses. Furthermore, we explored the benefits of XAS in obtaining data on the local geometric and electrical structure of active metal centers as a weighted average [2]. XPS, on the other hand, provided valuable information about the surface composition, valence, and atom coordination environment. EELS offered insights into the atomic arrangement, elemental composition, catalytic mechanisms, and interactions between the catalyst and reactants. Lastly, FTIR allowed for the identification of various functional groups within a material and facilitated the determination of molecular structure. Each technique brings up certain advantages and disadvantages, and in the majority of circumstances, a combination of these techniques is required to reach a reliable conclusion. The characterization of every single atom, as opposed to the statistical mean information of all SAs with varied spatial distributions and chemical bound states, is essential for a thorough comprehension of the structure–performance relationship. Furthermore, combining experimental and theoretical tools help us to reveal information about SACs, including their location, structure, composition, nature, bonding state, and much more. In upcoming chapters, the applications of SACs in a large number of electrocatalytic reactions such as oxygen reduction reaction (ORR), oxygen evolution reaction (OER), hydrogen evolution reaction (HER), carbon dioxide reduction reaction (CO_2RR), nitrogen reduction reaction (NRR), and alkali metal ion sulfur batteries were studied in detail. It was shown that they exhibit exclusive activity, selectivity, and higher stability compared to their bulk/nanoparticle counterparts. The common statement "Plenty of Room at the Bottom" is now in full bloom and being fully satisfied by SACs [3]. At the core of advancing electrocatalysis technology lies the essential aspect of thoughtfully designing and synthesizing catalysts capable of facilitating a diverse range of electrochemical reactions.

While notable advancements have been made in recent years in the development of exceptionally stable SACs along with their synthesis approaches, support selection, characterization techniques, theoretical investigations, and applications in energy conversion and storage. However, there are still a number of thorny challenges associated with SACs in terms of their synthesis, characterization, catalytic activity, and stability that restricts the industry-relevant applications and hence must be overcome and are described in the following section (Fig. 10.1):

1. **Low loading of SAs**—The loading amount of single-atom sites is closely linked to the catalytic performance. The higher metal loading is a driver to enhance the catalytic performance of SACs in various catalytic reactions. Higher loading will increase the number of active single sites where more and more reactant molecules reach. However, it is difficult to simultaneously balance the increased metal loading content and atoms in a singly dispersed state. It is believed that preparing SACs with high loading amounts ($>$ 5 wt.%) of SAs is still an open challenge, which is significantly less than the benchmark catalysts used in the industry (20 wt.% Ir/C or Pt/C). This is because a surplus

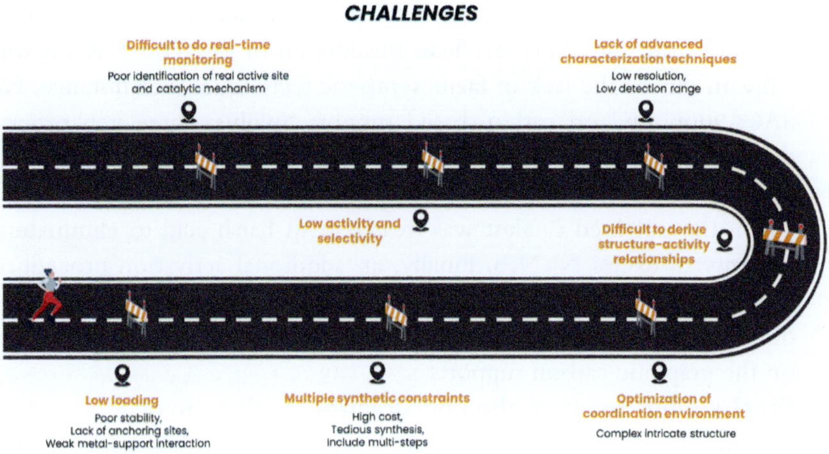

Figure 10.1 Various challenges associated with SACs.

of metal atom loading, however, would aggregate to form nanoparticles (NPs) owing to their high surface free energy and high agglomeration tendencies, thus, invariably resulting in metal leaching and reducing the stability. Moreover, employing traditional synthetic procedures to increase the loading amount of SAs seems to be a slow process (For instance, there is a high chance of agglomeration of SAs during their synthesis via high-temperature pyrolysis due to long heating exposure time). Therefore, further investigations are needed on how to improve the density of individual SAs on the support without affecting their stability. An in-depth comprehension of the preparation strategies of SACs and exploring feasible methods (e.g., heteroatom doping, defects, crystallinity, and appropriate precursors) is vital to enhance the active-site density of SACs. Moreover, during the design of SACs, supports play an important role in regulating the metal-support interactions and achieving durability. There are some selected supports that are employed to stabilize SACs, such as metal oxides, carbon-based materials, MXenes, TMDs, etc. Therefore, to prepare high-density and stable SACs, it is significantly needed to explore advanced support with coordinated atoms and plentiful anchoring sites that can limit the aggregation of active SAs and make them robust.

2. **Multiple synthetic constraints**—Despite huge progress in the synthetic technologies of SACs, still, most of the time, the preparation of SACs is expensive and usually involves tedious or complicated steps. Moreover, the synthesis process often consists of repeated experiments

that frequently led to material and energy consumption as well as a waste of time and expenses. Mass production of low-cost SACs is still a dream due to the lack of facile synthetic techniques. For instance, Ni SACs supported on carbon-based support involves three-step procedures [4]. Firstly, Ni-based MOFs were pyrolyzed at 700°C in the N_2 environment to produce Ni NPs incorporated in the graphene layers. Then, the obtained catalyst was treated with harsh acid to eliminate a majority of excess Ni NPs. Finally, an additional activation procedure using the electrochemical cycling potential is necessary to boost the dissolution of any remaining Ni NPs to obtain only Ni SAs supported on the graphitic carbon support.

3. **Precise adjustment of the coordination environment**—The SACs feature intricate structures that include single metal atoms, support, and functionalities/impurities. The active site of SACs does not consist of only a single metal atom but rather an active region comprising of a single metal atom and nearby atoms of the support. Hence, the type of active metal center, the nature of supporting material, and the local coordination environment (coordination number, coordinative atoms) of a single metal atom greatly alter the electronic properties and geometric structure of SACs and hence affects the overall catalytic performance. Therefore, it is desirable to precisely adjust the coordination environment surrounding the active metal center and build SACs with a desirable structure to improve the catalytic performance. However, it is challenging to accurately regulate the bonding environment and local atomic structure of every single atom anchored on the support due to the inability of existing synthetic procedures.

   Moreover, due to the complex coordination environment of SACs, it is very difficult to accurately describe the catalytic performance with coordination information. Theoretical research may assist in determining the best SAC structures according to the type of catalytic reaction in which they are employed. Moreover, synthetic strategies should be modified to precisely control the coordination environment via doping, ligand anchoring, structural defects, vacancies, etc.

4. **Identification of potential active sites**—The active single-atom site might vary from its synthesized initial structure under actual working conditions. Predicting the nature of the actual active center involved in the catalytic reaction, the nature of the reaction intermediates that emerge throughout the reaction process, and monitoring the complex stepwise elementary reaction mechanism are still challenging tasks.

As a result, it is quite difficult to derive the correlation between structure and activity. Therefore, it is critical to conduct a detailed analysis of the dynamic reconstruction of the active center in order to pinpoint the real active site and understand the reaction mechanisms by real-time monitoring [5–7].

5. **Lack of advanced characterization techniques**—Even if numerous catalytic mechanisms have been put forth, present characterization techniques are still not capable to examine in-depth the influence of support material, type of active metal center, modified support with heteroatoms, ligands, vacancies, etc., on the catalytic performance of SACs. For instance, Bi-coordinated N-doped carbon networks favor the CO_2 electroreduction process while inhibiting the HER process [8]. On the other hand, the HER is preferred to be carried out for the catalytic system of P-coordinated Mo_2C [9]. These examples demonstrate that the relationship between catalytic activity and structure is still far from being understood. In situ studies are crucial and important to reveal a deeper knowledge of the mechanisms and construct a relationship between structure and catalytic performance. Hence, building connections between catalyst structure and catalytic performance is crucial for fabricating catalysts with improved efficiency and adjustable product selectivity.
6. The quantitative relationship between metal–support interaction and the catalytic performance of SACs is not well established. For instance, it is generally predicted that the enhanced charge transfer between SACs and support will lead to stronger interactions; nevertheless, it is not obvious to what extent increased charge transfer will increase catalytic activity. Therefore, careful consideration is required in this aspect to fully understand the relationship.
7. The performance of SACs is sometimes inferior to their nanoparticle counterparts in certain reactions. For instance, it was discovered by scientists in 2015 that Pt nanoparticles serve as an active site instead of Pt SAs for CO oxidation and water-gas shift reactions [10].
8. Last but not least, present studies on SACs are still carried out on a lab scale and do not take into account the real-world factors that may drastically affect the SACs performance. For instance, testing SACs in ideal circumstances fail to account for the contaminants or harmful compounds present in industrial settings that might potentially render the SACs inactive. Similarly, when employing SACs as the electrocatalysts of fuel cells in electric vehicles, their operating conditions differ

slightly from those in laboratories. Therefore, more work still needs to be undertaken to establish working environments of SACs that are more realistic and to create guidelines that will direct future research. To address the mentioned challenges, it is crucial to find solutions that effectively tackle the issues at hand, allowing for the rational design and practical implementation of a new generation of SACs.

All of these challenges discussed earlier call for new ideas and contributions in synthetic methodologies, exploration of novel support materials, refined characterization techniques, thorough catalyst screening, detailed understanding of mechanisms, and application-oriented catalyst development. These efforts aim to resolve stability issues, increase the loading capacity of SAs, and improve catalytic efficiency, as elaborately discussed in the following (Fig. 10.2).

1. **Exploiting artificial intelligence (AI) and integration with density functional theory (DFT):** It is not viable to conduct trial-and-error experiments to choose potential combinations of metal and substrate due to time constraints and high economic costs. Hence, new advances in AI present excellent prospects to save time and money as well as improve catalytic performance in terms of activity, selectivity, and stability. This is because AI may drive and enable direct evolution. AI and data science can collaborate with the widely used DFT calculations to conduct a more thorough examination of the structure–performance relationships, which can serve as a foundation for developing high-performance SACs.

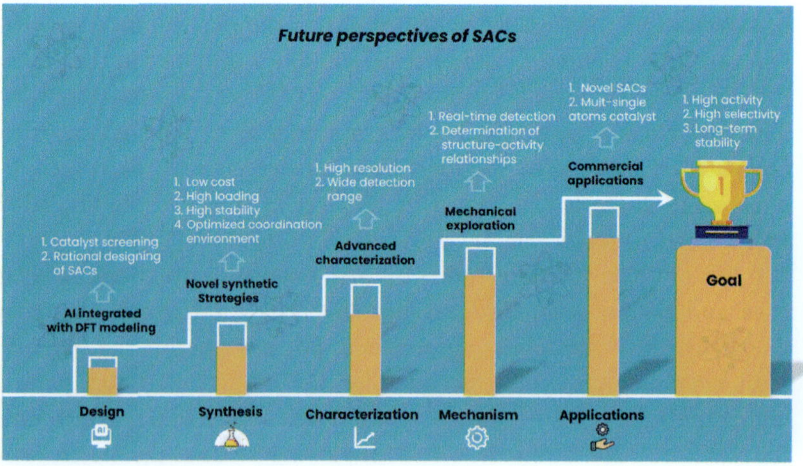

Figure 10.2 Future perspectives of SACs.

For instance, candidate crystal structures are pulled from a database containing vast sets of component arrangements by algorithms. Then, machine learning offers the potential for screening active SACs with a favorable coordination environment according to the database comprising adsorption energies of intermediates, rate-limiting steps, and reaction energy barriers. DFT is then used to establish the thermodynamic stability of these structures. Even though there have been many studies on DFT-predicted high-performance SACs, the majority of them are constrained by the size of the data set and cannot offer thorough explanations for the implementation of experiments. The data collection might be expanded by combining DFT calculations with AI. Therefore, more research efforts should be carried out by theoretical chemists to extend its application to other areas, which are recently applied in limited applications (mainly biomedical applications).

2. **Developing facile and low-cost novel synthetic strategies—** Various synthetic techniques have already been employed through the combined efforts of researchers to prepare SACs, such as pyrolysis, wet chemistry, ALD, ball milling, etching, hydrothermal/solvothermal, etc. Despite significant progress in synthetic procedures of SACs, there is still a lack of effective methods for producing cost-effective and high-density SACs on a wide scale along with outstanding electrochemical activities. Recently, the mechanochemical technique has shown the potential to fabricate SACs on a kilogram scale [11,12]. Additionally, although high metal loadings of 20 wt.% or even higher were recently achieved, the majority of reported SACs still feature a small number of active catalytic sites with an atom loading usually less than 5 wt.% [13,14]. This is because the synthesis of SACs is not easy due to their high tendency to agglomerate to form bigger clusters/nanoparticles, thus restricting the improvement of the electrocatalytic activity. Moreover, the maximum possible loading that can be achieved is currently unknown. Hence, it is extremely important to focus our future research on improving the original preparation procedure or developing cheap, easy, and universal synthetic strategies to produce large-scale and highly dispersed SAs that are firmly anchored to the supports to encourage the commercialization of SACs. The targeted high metal loading without affecting the SACs stability can be achieved via adopting two pathways (1). Exploring cheap, innovative, and simple synthetic techniques that can precisely modulate the coordination configuration (type of coordination atoms, coordination number) around

SAs to obtain specific SACs structure, favoring high loading content of SAs with high stability to excel catalytic performance toward targeted reaction. For instance, adopting low-temperature controlled synthesis (icing-assisted photocatalytic reduction) reduces the exposure of precursors in high-energy environments, thus avoiding agglomeration [15]. Also, using a novel synthetic technique based on the cross-linking and self-assembly of the graphene quantum dots, Ir SACs anchored on the carbon support were able to obtain an unusually high loading of Ir SAs up to 41.6 wt.% [16]. In this technique, amine-functionalized graphene quantum dots were utilized to undergo cross-linking and self-assembly, resulting in the formation of a layered carbon structure on a larger scale. This structure offers an abundance of anchoring sites for loading a dense concentration of SAs, while also maintaining appropriate spacing between the atoms to prevent aggregation. In another approach, SACs with a metal loading as high as 30 wt.% were achieved via pyrolyzing coordinated polymer strategy on a massive scale [17]. (2) exploring new metal precursors and searching for new modified substrates with enough anchoring sites to load high density of metal atoms. For instance, using organometallic complexes as metal precursors. These precursors have the benefit of providing abundant coordination sites to strongly bind metal atoms and therefore prevent the aggregation of metal atoms even during high-temperature pyrolysis. Recently, a simple and low-cost method for the mass production of robust Ni SACs for efficient and selective CO_2RR was established by using hydrogen-bonded organic frameworks (HOFs) as the precursors and utilizing effective and affordable guar gum. Massive structural pores in HOFs could effectively trap the Ni source [18]. Further, advanced support with a modified surface via heteroatom doping, defects, vacancies capture, functional groups/ligands, and anchoring should be developed. For instance, Ru SAs build a strong interaction with N, S co-doped graphene oxide support (N,S-Go) via strong Ru-N and Ru-S coordination bonds [19]. Hence, the Ru/N,S-Go catalyst showed outstanding stability of 50 h and no degradation in activity even after 10,000 CV cycles. However, understanding the defect effect is still in its initial stages. A large density of defects sometimes may have more negative than positive effects, such as the fall of electron conductivity or low selectivity. Hence, the influence of these modification parameters should be properly investigated in detail. Also, we can choose support with a special 3D morphology that has a high surface area along with a hierarchical porous structure that will boost the number of exposed active sites, resulting in high metal loading. These strategies are the key to proceeding to the next stage in the industrialization of SACs.

3. **Deciphering the single atom-support interaction:** For understanding the fundamentals of supported single metal atoms, understanding the interactions between the isolated atoms and the support is a key advancement. Although strong interactions are essential for activating SACs and preventing the SAs from agglomeration, excessively strong interactions can sometimes inactivate the atoms [20]. Moreover, it is still not understood whether support is able to interact with the inner atoms or not. Therefore, the impact of the support on core-shell SACs needs to be given additional consideration. Additionally, theoretical simulations and modeling based on data from microscopy and spectroscopy should be developed to obtain detailed structural knowledge of the impact of the support on core-shell SACs. The type of support is crucial in altering the single atom-support interaction and determining the overall catalysis mechanism and performance, and the development of a more precise structure—performance relationship linked to chemical and physical properties is highly advantageous for their design for various applications. Certain modifications in the form of elemental doping, introducing defects, structure engineering, etc., are done in the support to enhance the interaction of the active single site with the support. It is easy to modify carbon supports with defects and doping (O, S, N, and other elements). Hence, they have become the preferred option for SACs support. However, employing carbon-based materials as supports is severely constrained by their instability at high temperatures and oxygen concentrations. The scope of single-atom catalysis can be expanded by creating a wide range of SAC supports, such as functional carbon materials, and choosing appropriate preparation techniques for the support type, like UV irradiation for TiN.
4. **Utilizing advanced characterization techniques for catalyst monitoring:** It is essential to develop advanced characterization technologies with better device resolution, wide detection range, image reconstruction algorithms, and specialized sample holders for a detailed survey of catalytic reactions ex situ and in situ [21,22]. Currently, ex situ spectroscopic and microscopic methods have been widely used to identify the SACs structure (geometric and electronic structure) and monitor the vast catalytic processes in a range of applications.

However, ex situ techniques are incapable of unveiling the possible reaction mechanisms and identifying the actual active sites participating in the catalytic reaction in real time because the atomic structure of SACs might undergo transformation. Hence, advanced in situ/operando characterization techniques, such as TEM, XAS, XPS, FTIR, FESEM, etc., should be developed to reveal the real-time morphologies, directly monitor the active sites' evolution process during operating conditions, and observe the changes in the electronic structure of the active center, intermediate formation, reactant adsorption, and product desorption processes [23,24]. This will certainly help to unravel the complex single-atom reaction mechanism at an atomic level in real time and construct a true relationship between catalytic complex structure and catalytic performance. In addition, theoretical modeling is also a powerful tool and is urgently required to design catalysts with an optimized structure and study in-depth the catalytic kinetics and reaction mechanisms. Therefore, exploring the direct in situ/operando characterization techniques as well as theoretical simulation studies are the need of the hour to develop optimized catalysts with suitable coordination environment for a specific reaction and survey the catalytic kinetics, and understand the mechanism in detail [25]. These techniques or studies are still not developed fully but can be pursued in the future. To advance the field of SACs and unlock their full potential for diverse applications, it is essential to continue exploring and developing an expanded array of advanced characterization techniques with improved resolution along with computational studies. The acquired knowledge and information will set the foundation for the development of SACs with superior activity, high selectivity, and long-term stability. Moreover, a universal and straightforward SACs characterization method is highly demanding due to the shortage and high cost of current equipments.

5. **Discovering new SACs with dual or multiple SAs**: Further development of new SACs with high catalytic activity and selectivity is required. SACs struggle with their lack of synergistic active sites and structural simplicity, which prevents them from outperforming more complex electrocatalytic processes like ORR, CO_2RR, and NRR, containing several reaction steps and intermediates. Hence, a promising solution is to bring the concept of dual or multiple SACs, which can considerably increase the catalytic activity of the process and improve the selectivity of the catalytic reaction by utilizing the synergistic effect

of various single-atom centers. The wide tunability of the active centers provides more opportunities to optimize the binding strength of reaction intermediates [26]. For example, there are a large number of challenges when employing SACs for CO_2RR, such as competing reaction HER, the complexity of the CO_2RR mechanism, and the formation of abundant reaction intermediates and products that drastically affects the catalytic performance [27]. In contrast, dual-atom catalysts (DACs) where two isolated metal atoms are stably anchored on suitable support may be able to overcome these problems due to the flexible dual-metal centers and synergistic effect between two metal atoms [28]. They can provide the distinct C-affinity and O-affinity sites for reinforcing different intermediates binding in a complicated CO_2RR mechanism. Moreover, the synergistic effect of two individual active sites leads to the improvement of sophisticated photocatalytic reactions, which promotes the research direction and expands the SACs to a whole new train of thought in the future, even if the development of DACs is still in its infancy. Due to the synergistic effect, DACs (Cu_2@C_2N) display greater catalytic activity for CO oxidation reaction than SACs (Cu_1@C_2N) [29].

6. **Development of universal activity evaluation indicators:** To aid in the logical design of high-performance SACs, unified activity evaluation indicators for each type of SAC should be presented. It is realized that most of the current research papers do not reveal crucial information on the performance of the catalyst. Examples include statistics regarding the number of active sites, selectivity apparent quantum yield, etc. Furthermore, it is difficult to compare performance under various conditions for a similar experiment. Hence, standardizing testing parameters for each catalytic reaction is highly demanding [30].
7. In future research, it is essential to incorporate highly accurate structural models that reflect real-world conditions rather than relying on simplistic models in idealized stationary states for theoretical calculations pertaining to specific reactions. This approach will yield deeper insights into the nature of SACs and the mechanisms associated with the reactions, thereby unraveling potential relationships between structure and performance. Additionally, theoretical calculations facilitate the efficient screening of novel catalysts for targeted reactions. Extensive theoretical analysis has already indicated that certain transition metal SACs hold promise as catalysts for specific reactions, although their experimental exploration is still at an early stage.

Consequently, it is crucial to invest more effort in the development and design of such catalysts.

8. In the pursuit of highly efficient catalytic materials, it is imperative to not only prioritize the use of minimal quantities of precious metals and maximize the utilization of their atoms but also explore alternative catalytic materials based on nonprecious metals. The aim is to identify nonprecious metal-based catalysts capable of achieving comparable or even superior catalytic performance when compared to noble metals.

The widespread implementation of SACs would mark a monumental milestone in the field of catalysis. Nevertheless, challenges such as the risk of agglomeration, lack of specificity between preparation techniques and coordination environments, and the intricate relationship between coordination environments and catalytic activity highlight the need for substantial advancements and discoveries to enable their extensive application on a larger scale. In short, the era of SACs is undeniably just around the corner and encourages and promotes the further advancement of this emerging research field. We anticipate that additional research efforts will be devoted to advancing this fascinating subject so that it might finally realize the practical application and make significant contributions toward carbon neutrality. Research needs to be directed to explore in-depth the mysterious catalytic nature of SACs and develop statistically significant structure–property relationships for more novel design. This can be achieved by closely integrating theoretical approaches, including high-throughput screening and AI techniques with advanced experimental methodologies. Hence, rapid advancements made in their design and construction, as well as the thorough investigation of their mechanisms in vast electrocatalytic energy conversion and storage devices, will surely maximize the catalytic efficiency of these processes achieving high activity, selectivity, and long-term stability in the near future. The application of more efficient and effective SACs can also be expanded into emerging domains, including photoelectric conversion, electrosynthesis of organic compounds, dye degradation, and rechargeable batteries [31–33].

References

[1] P. Aggarwal, D. Sarkar, K. Awasthi, P.W. Menezes, Functional role of single-atom catalysts in electrocatalytic hydrogen evolution: current developments and future challenges, Coord. Chem. Rev. 452 (2022) 214289. Available from: https://doi.org/10.1016/j.ccr.2021.214289.

[2] P. Liu, R. Qin, G. Fu, N. Zheng, Surface coordination chemistry of metal nanomaterials, J. Am. Chem. Soc. 139 (2017) 2122–2131. Available from: https://doi.org/10.1021/jacs.6b10978.

[3] R.P. Feynman, There's plenty of room at the bottom, J. Microelectromech. Syst. 1 (1992) 60–66. Available from: https://doi.org/10.1109/84.128057.

[4] L. Fan, P.F. Liu, X. Yan, L. Gu, Z.Z. Yang, H.G. Yang, et al., Atomically isolated nickel species anchored on graphitized carbon for efficient hydrogen evolution electrocatalysis, Nat. Commun. 7 (2016) 1–7. Available from: https://doi.org/10.1038/ncomms10667.

[5] L. Reith, J.N. Hausmann, S. Mebs, I. Mondal, H. Dau, M. Driess, et al., In situ detection of iron in oxidation states ≥ IV in cobalt-iron oxyhydroxide reconstructed during oxygen evolution reaction, Adv. Energy Mater. 13 (2023) 2203886. Available from: https://doi.org/10.1002/aenm.202203886.

[6] Z. Chen, H. Yang, S. Mebs, H. Dau, M. Driess, Z. Wang, et al., Reviving oxygen evolution electrocatalysis of bulk La–Ni intermetallics via gaseous hydrogen engineering, Adv. Mater. 35 (2023) 2208337. Available from: https://doi.org/10.1002/adma.202208337.

[7] P.W. Menezes, C. Walter, B. Chakraborty, J.N. Hausmann, I. Zaharieva, A. Frick, et al., Combination of highly efficient electrocatalytic water oxidation with selective oxygenation of organic substrates using manganese borophosphates, Adv. Mater. 33 (2021). Available from: https://doi.org/10.1002/adma.202004098.

[8] J. Lin, S. Yan, C. Zhang, Q. Hu, Z. Cheng, Electroreduction of CO_2 toward high current density, Processes 10 (2022). Available from: https://doi.org/10.3390/pr10050826.

[9] X. Yin, J. Shao, W. Zhang, Q. Gao, Biomass-derived Mo_2C@N, P Co-doped carbon as an efficient electrocatalyst for hydrogen evolution reaction, Energy Fuels 36 (2022) 11261–11268. Available from: https://doi.org/10.1021/acs.energyfuels.2c02265.

[10] K. Ding, A. Gulec, A.M. Johnson, N.M. Schweitzer, G.D. Stucky, L.D. Marks, P.C. Stair, Identification of active sites in CO oxidation and water-gas shift over supported Pt catalysts, Science 350 (2015) 189–192. Available from: https://doi.org/10.1126/science.aac6368.

[11] H. Jin, S. Sultan, M. Ha, J.N. Tiwari, M.G. Kim, K.S. Kim, Simple and scalable mechanochemical synthesis of noble metal catalysts with single atoms toward highly efficient hydrogen evolution, Adv. Funct. Mater. 30 (2020) 4–9. Available from: https://doi.org/10.1002/adfm.202000531.

[12] X. He, Y. Deng, Y. Zhang, Q. He, D. Xiao, M. Peng, et al., Mechanochemical kilogram-scale synthesis of noble metal single-atom catalysts, Cell Rep. Phys. Sci. 1 (2020) 100004. Available from: https://doi.org/10.1016/j.xcrp.2019.100004.

[13] Z. Li, R. Hong, Z. Zhang, H. Wang, X. Wu, Z. Wu, Single-atom catalysts in environmental engineering: progress, outlook and challenges, Molecules 28 (2023). Available from: https://doi.org/10.3390/molecules28093865.

[14] H. Zhang, X.F. Lu, Z.-P. Wu, X.W.D. Lou, Emerging multifunctional single-atom catalysts/nanozymes, ACS Cent. Sci. 6 (2020) 1288–1301. Available from: https://doi.org/10.1021/acscentsci.0c00512.

[15] P. Zhou, X. Hou, Y. Chao, W. Yang, W. Zhang, Z. Mu, et al., Synergetic interaction between neighboring platinum and ruthenium monomers boosts CO oxidation, Chem. Sci. 10 (2019) 5898–5905. Available from: https://doi.org/10.1039/C9SC00658C.

[16] C. Xia, Y. Qiu, Y. Xia, P. Zhu, G. King, X. Zhang, et al., General synthesis of single-atom catalysts with high metal loading using graphene quantum dots, Nat. Chem. 13 (2021) 887–894. Available from: https://doi.org/10.1038/s41557-021-00734-x.

[17] Y. Xiong, W. Sun, P. Xin, W. Chen, X. Zheng, W. Yan, et al., Gram-scale synthesis of high-loading single-atomic-site Fe catalysts for effective epoxidation of styrene, Adv. Mater. 32 (34) (2020) e2000896. Available from: https://doi.org/10.1002/adma.202000896.

[18] S. Han, S. Jia, W. Xia, X. Xing, R. Qi, H. Wu, et al., A new strategy for mass production of single-atom catalysts for high performance of CO2 electrochemical reduction, Chem. Eng. J. 455 (2023) 140595. Available from: https://doi.org/10.1016/J.CEJ.2022.140595.

[19] M.J. Wang, M. Ji, X. Zheng, C. Jiang, H. Zhao, Z.X. Mao, et al., Enhanced hydrogen evolution of single-atom Ru sites via geometric and electronic engineering: N and S dual coordination, Appl. Surf. Sci. 551 (2021) 148742. Available from: https://doi.org/10.1016/j.apsusc.2020.148742.

[20] J. Liu, Catalysis by supported single metal atoms, ACS Catal. 7 (2017) 34–59. Available from: https://doi.org/10.1021/ACSCATAL.6B01534.

[21] J.N. Hausmann, S. Mebs, H. Dau, M. Driess, P.W. Menezes, Oxygen evolution activity of amorphous cobalt oxyhydroxides: interconnecting precatalyst reconstruction, long-range order, buffer-binding, morphology, mass transport, and operation temperature, Adv. Mater. 34 (2022) 2207494. Available from: https://doi.org/10.1002/adma.202207494.

[22] H. Yang, J.N. Hausmann, V. Hlukhyy, T. Braun, K. Laun, I. Zebger, et al., An intermetallic CaFe6Ge6 approach to unprecedented Ca − Fe − O electrocatalyst for efficient alkaline oxygen evolution reaction, ChemCatChem. (n.d.) e202200293. https://doi.org/10.1002/CCTC.202200293.

[23] J.N. Hausmann, P.W. Menezes, Effect of surface-adsorbed and intercalated (Oxy) anions on the oxygen evolution reaction, Angew. Chem. Int. Ed. (2022). Available from: https://doi.org/10.1002/ANIE.202207279.

[24] P.W. Menezes, C. Walter, J.N. Hausmann, R. Beltrán-Suito, C. Schlesiger, S. Praetz, et al., Boosting water oxidation through in situ electroconversion of manganese gallide: an intermetallic precursor approach, Angew. Chem.—Int. Ed. 58 (2019) 16569–16574. Available from: https://doi.org/10.1002/anie.201909904.

[25] I. Mondal, J.N. Hausmann, G. Vijaykumar, S. Mebs, H. Dau, M. Driess, et al., Nanostructured intermetallic nickel silicide (Pre)catalyst for anodic oxygen evolution reaction and selective dehydrogenation of primary amines, Adv. Energy Mater. (2022) 2200269. Available from: https://doi.org/10.1002/AENM.202200269.

[26] Y. Ying, X. Luo, J. Qiao, H. Huang, "More is Different": synergistic effect and structural engineering in double-atom catalysts, Adv. Funct. Mater. 31 (2021) 2007423. Available from: https://doi.org/10.1002/adfm.202007423.

[27] Q. Chen, P. Tsiakaras, P. Shen, Electrochemical reduction of carbon dioxide: recent advances on Au-based nanocatalysts, Catalysts 12 (2022). Available from: https://doi.org/10.3390/catal12111348.

[28] R. Li, D. Wang, Superiority of dual-atom catalysts in electrocatalysis: one step further than single-atom catalysts, Adv. Energy Mater. 12 (2022) 2103564. Available from: https://doi.org/10.1002/aenm.202103564.

[29] F. Li, Z. Chen, Cu dimer anchored on C_2N monolayer: low-cost and efficient Bi-atom catalyst for CO oxidation, Nanoscale 10 (2018) 15696–15705. Available from: https://doi.org/10.1039/C8NR03394C.

[30] J. Yang, X. Du, B. Qiao, Comprehensive activity evaluation of single-atom catalysts, Chem. Catal. 3 (2023) 100424. Available from: https://doi.org/10.1016/J.CHECAT.2022.09.049.

[31] H. Yang, P.V. Menezes, G. Dai, G. Vijaykumar, Z. Chen, M. Al-Shakran, et al., Activation of nickel foam through in-liquid plasma-induced phosphorus incorporation

for efficient quasi-industrial water oxidation and selective oxygenation of organics, Appl. Catal. B Environ. 324 (2023) 122249. Available from: https://doi.org/10.1016/j.apcatb.2022.122249.
[32] T. Kahlstorf, J.N. Hausmann, T. Sontheimer, P.W. Menezes, Challenges for hybrid water electrolysis to replace the oxygen evolution reaction on an industrial scale, Glob. Chall. (2023) 2200242. Available from: https://doi.org/10.1002/gch2.202200242.
[33] A. Indra, P.W. Menezes, M. Schwarze, M. Driess, Visible light driven non-sacrificial water oxidation and dye degradation with silver phosphates: multi-faceted morphology matters, N. J. Chem. 38 (2014) 1942–1945. Available from: https://doi.org/10.1039/C3NJ01012K.

Index

Note: Page numbers followed by "*f*" and "*t*" refer to figures and tables, respectively.

T

U

V

W

X

Z

CPI Antony Rowe
Eastbourne, UK
January 29, 2024